RURAL PLANNING AND DEVELOPMENT IN THE UNITED STATES

RURAL PLANNING AND DEVELOPMENT IN THE UNITED STATES

Mark B. Lapping
Thomas L. Daniels
John W. Keller

THE GUILFORD PRESS
New York London

A Division of Guilford Publications, Inc.
72 Spring Street, New York, NY 10012

Printed in the United States of America

This book is printed on acid-free paper.

Last digit is print number: 9 8 7 6 5 4 3 2 1

Library of Congress Cataloging-in-Publication Data

Lapping, Mark B.
Rural planning and development in the United States / by Mark Lapping, Tom Daniels, and John Keller.
p. cm.
Bibliography: p.
Includes index.
ISBN 0-89862-384-7
ISBN 0-89862-517-3 (pbk.)
1. Rural development—United States. 2. United States—Rural conditions. I. Daniels, Tom. II. Keller, John (John W.)
III. Title.
HN90.C6L36 1989
307.72′0973—dc20 89-2231
CIP

To Our Wives—Joyce, Kathy, and Kathy
To Our Children—Hans and Karin, Ethan and Jason, and Sean and Mary

Your support, love, and enthusiasm has never failed us.

Preface

This book is the second in a two-part strategy to address some of the needs, as we have seen them, in the literature of rural planning. This team published the *Small Town Planning Handbook* (Daniels, Keller, & Lapping, 1988), a straightforward, step-by-step approach for planners and the lay community. This present volume constitutes our attempt to write an informative and issues-and-problems text for rural planning practice. We hope that this book will be of genuine use both for teaching and as a practical reference.

Rural Planning and Development in The United States has taken a number of years to produce. And with each passing year, we have learned more about rural people and rural places. We wish to thank those who have had a large hand in helping us. In particular, we want to acknowledge contributions made by William Lassey, Frederic Sargent, James R. Coffman, Louis Swanson, Harvey Jacobs, Frederick Steiner, Jon Wefald, Jack Kartez, Cornelia and Jan Flora, George Penfold, James Hillestad, Robert Gray, Frank Popper, Charles Geisler, Jackie Wolfe, Fred Buttle, Pierre Clavel, Paul Windley, Sam Stokes, and Marty Zeller. At Guilford, a group of earnest and committed "book people" who stayed with us over the years, we want very much to acknowledge the support of Janet Crane, now with the AAG, Seymour Weingarten, Editor-In-Chief, who never gave up on us, and Pearl Weisinger, who has labored on this manuscript to make it a better one. At Kansas State University, which has provided us with a nurturing environment, we want to thank Jean Olson, Bertha Jackson, and Diane Potts for typing and for so much other support and help. Ditto to Gail Stecker at Iowa State University. Parris Riordan and Ann Traylor Petrushka have been very helpful as research assistants.

We wish to acknowledge with appreciation the permission of the following to reproduce valuable materials from previous publications:

National Association of Counties, University of Buffalo (SUNY) Law Review, John Wiley and Sons, Inc., Frederic O. Sargent, *Journal of Rural Studies* (Pergamon Press), Association of American Geographers, University of Minnesota Press, Johns Hopkins University Press, American Planning Association, *Science*, American Association for the Advancement of Science, *Scientific American*, and the Council of State Governments.

We have been teaching, researching and working as planners for nearly twenty years throughout North America. Innumerable good people living and working in small towns, in remote places, and in the open country have been our source of support and inspiration. We think the needs of rural people and rural places go to the very heart of a democratic, pluralist conception of America. It is our sincerest hope that this book will help to influence others and that they will be compelled, as we have been, to think seriously, critically, and strategically about what is at stake for our future in rural America.

Mark B. Lapping
Thomas L. Daniels
John W. Keller

Contents

CHAPTER ONE

An Overview of Rural America

Change, diversity, and complexity best describe contemporary rural America. To most people, the term "rural" suggests an image of open country, spacious vistas, and quaint villages, with farming the dominant way of life. But such an image applies only to a minority of rural places, and overlooks the far-reaching economic and social changes that have occurred in much of rural America since World War II. The most notable change has been the drop in farm population from 24.3 million people in 1945 to below 5 million in 1987; still, almost 60 million people now live beyond metropolitan boundaries. Today, "rural" is used to describe a diversity of landscapes, economies, and people. But change and diversity have made many rural areas more complex and difficult to understand. Therefore, a clear understanding of rural areas is essential for creating effective rural planning and development strategies.

The dominant rural land uses in the United States fall into three general categories: "open land," "production," and "settlement." Open lands are often remotely situated, and have few people and little economic activity. These areas include wetlands, tundra in Alaska, mountainous regions, desert and wilderness areas, and much of the vast high plains of the western United States. Rural production areas are characterized by a variety of economic activities and population sizes. Generally, counties dependent on natural resource industries will have smaller populations than counties with manufacturing plants. Rural production areas include (1) agricultural regions producing crops and livestock; (2) timber resource and commercial lumber areas; (3) mineral extraction and energy-producing regions; (4) fishing areas; and (5) manufacturing regions. Settlement areas are the small cities, towns, villages, rural–urban fringe areas, and unincorporated places that feature employment in service industries (wholesale and retail trade, finance, and government).

These industries serve not only the settled areas, but the adjacent hinterlands as well.

Three trends reflect the major changes underway in rural areas in recent years. First, between 1970 and 1980, more rural dwellers stayed in rural areas, and there was a net migration of about 4 million urban people to rural areas. Improved transportation and communication networks reduced the social and cultural isolation and homogeneity of rural inhabitants. Traditionally urban-based manufacturing plants have increasingly located in rural areas, along with new service, trade, and recreation industries. Government employment has also shown strong growth, and local public services have notably improved.

These new industries and employment opportunities have created a second trend: the economic restructuring of much of rural America. The importance of the natural resource industries—agriculture, fisheries, forestry, and mining—has declined in many communities, and these industries now employ only some 12 percent of the rural labor force. Moreover, the social and economic changes have altered the way rural people live and work and have narrowed the distinction between "urban" and "rural."

An important example of this narrowed distinction can be seen in the third trend: the greater impact of international economic and political events on rural areas. After the Soviet grain deal of 1972, American farm exports increased sharply, topping $40 billion in 1980, and accounting for about one-quarter of net farm income. But between 1980 and 1985, federal tight-money policy caused the U.S. dollar to rise in value some 40 percent against other world currencies, thus making American farm and manufactured products less competitive on world markets. Meanwhile, the world supply of wheat, soybeans, and corn increased, thanks in large part to greater output by Brazil (soybeans), Canada (wheat), Australia (wheat), Argentina (wheat), and the European Economic Community (wheat and corn). To make matters worse for U.S. farmers, the Soviet grain embargo of 1979–1981 merely forced the Soviets to look for suppliers other than the United States. The embargo also raised suspicions among other grain-buying countries as to the reliability of American grain supplies. This increased world competition has meant lower commodity prices for American farmers and financial stress in the major food-exporting states of the Midwest.

The greater world competition in agriculture is repeated in mining and manufacturing. World prices for minerals and energy are largely determined on New York and London markets. For example, an increase in the amount of copper supplied by Zambia or Chile can affect the world copper price and thus influence the profitability of operating a copper mine in Ajo, Arizona, or Butte, Montana.

Manufacturing employed a greater percentage of the rural work force than the urban work force (26 percent vs. 22 percent) in 1980. But manufacturing has since suffered from intense competition from foreign imports, as evidenced by America's $120 billion annual trade deficit in 1984. In addition, some rural manufacturing plants have been drawn overseas by lower labor costs and relaxed environmental standards. Since 1980, the United States has lost 400,000 manufacturing jobs, of which 150,000 jobs were lost in the textile industry concentrated in the rural Southeast.

The future growth of rural America is clouded with uncertainty. It is important to note that since 1980, urban areas have once again been growing at a faster rate than rural areas. The key question for rural America is this: "How will rural people earn a living?" As employment in agriculture and manufacturing continues to be vulnerable to foreign competition, the effects are quickly felt in the shops and stores of rural communities. Less money is available to maintain public services, and local banks are hard pressed to justify new loans, much less to retire old loans. Furthermore, the federal deregulation of banking has spurred a movement toward interstate banking, which could mean less investment in rural areas. Deregulation in the transportation industry has permitted railroad, trucking, and bus companies to establish freight rates that have increased transportation costs in many rural areas or have resulted in the abandonment of service to some small towns. Finally, the huge federal budget deficits of the 1980s and cuts in federal domestic programs suggest that, in the future, less federal money may be channeled toward rural America.

"Economic development" continues to be the catch phrase in much of rural America. But this supposed panacea to raise local incomes, employment, and tax bases is often a desperate cry for help. Many rural areas need new businesses and industries to diversify, if not to totally overhaul, the local economy. Yet such diversification is usually a slow and uncertain process. Although economic development and diversification are important tasks of rural planning, there are other important challenges. Not all rural areas need to promote growth; some areas must control growth to maintain the environmental and social qualities that make these communities desirable places to live. The first step a planner can take is to understand the differences between rural and urban areas. Many recently published books on rural and small-town planning have merely tried to apply urban planning techniques to rural areas. This is often wrong. For example, detailed comprehensive plans with a 20-year time horizon are often of little use in a sparsely settled region or small community dependent on the health of one or two industries. The use of sophisticated growth controls may run into opposition from rural landowners who feel that they have a right to do whatever they want to with their land. The planner must be sensitive to the economic needs and the deep-seated attitudes found in rural America. Education and compromise are likely to be more effective than a heavy-handed bureaucratic approach. Second, a planner must learn how rural areas differ from one another. The diversity in rural America, even within a single county, can make the use of a single model of rural planning wholly ineffective.

Rural Populations

The key features of rural life remain space and distance. Rural people live in fewer numbers and farther apart than urban dwellers. Rural residents must often travel farther to schools, hospitals, stores, and other services; some lack access to many of these services and amenities (Dillman & Hobbs, 1982). In recent years, there has been a trend toward the consolidation of public services (e.g., schools) and private services (e.g., regional shopping malls). This consolidation has made living in more remote areas increasingly difficult.

Rural people also have tended to differ from urbanites in their social and political attitudes and in their relationship to the land. Rural people have historically been viewed as being more politically conservative, more traditional, and slower to accept new social ideas; furthermore, land is seen as a resource for the production of food, fiber, and minerals, rather than as a commodity to be consumed for housing or recreation. The validity of these generalizations of rural attitudes certainly varies among communities and regions. For example, it is instructive to remember that rural America has been the wellspring of many indigenous radical movements, such as Populism, Progressivism, and the Industrial Workers of the World.

Nonetheless, there are certain quantifiable characteristics of rural populations that set them apart from urban people. Although the United States can be described as an urban society, more than 25 percent of the 238 million Americans were living in rural areas in 1987 (see Table 1.1). According to the U.S. Bureau of the Census, there are two ways to define "rural": (1) a nonurbanized area of less than 2,500 people (low population density); or (2) a county without a city of greater than 50,000 inhabitants, known as a "nonmetropolitan county." Of the 3,137 counties in the United States, only 798 are considered urban; moreover, rural counties comprise 85 percent of the nation's 2.3 billion acres and contain 30 percent of all townships and 70 percent of all municipalities. In fact, more than 70 percent of all Americans live on about 35 million acres, or 1.5 percent of the total national land base. (See Figure 1.1.)

TABLE 1.1. U.S. Rural Population by Size of Place, 1980

	Places	Population	Percent of total U.S.	Percent distribution
Total rural population	13,764	59,494,813	26.3	100.0
Places of 1,000–2,500 people	4,434	7,037,840	3.1	11.8
Places of less than 1,000 people	9,330	48,593,503	21.4	81.7
Other rural (living in unincorporated areas)	—	3,863,470	1.7	6.5

Note. From *1980 Census of Population United States Summary General Population Characteristics*, Part 1, B (pp. 1–19) by the U.S. Bureau of the Census, 1980, Washington, DC: U.S. Government Printing Office.

But population density alone may be a misleading criterion for determining rural and urban places. Rural pockets may be found within a "Standard Metropolitan Statistical Area" (SMSA), defined as a county with a city of greater than 50,000 inhabitants. Similarly, settlements of urban density may exist in nonmetropolitan counties. Communities of over 2,500 people may be found in rural areas, and jurisdictions of less than 2,500 adjacent to urban areas. The 1980 census indicates that of the

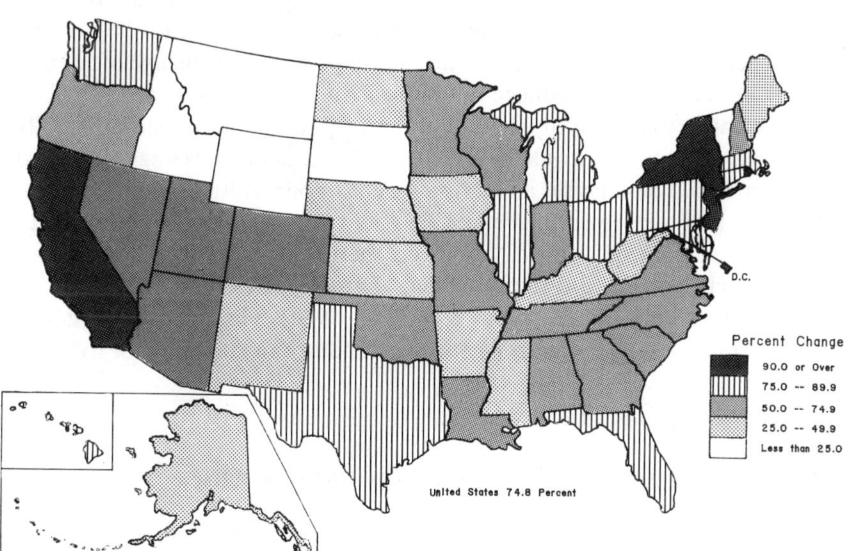

FIGURE 1.1. Percentage of population in Standard Metropolitan Statistical Areas (SMSAs) by states, 1980. Data from U.S. Bureau of the Census, 1980. *1980 Census of Population Number of Inhabitants, United States Summary*, Part 1, A, pp. 1–27.

169 million people living within SMSAs, 24 million were considered rural dwellers, and of the 57 million people living outside of SMSAs, 35.5 million were defined as rural (U.S. Bureau of the Census, 1980). In addition, within rural populations, a distinction can be made between farm and nonfarm residents. Since 1910–1920, when the farm population reached an estimated 32 million, the number of people living on farms has declined to about 4.9 million in 1987, or roughly 8 percent of the total rural population (U.S. Bureau of the Census, 1988). About half of all farm residents live in the Midwest; over one-quarter reside in the South; one-seventh live in the West; and only 6 percent are found in the North.

The U.S. Bureau of the Census classifies individuals or families as "rural nonfarm" if they live outside an incorporated community of 2,500 or more people and they do not earn at least some minimum level of income from farming. Obviously, a number of rural nonfarm residents live in rural–urban fringe areas as well as in more remote places. Rural nonfarm people in the fringe areas tend to commute to jobs in the urban centers or suburbs. In more isolated areas, nonfarm people often live in housing developments located along good transportation routes, rather than in small towns, to take advantage of scenic surroundings, lower property taxes, and commuting convenience.

A comparison of rural and urban population characteristics reveals some noteworthy social and economic differences. In rural areas, there are proportionately more males, fewer members of minority groups, more people under 18, more families, more people over the age of 55, and fewer high school and college graduates. Rural people also tend to have higher rates of unemployment and poverty, and lower median family incomes, than urban dwellers. These rural characteristics suggest a need for increased educational and employment opportunities in rural regions to raise incomes toward urban levels and to discourage migration to already overcrowded cities. Moreover, the urban–rural disparity in economic well-being and differing social characteristics has widened in recent years (Henry, Drabenstott, & Gibson, 1986). This growing disparity is in contrast to the convergence of rural and urban standards of living in the 1970s.

The Brief Population Turnaround in Rural America

The 1980 census revealed that for the first time since the census was begun in 1790, rural areas and small towns had achieved greater rates of population growth than metropolitan centers. Furthermore, this trend was observed in all regions of the nation (see Figures 1.2 and 1.3).

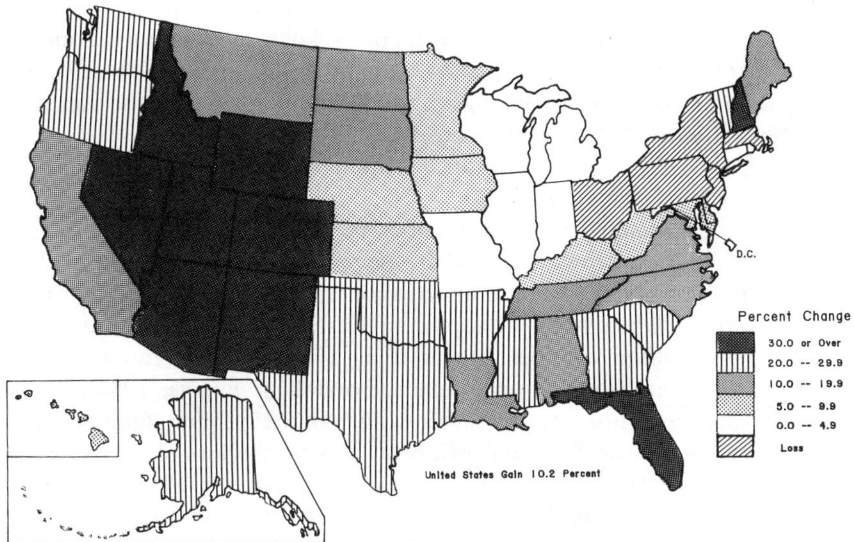

FIGURE 1.2. Percentage change in metropolitan population by states, 1970–1980. Data from U.S. Bureau of the Census, 1980. *1980 Census of Population Number of Inhabitants, United States Summary*, Part 1, A, pp. 1–28.

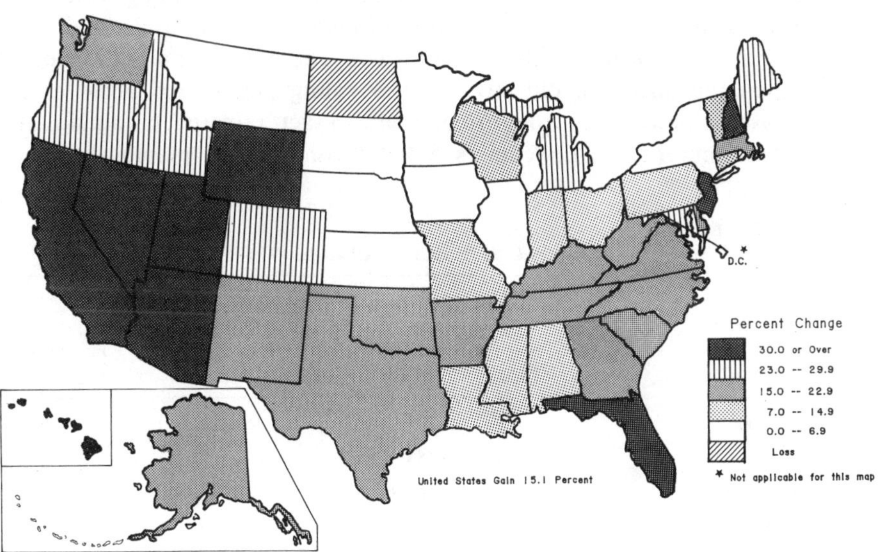

FIGURE 1.3. Percentage change in nonmetropolitan population by states, 1970–1980. Data from U.S. Bureau of the Census, 1980. *1980 Census of Population Number of Inhabitants, United States Summary*, Part 1, A, pp. 1–28.

Between 1970 and 1980, nonmetropolitan areas adjacent to metropolitan areas experienced average growth rates of 17.4 percent, and more remote nonmetropolitan areas grew at an average rate of 14 percent. These rural growth rates exceeded the average for SMSAs (10.2 percent), the average for central cities (0.1 percent), and the U.S. national average (11.4 percent). Only suburban areas gained population at a faster rate (18.2 percent). (See Table 1.2.)

In total, nonmetropolitan counties received 37 percent of the decade's population growth. This hefty increase differs sharply from the situation in the 1960s, when nonmetropolitan population grew by only 4.4 percent, well below the metropolitan average of 17 percent. But since 1980, metropolitan counties appear to be growing more rapidly than nonmetropolitan counties once again (Forstall & Engels, 1984). This change has been influenced by the downturn in the farming, mining, and energy sectors in the West and Midwest, and by the loss of many rural manufacturing jobs in the South. For example, between 1980 and 1987, the nation's farm population fell from over 6 million to just under 5 million, the lowest since the 1850s (U.S. Bureau of the Census, 1988). In short, it has become more difficult to earn a living in many parts of rural America in the 1980s.

The reasons for the 1970s trends of "urban flight" and increased indigenous rural population growth were both economic and social. Improved transportation and communication networks, along with federal programs, facilitated the location of manufacturing, trade, and service industries in previously isolated areas. These new industries helped to stem the historical outflow of rural young people and acted as a catalyst for newcomers from urban areas. In fact, researchers found that the vast majority of newcomers did not commute to previously held

TABLE 1.2. U.S. Population Change, 1970–1980

	Population (in millions)			
	1980	1970	Percent change 1970–1980	Percent of U.S. total
U.S. total	226.5	203.3	11.4	100.0
SMSAs	169.4	153.7	10.2	74.8
Central cities	67.9	67.8	0.1	30.0
Suburbs	101.5	85.6	18.2	44.8
Nonmetropolitan areas	59.4	49.6	15.1	25.2

Note. From Hauser, "The Census of 1980," by P. M. Hauser, 1981, *Scientific American, 245* (5), p. 57. Copyright 1981 by *Scientific American*. Reprinted by permission.

urban jobs; rather, they worked in the rural areas in which they resided (Dillman & Hobbs, 1982). In addition, some rural public services such as schools and sewer and water systems were upgraded, thus enhancing the quality of life. Perhaps more importantly, many rural newcomers appeared willing to sacrifice urban-level salaries and conveniences for a perceived superior living environment—clean air and water, open space, less crime, and greater community spirit—generally ascribed to rural areas.

But rural population increases have also produced some undesirable results. Environment impacts have featured polluted groundwater, overcrowded solid waste disposal sites, loss of open space, and greater traffic congestion. Some rural communities have grown so rapidly that local governments have not been able to keep pace in providing public services, and it has been found that newcomers tend to demand more services than long-term residents. Although newcomers may initially tolerate limited services, over time they are often reluctant to accept a level of services below that to which they were accustomed in urban areas. The cost of enhancing local services is often met by increasing property taxes; this tends to put pressure upon many long-term rural residents. Furthermore, much of the recent population growth has occurred outside of population centers. These dispersed settlement patterns have added to the cost of providing utilities and other services, and have often displaced farm and forest operations.

Disruptions in social life have also occurred with the influx of new inhabitants. Clashes of ideas, beliefs, and customs have created tensions between newcomers and long-time inhabitants. Permanence, stability, and tradition are frequently cited as rural traits. Rural people tend to cling longer to traditional views and resist new ideas. Social institutions, such as churches and clubs, have often served to reinforce these values. Despite the relative decline of natural-resource-based industries in many rural areas, most local residents still often consider land to be more valuable for production of food, fiber, and minerals than for more intensive uses.

Newcomers, on the other hand, represent change, which is often highly disruptive to the social, cultural, and economic status quo. They tend to strain social institutions with sheer numbers; their urban backgrounds and values may make them appear strange and even offensive; they generally have little affinity with a community's past; and they may earn their living in a newly located industry. Moreover, they may view land not as a productive resource, but as a commodity to be consumed for residential and recreational uses. Even so, since 1970, there has been a resurgence of part-time farming in the United States. Over 70 percent of the nation's farms produce only 12 percent of total farm output. Many of these farms are "hobby farms" of less than 50 acres that generate less than $10,000 a year in sales. In some regions, hobby farms have threatened the long-term viability of commercial farms by forcing up land prices and by precluding expansion through fragmenting landownership.

Making a Living in Rural Regions

The economic base of many rural areas has changed dramatically over the past 40 years. A major cause has been the change in the relative importance of agriculture in many communities. Between 1945 and 1982, the number of U.S. farms fell from 6 million to 2.2 million, and the average farm size more than doubled to 439 acres. Today, only a relatively small portion of the rural labor force is engaged in agriculture. In fact, a mere 6 percent of all farmers produce over 50% of the total farm output, and the majority of U.S. farmers now receive over half of their income from nonfarm employment!

The trend toward fewer but larger commercial farms has several explanations. First, the scale of technology—sophisticated machinery and chemicals—requires substantial amounts of capital and is designed for large farms with mass production as a means of lowering the cost of producing food. Second, many farm workers have been displaced by machinery and have been attracted to job opportunities in urban areas; some have simply become redundant to society. Third, and perhaps most importantly, the political economy of farming in the shape of federal programs—price supports, water subsidies, acreage-set-aside payments, credit, and research—has favored large farms.

Although the number of farmers and farm laborers has declined, new economic activities have emerged, enabling more people to earn a living in rural areas. Manufacturing, trade, and service industries have been drawn to nonmetropolitan regions by attractive land and labor costs; physical amenities; improved transportation, especially highways; better communications; and the lure of state, local, and federal assistance and incentive programs. As a result, the manufacturing, trade, and service sectors accounted for almost two-thirds of all jobs in rural areas in 1980 (see Table 1.3).

Although agriculture remains the dominant industry in several rural counties, particularly in the Midwest and Great Plains, most localities have experienced a shift in their economic base away from agriculture and other natural resource industries toward more urban types of employment. For example, between 1970 and 1979, nonmetropolitan counties had a 24 percent increase in manufacturing jobs, compared to a 4 percent increase in metropolitan counties. As of 1987, 29 percent of the

TABLE 1.3. Rural Employment, 1980

Industry	Number employed	Percent of total
Manufacturing	6,062,136	26.2
Trade	4,254,594	18.6
Services	5,601,190	24.6
Agriculture	2,023,326	9.2
Forestry and fisheries	83,444	0.4
Mining	485,395	2.3
Construction	1,828,266	8.8
Transportation, communications, and utilities	198,131	0.8
Finance	914,026	4.4
Government	1,034,779	4.7
Total	23,885,287	100.0

Note. From *1980 Census of Population United States Summary General Social and Economic Characteristics*, Part 1, C (p. 87) by the U.S. Bureau of the Census, 1980, Washington, DC: U.S. Government Printing Office.

rural labor force was employed in service industries and 23 percent in manufacturing, compared to urban employment of 18 percent and 34 percent in those industries, respectively (U.S. Bureau of the Census, 1988).

Employment in forestry, fisheries, and mining, while not nationally large, is important in certain rural regions. Forestry is concentrated in the Pacific Northwest, northern Maine, and parts of the Southeast. Major commercial fisheries are found in the north Pacific, off New England, and along the Gulf coast. Mining is based primarily in the mountain states and Appalachia. Since 1950, all three industries have experienced declines in employment, although the number of jobs in mining sharply increased in the 1970s, spurred by the search for new supplies of oil, natural gas, and coal. Although technological advances have been made in each industry, forestry and fisheries have faced additional changes. Forestry has suffered from reduced product demand, such as the substitution of plastics for wood, and from high interest rates in the 1980s, which have dampened the home construction market. Fisheries have been subject to government-imposed limits on fishing seasons and limits to the number of fishermen allowed in a fishery. These actions have been taken in response to reduced fish catches in many years. In sum, the employment opportunities in forestry and fisheries will probably continue to decline. The number of jobs in mining will depend upon the price of energy and minerals.

A Typology of Rural Areas[1]

Rural areas can be identified according to geographic location, the dominant economic activity, or social characteristics. How these three features fit together in different rural communities will have important implications for both planning needs and strategies.

Rural areas are often separated into two general geographic categories: nonmetropolitan counties adjacent to metropolitan counties, and nonmetropolitan counties that do not border on metropolitan counties. This first kind of rural area is often called the "rural–urban fringe"; this term has been used to describe the scattered development patterns or urban sprawl around major population centers. In fact, nearly all settlements are surrounded by a "town and country" region that increases in density closer to community boundaries. Fringe areas vary tremendously in size and economic activity. They may extend anywhere from a fraction of a mile beyond small towns to 40 miles beyond major metropolitan centers. Around small towns, the fringe area features scattered housing and hobby farms. In the fringe areas around large cities, agriculture and other traditional resource-based industries may be found; however, these land uses are rapidly giving way to residential, commercial, and industrial uses and high population densities. New inhabitants have the option of commuting to work in the urban center or working at an urban type of job in the fringe.

Rural counties away from the rural–urban fringe can be categorized according to the county's dominant economic activity or by a high percentage of poor or elderly residents. In "agricultural" counties, at least 20 percent of personal income or 20 percent of employment depends on farming or ranching. These counties are concentrated in the Midwest and Great Plains states, and feature populations well below 50,000 people. Agricultural counties also show a lack of economic diversity, and population change is slow. In recent years, these counties and their small towns have tended to lose population, coinciding with farm consolidation and the downturn in the farm economy (see Figure 1.4). Job opportunities for young people are likely to be limited, causing migration to urban and regional centers. About 700 counties, or one-quarter of all nonmetropolitan counties, can be classified as agricultural.

"Mining" counties show at least 20 percent of income and employment from mining activities. These counties exhibit high rates of population change, as one might expect, given the boom-and-bust conditions of

[1]This section draws heavily from *Procedures for Developing a Policy-Oriented Classification of Nonmetropolitan Counties* by P. J. Ross and B. L. Green, 1985, Washington, DC: U.S. Department of Agriculture, Economic Research Service.

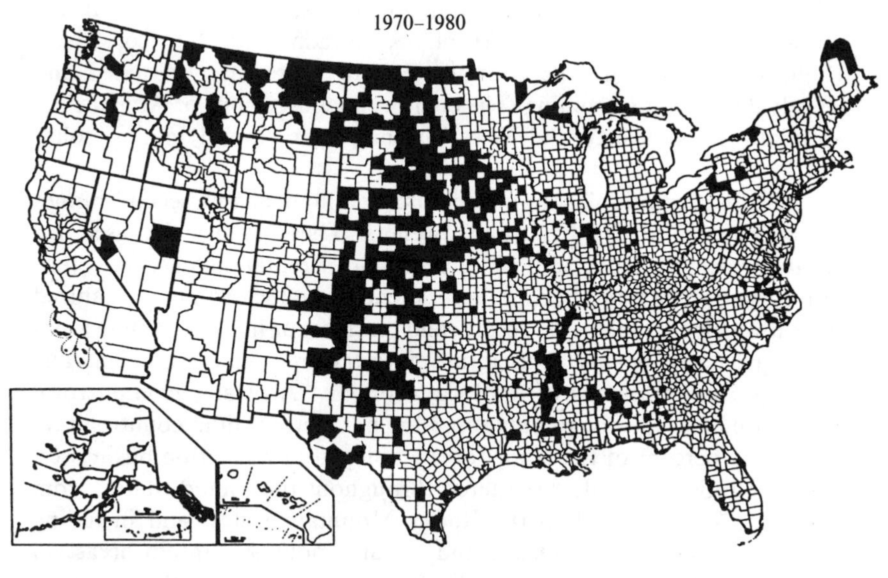

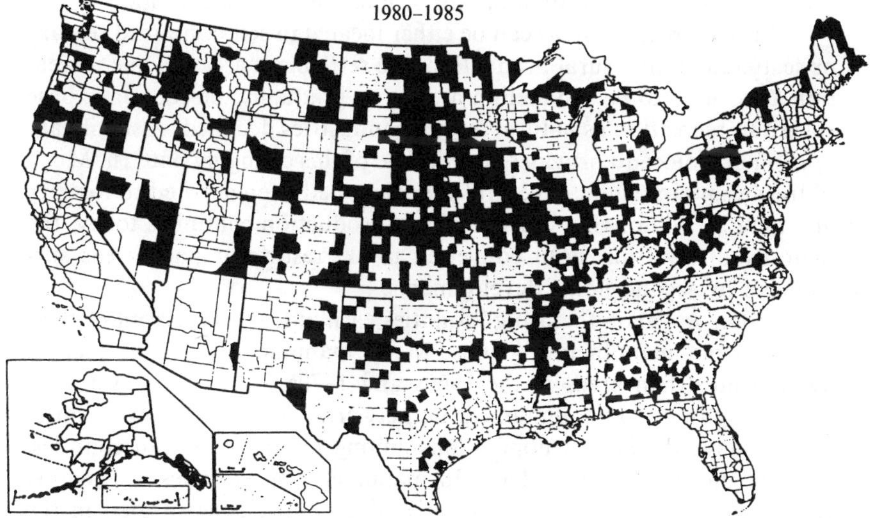

FIGURE 1.4. Nonmetropolitan counties with declining population, 1970–1980 and 1980–1985. From Brown and Deavers (1987) *Rural Economic Development in the 1980's*, p. 1–28. USDA Economic Research Service: Washington, D.C.

mining areas. Of the 200 or so mining counties, most are found in the states of Arizona, Colorado, Montana, Nevada, New Mexico, Texas, Utah, and Wyoming. The counties tend to be remotely situated and heavily dependent on mining. The coal-producing regions of Appalachia, especially in Kentucky and West Virginia, have several mining counties.

"Rural recreation" counties feature developments of second homes owned predominantly by urban dwellers. These homes are used for vacations, for weekend outdoor recreation, for rental income, or as part of full-time retirement. Second homes tend to be located near organized recreation areas, such as ski resorts and marinas, or in more remote areas of unique scenic beauty. Agriculture, forestry, and mining are generally not important segments of the local economy. All states contain rural recreation areas, although only some 150 nonmetropolitan counties have at least 10 percent of the labor force employed in recreation businesses. These counties are widely scattered throughout the United States, from northern New England to the Rocky Mountain states, and including coastal regions and the Ozark and Great Smoky Mountain areas. In recreation counties, population is likely to change significantly during the year because of the seasonal nature of most recreation activities.

Rural recreation areas can be either located in self-contained regions or scattered among rural production and settlement areas. Some recreation areas used to be production areas, but gradually the economic base shifted to recreation as natural resource industries declined. A significant portion of Colorado provides an example of this changing pattern. Much of the mountain region of central Colorado had been devoted to forestry and mining. Although a portion of this production continues today, the majority of land has been converted to ski resorts or park areas surrounded by second-home developments.

People in "manufacturing" counties earn at least 40 percent of their income from manufacturing jobs. Many manufacturing firms located away from urban areas in the 1960s and 1970s, thanks largely to improved highway and communication networks. Manufacturing counties tend to have the largest populations among nonmetropolitan counties, and are more urbanized. The 678 manufacturing counties are mainly found east of the Mississippi River and are most concentrated in the Southeastern states. These counties may be vulnerable to foreign manufacturing competition, as has occurred with the Southern textile industry. Manufacturing is the largest nonmetropolitan employer, accounting for 26 percent of the nonmetropolitan labor force.

"Government" counties derive at least one-quarter of their total personal income from federal, state, and local government jobs. Government counties are spread throughout the United States and have experienced rapid population growth in the last 20 years.

Almost all counties with at least one-third of the land base in federal ownership are found in the Western states and Alaska. There are 247 "federal lands" counties, and the economic base features mining and government employment along with some ranching and forestry.

"Poverty" counties show an average per capita income in the bottom 20 percent of all U.S. counties. These counties make up about one-tenth of all nonmetropolitan counties and tend to be concentrated in the Southeast and Appalachian states. Poverty counties have a higher percentage of blacks and female heads of households, but a lower proportion of participation in the work force than other nonmetropolitan counties.

"Retirement" counties feature a population comprised of new inmigration of people 60 years or older. About one-fifth of all nonmetropolitan counties can be classified as retirement areas. The majority of retirement counties are found in the South and Southwest. Retirement counties show a large service economy and a high proportion of income from transfer payments such as Social Security. On the whole, these counties grew quite rapidly during the 1960s and 1970s.

Some nonmetropolitan counties do not fall into any of the categories listed above, whereas other counties may contain elements of more than one category. The purpose of categorizing rural areas in this manner is to show the diversity of economic and social features that make up rural America. This diversity has several implications for public policy and planning. Most importantly, a single national policy on rural America is not likely to speak to the differing needs of the many kinds of rural areas. At the same time, a single planning approach to rural America is not likely to succeed because of the diversity of economic and social problems and personal attitudes.

Land and the Creation of Wealth in Rural Regions

Changes in the economic activities and social composition of rural communities have intensified the debate over the role of land as a source of wealth. People involved with traditional natural resource industries tend to value land according to its ability to yield food, fiber, and minerals. Farmers, in particular, see land as supporting a working rural landscape that has intrinsic value and cannot easily be replaced. Other rural people may value land for supporting factories, stores, and houses. These people may derive certain amenities from "consuming" the open space provided by natural resource industries, but may not see these traditional enterprises as vital to the growth and survival of a community.

As a rural community grows in population and economic wealth, the local real estate market is often distorted. The increased demand for land causes land prices to rise. Land is revalued above its worth as farmland or forestland, and farmers and forest owners may have difficulty in paying property taxes, expanding operations, or resisting lucrative offers to sell or convert land to other uses. Between 1950 and 1979, the value per acre of American farmland increased by more than 400 percent, after accounting for inflation. Between 1969 and 1979, the real value (after inflation) of rural land for recreation grew about 700 percent, and timberland registered average gains of 500 percent. Often prices for farmland and forestland were higher than the income the land could generate in farm or forest uses, and the high prices made it difficult to expand operations. The greater demand for rural land for second homes, primary residences, and commercial and industrial sites caused rural land to become more scarce in economic terms, as measured by the large increases in real land prices.

These changes in the rural land market continue to have far-reaching consequences for communities in the rural–urban fringe, in rural recreation areas, in small towns experiencing indigenous population growth or an influx of newcomers, and in remote communities with declining populations. Although rural land prices have leveled off or even fallen in the 1980s, rural land is a far scarcer commodity than it was 30 years ago. Above all, the higher land prices reflect a rising cost of living in rural areas, and more often this cost is affordable only through income sources outside of rural areas or in new rural manufacturing and service industries that pay near urban-level salaries. Particularly in recreation areas, long-term local inhabitants and their offspring are often priced out of the land market. The possibility of earning large (and often quick) profits from rising land values has attracted urban land speculators and investors. Land speculation may lead to patterns and rates of development that radically alter the appearance of a community. In addition, significant amounts of rural land may be purchased by absentee owners who enjoy an occasional recreational experience or who are holding land as an investment hedge against inflation. As absentee ownership proliferates, the future of a community increasingly resides in the hands of those with little connection or concern for the locale.

State, local, and federal programs have also tended to view land as a commodity for consumption. Federal rural policy has long been dominated by farm policy. But over the past 20 years, federal programs have also sought to promote nonfarm growth through infrastructure development and housing and business loans. Nonetheless, rural federal policy has lacked coherence, and development goals have been couched in highly abstract terms. Little has been said about the current and future use of resources, particularly the land base.

State and local governments have made efforts to attract industry to rural areas, both to provide jobs and to broaden tax bases. Often these efforts have succeeded, but have also caused strains on local services and environmental quality. In the past, local governments existed to provide services that individuals could not provide themselves. But lately, prodded largely by federal regulations and new inhabitants, local governments have begun to develop comprehensive plans concerning new service goals and, in some cases, the regulation of certain land uses. The traditional values of rugged individualism and the assumed freedom to use one's land however one chooses are deep-rooted in rural America. However, a major challenge facing many rural communities involves finding ways to minimize land use conflicts so that the rural economy can maintain a diversity of wealth-creating land uses. In this way, a community will become more economically self-sustaining without sacrificing those amenities that provide a satisfying quality of life.

A Look to the Future

Future growth in rural areas will depend primarily on the price and availability of energy, on labor costs and productivity, and on other resources such as water. Because of dispersed settlement patterns, rural inhabitants rely heavily on transportation for importing and exporting goods and services. Public transportation is very limited, and private cars and trucks are often the only means of travel. Thus, if energy becomes costly and scarce, population growth rates in nonmetropolitan areas could be curtailed. Still, in the 1970s many Americans absorbed sharp increases in the price of oil and occasional gasoline shortages to settle or stay in rural areas.

The quality of labor and its cost will determine how competitive rural labor is with urban labor and overseas workers. Increased productivity through improved training and greater investment in technology will be crucial.

Another major influence on rural growth will be whether rural areas can continue to provide those qualities that have attracted people in the past—namely, clean air and water, open space, and low crime rates. An increase in rural populations will mean a greater impact on public service and natural environments, and portends more intense land uses and greater population densities. For example, water supplies in some areas have been strained because of increased demands, and ultimately the extent of these supplies will act as a limit to growth. In addition, the number of farms and farmers appears certain to decline, and manufacturing faces continuing foreign competition. These prospects raise many questions about the future mix of economic activities and land uses in many communities. Above all, will people be able to earn a decent living in rural areas, or will they be forced to migrate to major urban centers?

Much of the post-World War II development in rural areas has been generously subsidized by taxpayers and utility ratepayers. Income tax deductions for mortgage interest payments; municipal bonding for roads, water, and sewer systems; property tax breaks for new industry; and cheap rates for utility extensions—all have combined to encourage the location of single-family residences, commercial development, and industrial plants in rural America. In the future, the cost of providing public services may rise, and state, local, and federal monies may not be as plentiful. This could result in the curtailment of services and the allocation of a greater share of development costs to developers and newcomers. Both of these outcomes would tend to discourage growth.

Nonetheless, some nonmetropolitan areas may continue to experience substantial population growth, especially those places not far from metropolitan areas. National surveys have consistently shown that most Americans would prefer to live in the country or a small town. The challenge for rural America is to build and maintain a diverse economic base, to mitigate negative environmental impacts, and to accommodate social change. In this way, many of the desirable features of small rural community life can be retained. Planning to determine the location, intensity, type, and timing of development is increasingly needed in order to accommodate growth in growing areas and to help stimulate growth in declining areas.

The Role of the Planner

Planning in rural and small-town America is not difficult to justify. Rather, it is a necessary ingredient in the fostering of sustainable communities, the development of greater self-reliance, and the promotion of economic development and environmental protection.

From a national perspective, rural planning and development can help close the economic gap between rural and urban dwellers. Planning can aid in slowing out-migration from rural areas to already crowded urban centers. In addition, planning can assist in the allocation of rural lands and resources among competing uses. Although the free market is efficient at placing private values on land and resources, public values in terms of health, safety, and welfare may be quite different, especially over the long run. Rural planners can seek only to modify or regulate the free market, not to replace it completely. That is, planning serves to guide the market toward socially desirable results while recognizing the key role of private entrepreneurs in rural economies.

An important initial task for rural planners is to look at rural areas realistically: How are wealth and power created and distributed? Who controls the local economic base? Who owns the land? What are the barriers to economic development or environmental protection? An objective assessment of a place and its people is necessary in order to identify what needs to be done, what can be done, and how the planner can help.

The planner has a diverse and demanding role to play in a rural setting. Belonging to a small planning staff, the planner, whether in a county, regional, or small-town office, will have to become knowledgeable about a variety of issues affecting the planning area. Such issues may include farming, forestry, natural areas, water, toxic waste, economic development, land markets, housing, the elderly, mining, and government programs, among others. This knowledge is essential for drafting and updating comprehensive plans, zoning and subdivision regulations, and capital improvements plans.[2] Furthermore, the planner will become better equipped in the review of development proposals, the search for outside grant funding, and the creation of new planning programs.

Because education is an important part of the planner's job, the more a planner knows, the better educator the planner is likely to be and the more likely it is that the planner will be able to "sell" planning to the local community.

In carrying out planning activities, the planner needs to place day-to-day actions and decisions within a long-term strategy. This is also important for the planner to convey to the planning commission and the local elected governing body. For example, how does a certain development proposal fit within the goals and objectives of the community? What are the likely economic, social, and environment impacts? Is it proposed for a site that can support the development? Will many newcomers be brought in? How many new jobs will be created, how much will be gained in property taxes, where will public funds be spent, and how will the structure of the local economy be affected?

Often such questions do not need to be raised over small development proposals. Still, many small developments can add up to major changes. Perhaps the overriding question that needs to be addressed is this: How will the needs and desires of the community be served by a

[2]For a thorough discussion of drafting comprehensive plans, zoning and subdivision regulations, and capital improvements programs, see Daniels, Keller, and Lapping (1988).

certain proposed development? The planner should provide information and advice that will enable the planning commission and the governing body to weigh the pros and cons of a proposal and to reach a reasonable decision.

There may seem to be little planning work for a number of rural areas, because many rural counties lack a basic comprehensive plan or zoning and subdivision regulations. But the rural planner must be able to demonstrate the value of planning. Economic and social circumstances can change quickly, even in rural areas. Planning is a way to anticipate change and establish contingency strategies for a community. The old rural attitude of "If it ain't broke, don't try to fix it" is outmoded in a global economy. Rural residents have less control over their lives than ever before, and planning is an attempt for them to gain more control and shape change.

A crisis mentality permeates much of rural America today. First, the farm crisis has caused the exodus of thousands of farmers from the land. Nearby towns have lost business and population. Many towns below 2,500 inhabitants are struggling to survive. Other rural industries, particularly mining, energy extraction, and manufacturing, are suffering from global competition.

There is also an environmental crisis focusing on the supply and quality of water, the source of life. Hazardous waste, solid waste, and the development of fragile lands near aquifers—all threaten the future availability of clean water for households, recreation, and industry. In many rural areas, water or the lack of it may soon become a very real limit to growth.

Factors contributing to the social crisis in rural America include the high poverty and unemployment rates of the 1980s, out-migration, and the high proportion of elderly people who live there. There is a need for increased training and job opportunities. Many rural areas, particularly in the Midwest, the Great Plains, and the intermountain West, are losing population. Often it is the young who leave in search of better jobs; in many rural counties more than 15 percent of the population is over 65, compared to 11 percent for the nation as a whole. Elderly people require special services, particularly in health care, and tend to live on fixed incomes. They are often not actively involved in the local labor force, and they may be reluctant to move to a congregate form of housing.

Another side of the social crisis in rural America involves the confrontation of urban and rural attitudes and tension over the control of rural land. Urban dwellers tend to view rural land as providing outdoor recreation and vacation experiences. Furthermore, urban dwellers often have a greater ability and willingness to pay for rural land than do rural dwellers. This means that rural land will first be put to those uses that the highest bidders choose. In short, rural dwellers are generally at a disadvantage in competing for the land base with urban dwellers.

This situation has created no end of friction and resentment. Urbanites have been a major force in bidding up the price of rural land for vacation/recreation uses. Vacation land is absentee-owned and used only occasionally; this may seem something of a waste to local residents. In addition, urban migrants to rural areas tend to be wealthier than local residents and have helped force up the price of land for rural residential uses.

Long-time rural dwellers tend to view land as providing productive services: farming, forestry, mining, and wildlife for hunting. The intrusion of vacation and rural residential uses works against natural resource industries by fragmenting the land base and driving up the price per acre. This can lead to far-reaching changes in the local economic base, as well as to major changes in the landscape and social structure of a community.

Finally, there is a political crisis of leadership and direction. The federal government continues to view agriculture as the dominant force in rural America. But farming is a main economic sector in only a minority of nonmetropolitan counties. Federal farm programs affect only a small proportion of the rural population and have not succeeded in keeping people on the farms. Federal programs do not reflect an awareness of the dispersed settlement patterns in rural America and the ways in which these settlement patterns make economic development difficult. Some of the billions of dollars in farm subsidies could be better spent on programs for developing rural growth centers with the critical mass to be sustainable communities.

Federal aid to rural America will probably be rather limited in the foreseeable future. The huge federal budget deficit is a major reason, along with reductions in domestic spending, exemplified by the end of federal revenue sharing to communities in 1987. The message from Washington is that rural areas are going to have to become more self-reliant in providing public services and generating economic development.

These four crises—farm, environmental, social, and political—simply mean that rural and small-town America will have to become better organized to provide decent housing, jobs, and overall quality of life. Planning is essential for inventorying and organizing resources and for taking action to guide dynamic economic and social circumstances toward desirable outcomes.

There is much a planner can do in rural areas. Education, information gathering and analysis, grant writing, and drafting and updating plans and regulations constitute more than a full-time job. Being useful, not just busy, is of course the goal.

CHAPTER TWO

A History of Federal Rural Planning and Development Programs

In just over 200 years, the United States has been transformed from a very largely rural and wilderness nation into an urban one. Rural Americans have gone from being the first majority to the last minority, to paraphrase historian John Shover (1976). All through this time, policies directly and indirectly oriented to rural areas have been enunciated by the federal government. Rarely has the coordination of policy been sought or achieved, however. Also, for generations, the easy assumption was made that agricultural policy constituted rural policy. Such a gross oversimplification has tended to mask some of the truly substantive problems of poor housing, unemployment, and malnutrition faced by rural populations.

Federal planning policy for rural America has featured four broad themes: the distribution and management of land; the development of human resources and physical infrastructure; financial support for farmers; and the alleviation of poverty in peripheral regions. The first two areas of concern tended to define the policy of the 19th and early 20th centuries, whereas the viability of agriculture and the persistence of rural poverty have been overriding concerns since the 1930s.

Settling the Continent

Federal Policies Promoting Settlement

Until the turn of this century, rural policy was mostly directed toward the distribution of nearly 1 billion acres of land from the federal government into private ownerships. Under a variety of grants, the federal government provided road systems, canals, bridges, and railroad corridors to open up the Western frontier. The goal of such programs was to stimulate population expansion and settlements as rapidly as possible, to complement the land grants made to individuals, groups, and corporations.

Up until the 1860s, the land grant system required the fee simple purchase of acreage from the government and its land agents. In 1862, the Homestead Act changed this by providing a system of free land grants of 160 acres for settlers willing to live on the land for at least five years (ch. 75, 12 Stat. 392). This departure in policy was influenced by a number of factors, especially the Civil War. President Abraham Lincoln sought to curry favor with "Western interests" that stood to gain from increased population settlement. In addition, the long-standing Southern opposition to Western development was rendered politically irrelevant by the war. Though subject to many abuses, the Homestead Act had a profound impact on American life. As land historian Paul Gates has noted, "few laws have so shaped American development, so accelerated its growth, so intimately touched the lives of millions of people" (quoted in Ebeling, 1979, p. 163). Also in 1862, a new series of grants was implemented for the railroads, thus guaranteeing that they would remain a dominant force in the shaping of the West.

The changing landscape of America was, then, directly attributable to federal policies and programs that sought to settle people and create communities; to establish transportation and communications networks; and to develop wealth through agriculture, forestry, and mining (Carney, 1912). The latter activities were stimulated by an elaborate system of irrigation subsidy programs (such as the Carey Act of 1894), mining and mineral laws favoring exploitation, and timber-harvesting supports. But with a tight credit policy based upon the "gold standard" of currency, expensive railroad shipping rates, and a high-tariff policy, agriculture's position was always vulnerable. This fueled much of the agrarian radicalism of the era. And even though conservation laws were later promulgated to protect the lands remaining in the public domain, the clear emphasis of this period's policy was the promotion of settlement and bringing more and more land into active income-generating pursuits. Furthermore, this was a period typified by a convergence between public and private development interests, especially the railroads, the banks, the mining companies, and the new state governments.

The Land Grant Universities

The year 1862 was significant in still other ways for rural America. If much of the emphasis of federal policy was directed toward land settlement and the provision of the physical infrastructure to promote such development, the Morrill Act of 1862 attempted to build up the human resource capital of the frontier societies, as well as the long-established agricultural economies of the East and South (12 Stat. L. 503). The Morrill Act authorized additional federal land sales, with the receipts directed toward the establishment of state colleges and universities where agriculture, the mechanical arts, and home economics instruction would be provided. The universities created under this program have become known as the "land grant universities," and they are among the largest in the nation. As with the legislation to stimulate the expansion of Western land settlement, the land grant program was also aided by the Civil War. Senator Justin Morrill of Vermont had sought to pass such legislation in the 1850s, but his efforts were consistently rebuffed by a powerful Southern congressional leadership opposed to public, and especially federal, support for higher education. Once this opposition ceased to be relevant, congressional passage and presidential support were achieved.

The law granted each state 30,000 acres of land per member of Congress for sale on the open market. The result of this provision was that the more populous states received more land to sell and hence could invest more money in their land grant institutions. Although this worked against the younger and less populous states, such a formula was politically necessary to secure the support of the larger Eastern states for a program that ultimately benefited the agrarian states in a disproportionate way. Only 17 states turned their land grants over to established universities, as New York did, for example, in the case of Cornell. The overwhelming majority chartered new universities with a mission to provide "practical" instruction in agriculture, science, and home economics to the expanding working and middle classes. In 1890, the Southern states took advantage of a new land grant act to create new state universities for black students. Although this perpetuated the "Jim Crowism" and racial segregation of the time, such state supported colleges did provide blacks with some public higher education options, as well as a means to address some concerns of rural blacks.

The land grant system continued to grow in importance as a provider of trained talent for rural America as it matured. In 1887 Congress passed the Hatch Act, which created a system of agricultural experiment stations at each land grant university (unless a state had previously established one elsewhere, as in the case of Connecticut at the private Yale University) (24 Stat. L. 440). The monies to fund the experiment stations came, once again, from receipts of land sales. The purpose of the Hatch Act was to direct university faculty to conduct scientific and technical research on specific agricultural problems. Suffice it to say that a long-standing struggle ensued at nearly every land grant university between those who sought to define a "practical" research agenda geared to solving everyday problems in the countryside and those who, imbued with the German university tradition of academic scientific research, sought to ask more theoretical and philosophical questions in search of pure knowledge. These were not matters solely of academic politics, for they often went right to the heart of the mission of the land grant universities and the agricultural experiment stations. Often these apparently ideological battles became matters of great public debate, with the Grange and the Farm Bureau, important farmer groups, becoming intimately involved in university issues (Carey, 1977; Nordin, 1974). Whole faculties were literally purged as one side or another became dominant. All of this underscored the growing importance of research to, and its role in, the emerging commercial agriculture of the United States.

The Smith–Lever Act of 1914 formally established the cooperative extension service, the third element of the land grant tradition (38 Stat. L. 372). Although the federal government had been supporting some extension activities in many counties, Smith–Lever forged a new partnership between the states and the federal government and brought a far more substantial array of services to rural people through land grant universities. Each state received a sum of money from the federal government, together with a differential grant based upon the proportion of the state's rural population, to be matched by state funding to support a cooperative extension service. Additional funds, from counties, organizations and individuals, could be used to match federal support. An agreement between the U.S. Department of Agriculture (USDA) and the agricultural colleges soon achieved some degree of uniformity and convergence of programming. By 1917, the cooperative extension service began to offer rural Americans a way to communicate their needs to the land grant universities, which, in turn, would execute research or provide instruction to solve those problems. Extension agents became the liaison between government/university support for agricultural activities on the one hand, and for matters of the home, farmstead, and community on the other. The land grant university provided an imaginative and unique structure for a system of research institutions to address real problems; the extension service helped universities understand what people wished to be taught and what problems needed to be researched. Cooperative extension agents also disseminated the results of research throughout the countryside. In this way, information and advanced technology could be transferred to potential users, and instruction, formal and informal, became an everyday occurrence. The land grant university, together with the federal policies that brought people into the countryside, helped to create a rural sector of unparalleled prosperity and vitality, though problems persisted.

The Push for Modernization

Toward the latter part of the 19th century, a great transformation was taking place throughout the United States. The proportion of rural population was decreasing and the cities were growing. Rural incomes were beginning to lag behind urban incomes, and the political power of rural America was beginning to wane. Much of the redundant labor in the rural areas was shifting to the cities, where industrialization and the promise of a better life enticed many. Most of the millions of new immigrants also saw their fortunes in the cities rather than in the hinterland. Still, Americans continued to cling to a populist kind of warmed-over Jeffersonianism, which romanticized life in rural areas and exalted the values of small-town America. This "rural fundamentalism," which attributed a certain near-saintliness to rural people, became a dominant theme of rural policy as well as the moral thrust of the agrarian radicals of the 1890s (Holmes, 1912). But what was stalling the progress of rural America was its failure to modernize, to take advantage of the efficiencies promised by new technologies, to realize its potential through scientific management and reformed institutions, according to some progressive leaders.

A reform movement that sought to speak for rural America and its interests emerged in the early 20th century. Known as the Country Life movement, this widespread social phenomenon came to define much of the debate over rural development in the pre-World War I era. It was rooted in a type of agrarianism that extolled the virtues of farming and manifested a deep suspicion of cities and urban dwellers. Country Lifers understood the impact of emerging technology and science and the greater concentration of population and production. They aimed, however, to secure for rural people an appropriate niche in the unfolding urban/industrial order. Although the Country Life movement took its name from the Country Life Commission initiated by President Theodore Roosevelt in 1908, it existed in Canada as well as the United States. Indeed, the Country Life ideology swept throughout the rural North American landscape as it sought to organize rural people for the great task ahead—modernization of the agricultural and rural sector (Roosevelt, 1917). Only through adopting a modern and rational approach could survival be assured.

Roosevelt, long concerned with natural resource issues and public lands policy, established the Country Life Commission for a number of reasons, not the least of which was his fear that farm depopulation could create vulnerability and insecurity within the American food system. Believing that agriculture was the basis of all economic prosperity, Roosevelt wanted to upgrade rural institutional life and create an efficient food production system. In this way, rural populations would be stabilized, the quality of rural life would be improved, and agriculture would assume its rightful place as the foundation for urban and industrial expansion.

Largely at the urging of Gifford Pinchot, America's premier scientific forester and a member of the USDA's staff, Roosevelt appointed Liberty Hyde Bailey to head the commission. As the dean of the college of agriculture at Cornell University, then America's most important center for agricultural research and extension, Bailey was clearly the foremost agricultural educator in North America. An accomplished plant scientist and an exponent of simplicity in living, Bailey was a prolific writer and speaker (Bowers, 1974). He was very much convinced of the wisdom of evolutionary science, pragmatism in education and politics, and the need to introduce technology into farming so that agriculture could conform to the emerging pattern of modern life.

In nearly every state, Country Life Commissions were formed on the local level to evaluate rural problems. Although the recommendations of the Country Life Commission failed to garner very much support from Congress and William Howard Taft, who had succeeded Roosevelt as president at the time of the commission's final report, the findings were critical to the future of rural policy in North America. Generally speaking, the commission and the movement it spawned saw agriculture in an extremely favorable light. As typical with rural fundamentalism, the Country Lifers ascribed to rural people an honesty, a simplicity, and a moral and emotional tone far superior to urbanites. As Bailey wrote, echoing the sentiments of Thomas Jefferson, farmers were the "fundamental fact of democracy" (1915, p. 131).

But many problems existed within the fabric of rural living that demanded attention. The litany of rural concerns emphasized the drudgery of farm life, the lack of culture and amenities, the poverty and ineffectiveness of country institutional life, and the failure of government policy. Among those institutions singled out for particular criticism and concern was the rural school. The failure of the rural school was often seen as the impetus for the depopulation of rural areas. The schools were primarily criticized for a perceived urban bias in the curriculum, the texts employed, and teaching approaches. Teachers were accused of steering the brightest rural youths to the cities rather than into farming; this "brain drain" seriously eroded the very foundations of rural life. Superintendents of rural schools hired poorly trained teachers, seldom placed emphasis upon continuing professional development, frequently failed to enforce laws related to attendance, and supervision was carried out by local farmers and not by professional educators—a condition that appalled the Country Lifers, who wished to see a greater level of "professionalism" in all areas of rural life, including the schools.

While the quality of professional leadership and teaching came under attack, so too did the very size of the typical country school. It was simply too small, too independent, and too informal to be effective. To the Country Life reformers, the one-room schoolhouse was the very symbol of rural inefficiency and institutional poverty. The answer to these and other related problems was school consolidation, the "best solution to the country school problem yet devised," according to Mabel Carney (1912, p. 187), one of the leaders of the movement. Although consolidation of schools was advocated several times prior to the Country Life movement, this reform became one of its hallmarks. The consolidated school would become the focus of a region's social and culture life. Grading, platooning, curriculum standardization, and the professionalization of teaching and supervision were all supposed advantages of the consolidated school system. Such an approach would almost immediately lead to an improvement in educational attainment and would permit instruction in scientific agriculture, domestic science, and industrial arts—all deemed essential to a modern rural education. Schools would instruct children "to do in a perfect way, the things their fathers and mothers are doing in a imperfect way, in the home, in the shop and on the farm" (Gates, 1912, p. 464), thereby helping them to achieve new levels of productivity and efficiency, the standards by which all policy reforms were to be evaluated.

The issue of the consolidation of rural school districts still remains a hotly debated issue in the present day. On the one hand, costs per pupil are on average higher in rural than in urban districts, so that there is some degree of economic advantage from consolidation. On the other hand, several studies suggest that students in rural schools perform academically as well as, if not better than, their urban counterparts. This implies that some quality in education may be lost through consolidation (Rasmussen, 1985). Rural communities are often very reluctant to give up their local schools to consolidation because of a perceived loss of control and definition. A community without a local school ceases to be a community, many feel.

Though other institutions, such as the rural church, were criticized by the Country Lifers, the movement's agenda stressed education and training as the true levers of social change and the modernization of the countryside. The reformers were absolutely convinced that farmers were not using scientific principles and the technological advances available to them. Thus, the role of Cooperative Extension Services, leadership development, and increased agricultural efficiency were all stressed. Leadership in this area was provided by Seaman Knapp, the nation's foremost advocate of demonstration work and agricultural extension. The Country Life movement stressed the economic benefits of agricultural development; this was clearly consistent with the objectives of the extension service system. Indeed, as one Country Lifer noted, "the small farm of today is similar in its organization to the shop of yesterday, and must surely give way" (Holmes, 1912, p. 523). Larger, highly specialized, and scientifically oriented agricultural units had to be created, and the farmer should come to see himself as both a scientist and a businessman. Theodore Roosevelt (1917, p. 13) argued that food shortages and high prices were the result of the fact that the farmer "still works by methods belonging to the day of the stage coach and the horse canal boat, while every other brain or hard worker in the country has been obliged to shape his methods into more or less conformity to those required by an age of steam and electricity." How then could agriculture, the base upon which all prosperity depended, be permitted to be so retrograde, different, and inefficient? The answer was simple. It could not.

The arguments and recommendations of the Country Lifers did not meet with the overwhelming approval of rural people, even though the movement stimulated thousands of meetings in the countryside, many state-level conferences, and much comment in the rural press of the day. It really remained for World War I to organize North American agriculture along the lines advocated by the Country Life Commission's leadership. As David Danbom (1979) has astutely noted,

> [T]he war, then, was the pivotal occurrence in the industrialization of agriculture. It unnaturally speeded the attainment of agricultural efficiency to the point where depression resulted, which in turn decreed that the gains of the war could not be lost. Not only did the depression [of 1919–1920] assure the cities of cheap food that had been their primary goal all along, it also stimulated the sharp competition in agriculture which assured that it would become still more productive. And finally, then, war accelerated those trends which were breaking down rural school institutions leading to increased realization of the Country Lifer's social goals. (p. 104)

The Country Life movement established both the mentality and the direction of rural development policy during the 20th century. The Country Life paradigm saw technology and modernization as the solutions to rural and agricultural problems. The dependence upon education, research, and extension for rural change further underscored the importance of technology transfer, mass production, and functionalism, all manifestations of the emerging urban industrial pattern of life. The Country Life movement required that farmers and other rural people see that their futures had to be directed toward specialization, efficiency, and "rational" production, as opposed to the model of small-scale, diversified production. Agricultural development was a top priority, and farm programs have continued to dominate federal rural policy, even though farming now provides less than 10 percent of rural employment. The electricity development of the Rural Electrical Administration (REA) and the Tennessee Valley Authority (TVA) are prime examples of this industrial emphasis.

The Country Life movement also stressed the need to consolidate the delivery of human services in rural areas. The emphasis was toward integration with the larger and stronger urban/industrial system, rather than greater cultural autonomy or self-reliance. For instance, highway construction, the central program of the Appalachian Regional Commission (ARC), attempted to make rural areas more accessible. The Economic Development Administration's (EDA's) sewer and water programs for industrial parks are other examples.

The Depression and the New Deal

The prosperity of the 1920s bypassed much of rural America, especially the Deep South, as a depression gripped the agricultural sector. Prices for many farm commodities were low, and it was during this period that the concept of "parity" emerged as a consistent demand of the agricultural community. Parity was based on the buying power of farm commodity prices in 1910, "the golden age of American agriculture," and would secure for farmers an income with a purchasing power equal to that of producers of manufactured goods. Throughout the 1920s, various parity formulas were introduced as part of five separate bills in Congress known as the McNary–Haugen bills. All were either rejected by Congress or were vetoed by President Calvin Coolidge. The concept of parity as embodied in McNary–Haugen re-emerged during the "New Deal" years of President Franklin Delano Roosevelt, however.

In 1929, President Herbert Hoover sought to increase farm prices by establishing the Federal Farm Board. Supported by a $500 million fund, the board made loans to farm cooperatives to reduce surpluses through better storage facilities and new marketing initiatives. By reducing commodity supplies, or withholding some from the market, prices would rise as demand, both domestic and international, increased; so the theory went. But with the advent of the Great Depression, both domestic and international demand actually diminished further, and the prices farmers received for their produce continued on a downward spiral. There simply was very little capital in rural America either for the farmers or for "Main Street" businesses, which depended upon the patronage of the agricultural community.

The impact of the New Deal upon rural America was substantial, if variable. Federal policies and programs were put together with great rapidity and sometimes very little coherence. Some programs were highly experimental, while others were very traditional. Modernization of farm production, infrastructure, and the exploitation of water power were concerns that dominated federal policy and employed hundreds of planners.

At the onset of the Great Depression, 44 percent of the nation's people lived in nonmetropolitan areas, and 22 percent were farmers. Farmers suffered from low crop prices, surplus production, and low incomes. Roosevelt's administration implemented an array of programs with a view to increasing prices farmers received for their crops and livestock, under the broad heading of the Agricultural Adjustment Act of 1933 (7 U.S.C. §§ 601–674). In 1934, additional support was provided to Southern cotton and tobacco farmers by the Bankhead Act (7 U.S.C. §§ 1010–1012) and the Kerr Tobacco Control Acts (7 U.S.C. §§ 751–756), respectively. But by 1936 these supply/management programs were partially withdrawn because of a U.S. Supreme Court decision that found them unconstitutional. Because the *Hoosac Mills* decision did not invalidate the entire market agreement provision of the existing law (U.S. v. Butler, 297 U.S. Reports 1), the Market Agreement Act of 1937 took care to avoid those specific problems of the program that the Supreme Court found objectionable (7 U.S.C. §§ 601–674). An even more sophisticated program emerged with the Agricultural Adjustment Act of 1938, which established a complicated system of market quotas and acreage allotments for specific commodities (52 Stat. L. 31). Although farm prices still fell, these programs managed to prevent further drastic declines in rural incomes and brought about some degree of stability. Additional farm-related programs were put into place during the New Deal years. These included federal crop insurance to reduce the financial loss of crop damage; the Soil Conservation and Domestic Allotment Act, to curtail farming on highly erodible lands (16 U.S.C. §§ 590 *et. seq.*); the Farm Mortgage Corporation, to provide low-cost loans on farm land; the Farm Credit Act (48 U.S. Statutes at Large 31; 1933), to support marketing initiatives as well as the extension of credit to tenant farmers and sharecroppers; and the Farm Security Administration, to establish small cooperatives to help low-income farm families.

Rural infrastructures was enhanced through the programs of the Public Works Administration and the Civilian Conservation Corps. These "make-work" programs provided bridges, roads, public buildings, hospitals, and schools, sewer systems, conservation projects, and parks for hundreds of rural communities. But perhaps no New Deal program changed rural America more than the introduction of electricity through the REA, established in 1935. Fully 90 percent of American farms were without electricity in that year; 15 years later, 90 percent of all American farms had power and light! Under the dynamic leadership of Morris Llewellyn Cooke, the REA made low-cost loans to private utilities to bring power to rural farms, homes, and businesses. Where private utilities failed to provide electricity, Cooke helped to establish many nonprofit cooperatives that brought power and light to even the most remote areas. As New Deal historian William E. Leuchtenburg (1963) has concluded, "no single act of the Roosevelt years changed more directly the way people lived than the President's creation of the Rural Electrification Administration" (p. 157). With electricity, farmers could be even more productive; the healthfulness of foodstuffs would be enhanced through refrigeration and better sanitation; and the quality of life in rural areas would more closely approximate the standards of the cities.

Along with rural electrification, the most famous New Deal rural program was the TVA, which sought to address the problems of an entire multistate rural region historically suffering from chronic unemployment and the out-migration of its youths. Organized along the river basin planning philosophy—wherein planning was adapted to natural watershed boundaries rather than political jurisdictions—TVA became a model for many developing nations in their assaults against rural deprivation (Friedmann & Weaver, 1979).

The roots of the TVA can be found in the World War I era, when the government sought to provide fertilizers for the domestic market by harnessing water power in the lower Tennessee River to produce the necessary nitrates. Roosevelt and his planners, notably Rexford Tugwell, Arthur Morgan, and David Lilienthal, saw in the TVA an institution of far greater potential than merely the production of agricultural fertilizer. In short, the production of electricity under public authority became the catalyst for change in the region. Associated benefits of harnessing the great river for power included flood control, recreational development, and enhanced navigation. By providing the valley with the cheapest industrial or bulk electricity prices in the nation, TVA helped to induce the industrialization of the region. But in doing so, TVA stimulated further population displacement. Whereas labor heretofore had left the region, now migration occurred within the region from the small towns and rural areas into the expanding cities, such as Knoxville. The end results were often one and the same: The cities grew, and the rural areas provided the necessary labor and natural resources for further urban development. TVA, then, actualized "growth pole" development theories, wherein development was concentrated in the larger communities and the rural areas became organized around these development poles or centers.

Opportunity and wealth supposedly "trickled down" from the cities into the hinterland, and in this manner rural incomes should rise and the quality of life should be enhanced. Although this did take place (at least to some degree), the region as a whole remained poor and subject to further environmental degradation and economic dislocation.

The TVA also attempted to upgrade and modernize the agriculture and forestry of the region. It tended to work through the existing framework of rural organizations, such as the land grant universities and cooperative extension. In this way, much of the experimental nature of the organization got sidetracked into more traditional patterns of rural change. Throughout its history TVA remained a target for conservatives, who were opposed both to public power and to the planning ethic that tended to define so much of TVA's programming. In terms of rural America, TVA sustained "modernization" approaches consistent with the Country Life movement's strategy for rural development. Further, it worked to promote urbanization and industrialization as the essential remedies for rural poverty.

Like the TVA, the damming of the Columbia River was one of America's greatest engineering feats and public works projects. In 1930, only 3 million people lived in the Pacific Northwest, and 70 percent lacked electricity (Reisner, 1986). In 1936, the Army Corps of Engineers built the Bonneville Dam, and the federal Bureau of Reclamation erected the world's largest concrete dam, the Grand Coulee. The cheap hydroelectricity from these dams enabled the aluminum and airplane industries to locate in the region, boosted irrigation for wheat production in the deserts of eastern Oregon and Washington, and helped to attract hundreds of thousands of new residents into the region. More dams followed on the Colorado, the Missouri, the Columbia, and throughout the nation. By the 1950s, dam building and irrigation projects involving hundreds of millions of dollars had become a favorite congressional "pork barrel." The rivalry between the Army Corps of Engineers and the Bureau of Reclamation often resulted in expensive projects with marginally productive benefits, significantly negative impact on wildlife, and the flooding of millions of acres.

The TVA was not the only New Deal agency that sought to spread the gospel of planning. The Natural Resources Planning Board (NRPB) undertook a number of regional studies aimed at better understanding the linkage between natural resources and economic development (Clawson, 1981). Like the TVA, it adopted a river basin planning orientation and also sought to relieve rural poverty through changing the countryside. The NRPB was an extremely valuable institution for the dissemination of planning ideas and concepts. And given that the private sector had produced unemployment rates of up to 25 percent of the labor force, the notion of a "planned economy" was not without its attractions. The NRPB served as a forum for discussions on the role of planning in society, though it was far from a neutral forum. The NRPB advocated only limited structural reform of society and the economy and tended to dismiss a rural perspective to regional planning, which had been advanced by Howard Odum and his associates at the University of North Carolina, and by the Regional Planning Association of America during the 1920s and 1930s (Clawson, 1981).

The New Deal also did some further experimenting with development of "new towns," and even with rural land reform through the Resettlement Administration, which later became the Farm Security Administration. The latter attempted to better the lot of tenant farmers and rural blacks in the Deep South (Baldwin, 1968). But once again, as in so many cases, the voices of conservative leadership throughout the nation successfully blunted attempts at aiding the rural underclass. In time the Farm Security Administration became the Farmers Home Administration (FmHA), which had fundamentally different goals and objectives as a lender of last resort for farmers, and as a source of funds for rural water and sewage systems and for small businesses and housing projects.

New Deal programs and policies were ultimately of great significance for rural planning, primarily because they expanded the role of the federal government in rural affairs. First, considerable policy and funds were directed toward increasing rural incomes by manipulating the prices farmers received for their produce. In this the New Deal sought to enlarge the federal government's role through idling land, reducing commodity surpluses, and giving price and income supports to farmers. This approach is very much in evidence in today's multibillion-dollar-a-year farm programs. Many people still believe that farm and rural problems can best be solved by maintaining farmers' incomes. Yet it is important to note that farmers currently make up only 2.4 percent of the nation's population and only 10 percent of all nonmetropolitan inhabitants. Although farm policy continues to dominate rural policy, in fairness it can only be concluded that farmers are being aided at the expense of rural communities.

Second, the New Deal was largely responsible for building a great deal of the physical infrastructure that made rural life better than it had been. Even though the impetus for these investments lay in a larger concern for the massive numbers of the unemployed, the results were important for rural people and places. Of special significance, of course, was the commitment to rural electrification, which altered so many dimensions of rural life in the United States.

Third, the New Deal introduced a serious national commitment to the concept of planning and societal guidance. This philosophy of planning did not treat seriously the concerns of rural areas, however. Rather, regional planning under the New Deal emphasized urbanization and industrialization as positive sources of change. New Deal planners advocated a modernization approach consistent with much of the academic thinking of the age. They stressed the need to rationalize agriculture through bigger farm units utilizing more advanced technologies, better technology transfer systems, the mobility of redundant rural labor into the cities, and the creation of more growth centers throughout the nation by spreading technology and with it, opportunity. With the coming of World War II, everything changed. The war years sidetracked much of the experimentation of the New Deal.

The Federal Role since World War II

The postwar years brought great prosperity to much of rural America. As incomes rose, pent-up domestic demand increased, and many foreign markets generated substantial new demands for agricultural, mining, and forestry products. America was beginning to "feed the world," and rural-based industries were among the first to recover from the war economy. Still, the farm population continued to decline—not only because of alternative employment opportunities in the cities and the emerging suburbs, but also because of greater productivity in agriculture as the result of more and better machinery, hybrids, fertilizers, pesticides, and herbicides.

During the Eisenhower years, new concerns were voiced over the plight of many low-income family farmers who were not enjoying the new prosperity. One of the responses to these problems was the first attempt by a federal government at a coordinated and cohesive rural development policy. The Eisenhower administration formed an interdepartmental Committee on Rural Development headed by True Morse, an undersecretary to Secretary of Agriculture Ezra Taft Benson, a leading Eisenhower advisor. Through Morse's efforts, a highly decentralized approach was adopted wherein state- and county-level rural development committees were established. These groups were charged with examining rural problems in their specific areas; through the efforts of many of these groups, pilot projects were often initiated that promised transferability to other locales. The federal government supported these efforts and also relied upon existing organizations, such as the cooperative extension services and the churches, to implement programs (Traylor, 1988). Perhaps not surprisingly, many of these local studies, including the administration's own *Development of Agriculture's Human Resources* (USDA, 1955) policy statement, strongly indicated that there were too many family farmers and that some attempt at "thinning out the numbers" was justified. The belief was strong that this would increase the economic viability of those farms left in production. Moreover, newly released rural labor either would migrate to the expanding metropolitan areas or would be employed in local industry. A major emphasis of the rural programming during these years was, then, the advocacy of rural nonfarm employment opportunities. Job training and health programs were initiated, and local governments, institutions, and industries were encouraged to support these and other initiatives.

The decentralization of the defense industry and the massive expansion of the federal interstate highway system under Eisenhower also generated change in the countryside. Lucky was the community with an interchange on the interstate! Many towns bypassed by the new transportation network saw their business districts dry up. Competition for these linkages was fierce, because rural people understood that roads of opportunity for some became roads of exit and decline for others. Eisenhower's commitment to rural America was genuine, for he, like so many millions of others, held to deeply cherished notions and myths about rural America. And growing up as he did in the small town of Abilene, Kansas, Eisenhower saw his own triumphs and successes as vindication and support of rural values (Traylor, 1988).

The Kennedy administration, short-lived as it was, attempted to institutionalize the rural nonfarm employment effort which was largely informal during the Eisenhower years. The USDA established an Office of Rural Area Development and brought the existing rural development committees under its authority. New committees were also nurtured, and the vast majority of states were now participating in some form of rural development planning. Although numerous federal agencies had programming responsibilities in rural areas, the USDA really became the lead agency, and a package of new low-interest loan programs administered through the USDA was initiated to support the construction of rural health facilities, recreational areas, industrial parks, schools, and community colleges. The rural focus during the Kennedy years was upon the rural poor. It was during this period that "the other America"—made up of those who lived in poverty and despair—was identified (Harrington, 1962).

If there was a further departure from existing tradition during the Kennedy years, it was the strong desire to help rural people to stay in rural areas as much as possible. It became increasingly obvious to policy planners that inducing the rural unemployed and underemployed to migrate into urban areas simply was not effective. Many of these people left rural poverty only to join the ranks of the unemployed in the urban setting. True, the support and welfare systems available for the urban poor were better, but this was not a solution. It simply transferred the problem from one landscape to another, and led in all probability to greater disaffection and hopelessness. This may have appeared initially as a subtle distinction in policy and thinking; it really was a significant change, however. The cities were no longer seen as huge, absorptive sponges for redundant rural labor. Rather, the key lay in making rural communities more economically diverse, with stronger institutions from which rural populations could draw strength and support. Although this reorientation in policy was important, programming lagged and few concrete results could be seen.

The alleviation of rural poverty became one of the hallmarks of the domestic policy of President Lyndon Johnson's administration. The Great Society could not countenance significant pockets of rural poverty amidst general prosperity. The War on Poverty was initiated under the terms of the Economic Opportunity Act of 1964 (7 U.S.C. § 2701 *et seq.*). The Job Corps was one of the law's major results. Many rural youths who lacked prospects for employment flocked into the corps and were trained under this more recent version of the New Deal's Civilian Conservation Corps. But job skills were not the only deficiency of rural youths. Basic literacy also became a focus of the program, as more and more policy makers came to understand that many rural youths were simply ill-equipped to deal with a changing and ever more complex world. The old Office of Rural Area Development was replaced during these years by the Rural Community Development Service, and an Interagency Task Force on Agriculture and Rural Life was also established. The former attempted, once again, to strengthen and target the federal government's rural development programs; the latter made recommendations specific to the need to upgrade the nutritional status of rural and older Americans, the extension of Medicare to the rural elderly, and even a guaranteed minimum income for all. Although some efforts were made to implement these policy directions, "the escalating costs of the Vietnam War forced the President to recommend sharp reductions in the proposals as part of his drive to cut Federal expenditures," according to USDA historian Wayne Rasmussen (1985, p. 4). It was a great paradox of the Johnson years that the federal government gave sustained attention to the analysis of rural problems, only to be forced to cut back on implemented poverty programs because of the war in Asia. In 1966, for example, the president persisted in addressing rural problems by establishing a National Advisory Commission on Rural Poverty. Its eventual report, *The People Left Behind* (President's National Advisory Committee on Rural Poverty, 1978), is a moving, powerful, and still relevant assessment of rural poverty in the United States. Very importantly, the study established that rural poverty was not the result of just one problem (e.g., chronic unemployment), but rather was a web of issues related to poor health and nutrition; inadequate housing and sanitation services; marginal educational and vocational training programs; the lack of accessibility to services and economic opportunites; and, in many cases, racism, sexism, and age discrimination. Only by addressing all of these matters could rural poverty be conquered.

Other important efforts were undertaken in these years in regional development planning for rural areas. Within the Department of Commerce, EDA was established as a result of the Public Works and Economic Development Act of 1965 (Pub. L. No. 89-136). While also attempting to address the needs of the inner cities, the EDA operated in rural areas where high rates of unemployment and chronic out-migration were typical. Grants for sewer and water lines, roads, and public buildings; grants to Indian reservations to aid Native Americans; and loans to new and expanding rural businesses were favored programs. Furthermore, the EDA offered a relatively new service in the form of "technical assistance." Often, what those in the countryside lacked was expertise in marketing, grantsmanship, infrastructure development, and basic business practices. The technical assistance activities of EDA sought to organize and provide some of this support. Rather than defining its service area along state or substate lines, the EDA innovated by creating multistate planning agencies where counties exhibiting high levels of rural poverty were contiguous. Thus, organizations such as the Ozark Regional Commission, the New England Regional Commission, and the Old West Regional Commission were established in the manner of the TVA and the independent ARC.

The ARC served a 13-state region, though only one, West Virginia, was totally within the ARC service area. ARC was given control and coordination responsibility for an array of federal programs in the areas of health, housing assistance, vocational education, soil conservation, timber development, mineland restoration, water resources, and highway development. Strictly speaking, ARC was not a federal agency per se, but a cooperative federal–state partnership. The commission was composed of the governor of each participating state (or the governor's representative), and a federal cochair who was named by the president. From that point on, ARC was subdivided into regional, state, and local development districts. At the regional level, ARC attempted to assess Appalachia's future role in the national economy, its needs in terms of regional public facilities (e.g., highways and airports), and other matters related to the socioeconomic well-being of the population. At the state level, ARC attempted to identify those specific areas where long-term investments might stimulate growth and development. On the local level, ARC was organized along multicounty lines, and here the goal was to provide some mechanism for citizen participation.

The overriding theme of the ARC, and also of the regional commissions organized by the EDA, was to target investment and to "concentrate in areas where there is the greatest potential for further growth and where the expected return on public dollars will be the greatest" (Appalachian Regional Act, 1965). In practice, ARC and the EDA followed a "growth pole" orientation, which tended to favor urban settlements over the open countryside or small towns. Indeed, evaluations of ARC spending patterns clearly indicate that the vast majority of federal funds were spent in the dominant growth centers of each state that lay within the ARC boundaries. Moreover, some of the largest investments were made in highway construction, since many argued that it was the lack of accessibility to other areas, regions, and urban markets that held Appalachia back.

The Nixon years involved further government activity on the rural front. A Presidential Task Force on Rural Development made a number of recommendations for a national rural policy, and while the Rural Community Development Service was phased out, it re-emerged elsewhere within the USDA. Perhaps the capstone of the Nixon administration's rural agenda was the passage in 1972 of the Rural Development Act (7 U.S.C. §§ 1013 a, 1921–1995). Like many complicated pieces of federal legislation, the act had several parts or titles. Some of these dealt with the expansion of water systems in rural areas; funds for the land grant universities to establish rural development research centers; enhanced support for cooperative extension activities; and additional financial help for conservation projects. By far the most important part of the law dealt with new federal incentives for rural industrialization. Funds were provided for the creation or improvement of industrial parks, loan programs for small business development, and water and pollution control systems for rural industries and communities. The emphasis of the 1972 act was clearly upon job creation in the rural nonfarm sector, especially through the efforts of the FmHA within the USDA. Additional support was provided to maintain the EDA-derived regional commission. Also, the 1972 Rural Development Act directed the Secretary of Agriculture to prepare an annual report to Congress on the progress and problems in rural development. This led, in turn, to the appointment of a new Assistant Secretary of Agriculture for Rural Development, who would have responsibility for this assessment. In this way, rural development policy planning was elevated to a far more important level than in some previous administrations. What had initially been a pilot project was rapidly assuming the proportions of a permanent part of the federal bureaucracy. And in 1972 a congressional "rural caucus" was organized, representing members from both political parties, to further push a rural agenda in the Congress and with the executive branch.

With the election of President Jimmy Carter, Americans selected a leader who celebrated his small-town and rural roots. Coming from a farming background in Plains, Georgia, Carter cast himself as the quintessential "Washington outsider." In his administration, the president issued a comprehensive Rural Development Policy (Carter Administration, 1980). Once again, the law reaffirmed the leadership role of the USDA. The Assistant Secretary for Rural Development was replaced by an Undersecretary for Rural Development. Under the direction of Alex Mercure, a Hispanic American from rural New Mexico, the new post marked an important departure for the USDA. Although it had long been recognized that the USDA was to have the central federal role in rural development, the new law and the new undersecretary guaranteed that, for the first time, rural development would be a major responsibility of the USDA. Until this time, rural development activities were of secondary or peripheral importance to the more traditional farm commodity and loan programs. Under the Carter administration, and through the leadership of Mercure and Secretary of Agriculture Robert Berglund, the USDA really became the U.S. Department of Agriculture *and* Rural Development, even though an official name change did not take place.

The Carter administration emphasized the need for greater coherence and coordination among different agency programs. The federal government has long had the problem that "the right hand does not know what the left is doing" when it comes to rural development policy and programming. The Carter administration sought to end this through intensive interagency conferences and working groups on various rural problems. Also, the administration hoped to achieve a greater level of equity between urban and rural areas in the allotment of federal funds and projects.

The 1980 Rural Development Policy recognized the importance of planning in a "grants economy," as well as the lack of planning expertise in the nation's rural areas to leverage federal funds for important projects and programs. Carter moved to overcome this by supporting, through FmHA and EDA, a system of "circuit-riding" planners who would work for a number of small rural communities and assist them in their efforts to secure adequate and fair treatment from the federal government. Philosophically, the Carter policy continued to rely upon a "growth pole" orientation featuring rural industrial development. In addition, new programs were established in rural housing, rural and alternative energy systems, and more and varied transportation options for rural people. Through the regulation of some industries and direct subsidies to specific firms with a commitment to serve rural markets, the government sought to overcome the problems of distance and scale in transportation and communications.

Finally, the Carter administration argued that the real goal of federal rural development policy and programming should be to build local "community capacity," so that continued reliance upon the federal government and federal resources might not become a genuine dependency. It was ironic, though, that increased federal activity and a renewed federal commitment were seen as the only means by which this reliance could be broken. Carter's government was perhaps attempting to resurrect the image of rural communities and people as truly self-reliant, independent, and fully capable of solving all of their problems. If the history of rural policy indicates any one thing, it is that rural development has always been a function of federal intervention and that the federal government has been the key "actor," along with rural people themselves, in shaping the history of the American countryside. The America the administration sought to create was an America that never truly existed.

Federal rural development policy in the Reagan years has featured a disengagement from the rural scene. The administration has been committed to the deregulation of the banking and trucking industries, as well as to the transfer of some social and economic responsibilities from the federal level to the state and government levels, and to the private and voluntary sectors. Furthermore, the Reagan administration has made relatively greater cuts in fiscal support to rural areas than to cities.

Generally speaking, the Reagan administration has tried to address rural needs and problems through the USDA's Office of Rural Development Policy, created in 1981 under Secretary of Agriculture John Block, himself a wealthy farmer from Illinois. The core of contemporary rural policy can be found in the administration's *Better Country: A Strategy for Rural Development in the 1980's* (USDA, 1983). Four key areas are given emphasis: improving facilities and services by returning authority to state and local areas; assisting local government by reducing regulatory control and enhancing technical assistance; increasing support for rural housing; and stimulating nonfarm job creation in the private sector through the promotion of exports, the creation of rural "enterprise zones," and an increase in rural credit.

The Reagan administration ended the federal revenue-sharing programs that dispersed billions of dollars to rural America each year. In addition, the activities of the U.S. Department of Housing and Urban Development were curtailed, especially the "701" program, which provided funds to support local and regional planning activities. The administration also proposed though unsuccessfully, the elimination of the ARC, the EDA, and the Small Business Administration—all of which have important rural development programs. From 1980 to 1986, for example, the staff of the TVA was cut in half, and other federally funded positions have also been reduced.

The reliance of "enterprise zones," initially implemented in the United Kingdom, is a reflection of the administration's emphasis on nonfarm job creation. While the government is committed to establishing only a small number of such zones as experiments, its reasoning is that many farm families depend upon some form of off-farm employment to make ends meet. If more such employment opportunities are created, it has been argued, more families will be able to stay on their farms during a period of profound economic dislocation for American agriculture. This was one of the findings of a major USDA study entitled *Rural Community and the American Farm* (USDA, 1984). And few would argue that in the middle and late 1980s, American agriculture has been suffering from a level of economic distress unknown since the years of the Great Depression.

The administration was also pinning many of its hopes on the indirect benefits of a reformed fiscal policy for the nation. By bringing down interest rates and lowering the value of the U.S. dollar relative to other currencies, the government argued that credit for farmers would be more readily available at lower rates than in the Carter years and that a cheaper dollar would significantly increase farm exports. Instead, largely because of increased production elsewhere in the world, U.S. farm exports fell to less than $30 billion a year in 1986 from the high of $44 billion in 1981.

Like previous administrations, the Reagan government had allowed farm policy to dominate rural policy. The plunge in farm commodity prices in the 1980s, huge crop surpluses, the decline in farm exports, and thousands of farm bankruptcies forced both the Congress and the executive branch into increasing expenditures on farm programs from $3.6 billion in 1980 to $26 billion in 1987 (U.S. Bureau of the Census, 1986). In 1983, Congress enacted the Reagan Payment-in-Kind (PIK) program to induce farmers to idle cropland in exchange for grain from government storage. The PIK program cost $10.7 billion, and farmers idled a record 78 million acres. But the farm economy continued to worsen. The 1985 Farm Bill reaffirmed a willingness to provide generous subsidies to farmers, even though the administration tried to reduce price supports toward market price levels (Pub. L. No. 99-198, 16 U.S.C. § 3801 *et seq.*). Such a reduction in farm commodity price supports would make American farm products more attractive for export and lessen the cost of federal farm programs.

The 1985 Farm Bill also included a major effort to encourage soil conservation on highly erodible cropland and to restrict the plowing up of fragile grasslands and the filling of wetlands for agricultural uses. The cost of these programs will probably exceed $1 billion a year for ten years—another indication of the perception of Washington politicians that farm programs are the most important rural problems.

A Look Behind and a Look Ahead

History occurs in the past, the present, and the future. The Roman god Janus had the enviable advantage of having two heads, thus giving him both hindsight and foresight. Would that planners were so fortunate! An overview of rural America's past shows that the federal government has played a role in rural development since the founding of the nation. This federal role first aimed at encouraging the settlement of the continent through privatizing government lands. The Great Depression and the response of the New Deal gave legitimacy to expanded federal involvement in modernizing rural economic and social development, according to the themes of the Country Life movement. Yet these federal programs and "free money" often have had the perverse effect of making rural areas even more dependent upon the national government than before, instead of creating more self-sufficient communities. The Reagan years have, in some ways, heralded a return to the privatizing of rural development efforts; thus far, only federal farm programs have retained "sacred cow" status.

The overriding question now is this: "How can rural communities become sustainable places in which to live and work?" The enormous federal budget deficits of the Reagan era suggest that federal subsidies to rural areas will be reduced in the near future. The continued crisis in agriculture suggests that farming alone will not be able to support many small towns, or, indeed, entire regions. The themes of modernization and consolidation championed by the Country Life movement, if followed further, would most likely lead to increased depopulation of the Great Plains, the Corn Belt, and the rural South. Meanwhile, no federal policy has been formulated to address the problems of the rural–urban fringe, where rapid population growth is bringing about significant changes in local and regional economies, societies, and environmental quality.

In the 19th century, America had a vision of the countryside as a vast void to be settled and civilized, even transformed into some bucolic paradise. The Jeffersonian ideal of a nation of small farmers seemed possible; indeed, the small farmers were the primary settlers of the Midwest, the Great Plains, and the West Coast. In the 20th century, the

Jeffersonian vision has lost relevance as agriculture has come to be dominated by a few hundred thousand farms, and farmers are now a small minority of rural residents.

A new vision for rural America has yet to be developed. This is a frustrating situation for rural inhabitants, political leaders, and planners. Visions provide direction for action. There is little disagreement that rural American should offer a favorable environment in which to live and work. But there is considerable uncertainty about how to bring this about. Should the federal government promote "growth poles" and ignore the small towns below 1,000 people? Should counties with fewer than 15,000 inhabitants be consolidated into fewer but larger counties? Should wilderness areas be fully exploited for their natural resources? Should population growth in the rural–urban fringe be discouraged and the countryside preserved for farms, forests, wildlife habitat, and mineral production? What will be the impact on rural America if energy prices rise sharply?

The future of rural America will ultimately depend on the interaction of several groups both within and outside of rural areas. Whether the federal government devises new rural programs or continues to withdraw funding from all but farm programs will be a major factor in shaping rural areas. Whether individual states will target spending programs to rural areas is another. But government programs will be secondary to the ability of the private sector to provide adequate employment, salaries, and profits. The decisions of individuals and corporations will determine how much private investment is made in rural America. At the same time, there are strong urban interests in the countryside, as urbanites like to escape the cities to engage in tourism and recreation as well. The spreading out of population from urban centers into the countrified cities of the rural–urban fringe represents an additional demand on rural land (Doherty, 1984).

Rural America today is the subject of a variety of dreams, demands, and expectations. Choices will have to be made by individuals, businesses, and governments alike. The planner can serve an important role by articulating the choices; predicting the likely consequences of an action; and showing how the planning of land use, infrastructure, human services, and economic base can help bring selected strategies to fruition.

CHAPTER THREE

The Planner and the Rural Community

The planning profession in America originated as a response to urban housing and sanitation needs at the turn of the 20th century. It was only a few years, however, before the "civic betterment" movement spread to rural areas and small towns. The difference between rural and urban planners is not only one of methodology, process, or plan elements. What sets rural planning apart from urban planning are the following: (1) the marked dissimilarity in personal values; (2) the resources available to the planner; (3) the lack of pluralism; and (4) communication networks.

Rural inhabitants are often characterized as conservative and resistant to change and progressive ideas. Rural planners have found that new ideas are sometimes late in arriving to rural areas, but that resistance to these ideas is probably no more prevalent than in urban areas. Rural dwellers, however, expect to have access to rural government. Such access is often difficult in an urban setting. Issues generate various groups seeking access to public decision makers. Urban officials continually face a diversity of groups and opinions. These officials operate under the assumption that in planning issues the access of any one group is limited.

Rural and small-town decision makers, on the other hand, face a deep-seated skepticism about the role of government. The most important value is that local government should be ready available. Nearly any group, ranging from the radical right to liberal progressives, can gain access to officials at regular meetings, on the streets of small towns, at morning coffee, at farm markets, and at their homes or businesses. Public agendas are frequently open, with few time limits placed u, on discussion. Another rural value is that local officials should basically keep to "housekeeping." Rural residents frequently reserve to themselves the right to vote on issues they view as "critical." Often "critical" issues include zoning ordinances, comprehensive plans, levies for economic development, and public services.

Rural and Small-Town Government Structure

Rural local governments vary widely, both in structure and in legal powers, throughout the United States and even within a single state. To the rural planner trying to work with local governments, the variations seem endless. Legally, local governments can be classified as either "municipal corporations" or "quasi-municipal corporations." This distinction is useful for the planner in determining whether or not a particular government is exempt from certain regulations. On the basis of function, local governments may be classified as either "general governments" or "special governments." General governments include counties, towns, and townships; special governments units consist of water districts, school districts, and rural fire districts. Rural planners mainly work with four main governmental units: (1) counties; (2) municipalities; (3) towns and townships; and (4) special districts.

Counties

Counties or county-type governments are the primary political government jurisdictions within 48 states (see Table 3.1). In Louisiana, the primary unit is the parish, which is very similar to a county. Alaska uses boroughs, and these have enough dissimilarities that they should be considered separately by planners.

Currently, the National Association of Counties lists 3,104 county-type governments in the United States, representing about 4 percent of the nation's 80,000 local governments. Counties display tremendous variations in both physical and demographic characteristics. They range in size from 3 square miles to 88,281 square miles (larger than the state of Nebraska). Total population ranges from approximately 114 people to 7 million people,[1] and population density varies from 2 people per 10 square miles to 5,000 people per square mile (Duncombe 1977, pp. 4–5). About 80 million people, or 35 percent of the U.S. population, live outside of incorporated communities and are governed solely by counties. Although counties

[1]Bullfrog County, Nevada, formed in June 1987, has a population of zero inhabitants. The county was created to forestall the federal government from using the area as a radioactive waste dump (Knudson, 1987).

TABLE 3.1. County-Type Governments in the United States: 1978

State[a]	Total	Counties	City–county consolidations	Independent cities
Total, all units	3,104	3,040	25	39
Alabama	67	67	...	...
Alaska	11	8	3	...
Arizona	14	14	...	...
Arkansas	75	75	...	...
California	58	57	1	...
Colorado	63	62	1	...
Delaware	3	3	...	...
District of Columbia	1	...	...	1
Florida	67	66	1	...
Georgia	159	158	1	...
Hawaii	4	3	1	...
Idaho	44	44	...	...
Illinois	102	102	...	...
Indiana	92	91	1	...
Iowa	99	99	...	...
Kansas	105	105	...	...
Kentucky	120	119	1	...
Louisiana	64	62	2	...
Maine	16	16	...	...
Maryland	24	23	...	1
Massachusetts	14	12	2	...
Michigan	83	83	...	...
Minnesota	87	87	...	...
Mississippi	82	82	...	...
Missouri	115	114	...	1
Montana	56	54	2	...
Nebraska	93	93	...	...
Nevada	17	16	1	...
New Hampshire	10	10	...	...
New Jersey	21	21	...	...
New Mexico	32	32	...	...
New York	58	57	1	...
North Carolina	100	100	...	...
North Dakota	53	53	...	...

TABLE 3.1. (Continued)

State[a]	Total	Counties	City–county consolidations	Independent cities
Ohio	88	88	. . .	. . .
Oklahoma	77	77	. . .	. . .
Oregon	36	36	. . .	. . .
Pennsylvania	67	66	1	. . .
South Carolina	46	46	. . .	. . .
South Dakota	64	64	. . .	. . .
Tennessee	95	94	1	. . .
Texas	254	254	. . .	. . .
Utah	29	29	. . .	. . .
Vermont	14	14	. . .	. . .
Virginia	136	95	5	36
Washington	39	39	. . .	. . .
West Virginia	55	55	. . .	. . .
Wisconsin	72	72	. . .	. . .
Wyoming	23	23	. . .	. . .

Note. Leaders (. . .) indicate not applicable. From National Association of Counties, *County Year Book, 1978*. Copyright © National Association of Counties. Reprinted by permission.

[a]Connecticut and Rhode Island do not have organized county governments.

typically invoke a perception of rural, agrarian government (and the majority of counties actually are such governments), in actuality the percentage of urban population ranges from 0 to 100 percent (Duncombe, 1977, p. 9).

Counties can be considered as functional, as nonfunctioning, or as independent cities. Functional counties have a defined geographical jurisdiction, specific powers, and elected offices. There are a number of nonfunctioning counties in the United States, mainly in the New England states. For example, as Table 3.1 indicates, Rhode Island's five counties do not have governing bodies, and Connecticut's eight counties have officially been abolished. In South Dakota, three counties are classified as "unorganized" and are attached to adjoining counties for governmental purposes. More than 20 counties in the United States no longer have separate corporate existence because of consolidation with other local units of government (e.g., the five counties within New York City; the parish of Orleans within New Orleans; or Dade County within Miami, Florida). Finally, there are nearly 50 county-type governments known as "independent cities," mostly in Virginia. These cities are independent of the counties in which they are located, but exercise both county- and city-type powers. Other than in Virginia, where cities may be designated as independent when they reach a population of 5,000, the best-known examples of independent cities are St. Louis, Baltimore, and the District of Columbia (Hallman, 1977, p. 222).

Functioning counties have remained remarkably stable in physical composition. The county seat still serves as the focal point of county government, even though the county seat may not be the largest, most conveniently located, or most efficiently run community in the county. Late 19th-century American folklore is rife with stories of community leaders sneaking to the county seat in the dead of night to make off with courthouse documents, so that they could declare the next day that *their* town was the county seat. Likewise, county boundaries, although they may be changed with ease by the state legislature, remain essentially fixed. Few events in county political life can cause as much uproar as a proposed boundary change. Counties, especially rural ones, often band together against "aggressor" counties seeking boundary changes, under the theory that "If it can happen to them it can happen to us." For example, the rural Midwest has dozens of counties with fewer than 15,000 inhabitants. Yet these counties have staunchly resisted attempts at consolidation into fewer but larger counties that would probably be cheaper to operate.

County governments perform a number of important public functions in the United States: law enforcement; judicial administration; road and bridge construction and maintenance; the supervision of legal documents; and social welfare. Unlike county boundaries, county functions have often expanded in the 20th century. State legislators have tended to grant counties relatively broad powers, with the hope that they would supervise smaller local governments. As towns and cities won greater "home rule" power from state supervision, rural legislators (still in the majority in many state legislatures) gave equal or counterbalancing powers to counties. This explains why many counties were among the first local governments to receive planning and zoning powers, laws that regulate annexation, and a great deal of discretion in economic development projects.

The diverse administrative structures found in American county governments can tax the imagination of even the most adaptable planner. Lines of authority are often confused because of the assortment of elected officials, deputies, supervisors, and appointed officers. Counties may be governed by judges (in Missouri), commissioners, selectmen, supervisors, freeholders, and other elected officials. But the most common form of county government is the "commission." Commissioners, usually three to seven in number, are elected from specific districts for a term of three or four years. Alternatively, sometimes one or more commissioners are elected at large. The commission serves as a legislative and policy-making body as well as an administrative authority. In some commissions, the commissioners sit as a group to perform administrative functions; in other commissions, each commissioner serves as an administrator/executive of individual county departments or functions. Under the unitary commission form, all county departments or functions are centrally administered. Under the fragmented form, one commissioner, for example, serves as the administrator for parks and recreation; another for roads and sanitary landfill; and yet another for planning, zoning, and noxious weed control (Wagner, 1974, p. 19).

A second form of county government is the "single-executive" or the "council-elected executive" government, which features the separation of executive and legislative functions. The county executive (similar to a mayor) is elected at large and often has some veto power over legislative matters. In some county governments, the executive is a member of the board of commissioners; in others, the offices are separate. The board of commissioners serves as a law- and policy-making body, and the elected executive is in charge of the county's administrative departments.

The third form of county government, the "county administrator" type, has become increasingly popular in recent years, especially among larger, urbanizing counties. In the "weak" form of this government, the county administrator is appointed by the board of commissioners, coordinates staff functions (e.g., budgeting or personnel), and may supervise some operational functions. Under the "strong" form of this government, the administrator (usually referred to as a "county manager") is appointed by the board of commissioners to oversee all administrative duties of the county, to hire and dismiss department heads, to propose legislation, and to prepare budgets (Duncombe, 1977, p. 41).

Although the three standard forms of government are available to counties, either under "home rule" provisions or by direct enabling legislation of the state, many counties have evolved a fourth government structure. A number of remote counties have never taken the initiative to reorganize their government structure, and thus have created an administrative vacuum. Since these counties lack formal administrative leadership, the county clerk becomes the chief administrative officer, courthouse manager, personnel director, and budgetary officer. The power of the county clerk in these counties far exceeds that granted to county managers or even elected officials. Though an elected official, the county clerk may well hold the position for life; few state statutes name mandatory retirement ages for elected officials, and many counties in effect have only one political party. Because many county commissions are noted more for their absence than their presence, the clerk may in fact run the entire government operation.

The different forms of county governments are summarized in Table 3.2.

Townships

The township plays an important role in America's local government and political scene. Yet townships defy systematic analysis. Some states have townships and some do not. In some states, townships are granted all powers of local governance; in other states, they appear to have no purpose or power whatsoever. All of the territory of the six New England states (save certain "wildland tracts"), for instance, is divided into townships of about 20,000 acres; each township is the primary unit of local government and performs most traditional county functions. The government structure of the New England townships varies from direct democracy via the town meeting, through a limited form of "representative" town meeting, to a council or council manager (Hallman, 1977, p. 23).

New Jersey provides another example of township government. Here, as in New England, the county has lost its importance and the townships have assumed regional governance under elected boards of "freeholders" or "selectmen." Within the irregularly shaped townships (usually 15–45 square miles) are incorporated towns, with their own governing bodies. Often the township will be composed of small unincorporated rural settlements, islands of urban growth, and unincorporated

TABLE 3.2. County Governments

Governing body	Authority
1. Board of commissioners	Legislation and administration
a. Unitary commission	All county departments/functions
b. Fragmented commission	Single departments/functions for each commissioner
2. Single executive (elected)	Administration
with board of commissioners	Legislation and policy making
3. County administrator (appointed) (weak)	Budgeting, personnel, some functions
with Board of commissioners	Legislation and most administration
County manager (appointed) (strong)	Administration, proposes legislation
with board of commissioners	Legislation
4. County clerk	Administration

"nameplaces" that are virtually indistinguishable from small cities but directly managed by the township board of freeholders.

In the central states, from Ohio to Colorado, townships generally do not perform the traditional functions of the town or county. Many of these townships were formed according to the Northwest Ordinance of 1785. This survey system divided each state into regularly shaped areas of six miles square (36 square miles), which are still referred to as "geographical townships." When the territorial legislatures created counties, each was divided into a number of "political townships," generally following well-defined "geographical township" boundaries. These Midwest townships typically maintain the small local roads, serve as voting districts, and often provide fire protection. They are governed by a clerk or alderman and two other township board members appointed by the county commission. Township budgets are prepared and approved by the county commission, but some autonomy over the administration of local functions is exercised by the township board. Two critical planning authorities often found in the political township are the signing of subdivision plats and the dedication of local roads. In a few states, townships have been given local planning and zoning powers in addition to subdivision duties. In Kansas, for instance, counties are empowered to establish a "zoning board" in any or all townships having a population of 13,000 or more. The zoning board has all the normal functions of a planning commission and is directly responsible to the county commission.

Rural planners working for New England-type townships frequently find structural situations similar to that of any other town. Planners employed by the few political townships having planning and regulatory functions find that a great deal of adjustment is necessary. Two masters must be served: (1) township supervisors and the planning commission; and (2) county commissioners, who have the ultimate authority over township zoning and planning decisions.

Towns and Small Cities

In all states, the predominant form of local government is the small town. The meaning of "small town" is not always clear. Researchers, census statisticians, scholars, and planners frequently use interchangeable (and conflicting) terms such as "rural community," "small town," "small city," and "nonmetropolitan area." Size, density, and population categories used to distinguish the small town from its larger counterparts are equally divergent. Threshold populations of 1,000, 2,500, 5,000, or even 50,000 are often used. In a sense, each population threshold can serve to designate a particular rural community (Seroka, 1986).

Both the style and scale of public planning activities differ sharply between rural communities and "large towns." First, in a small town, planning occurs mainly after dark, when official and quasi-official community meeting are held. Town meetings, city council and planning commission meetings, boards, and general community decision making (e.g., school board functions) are strictly night-oriented. Because of this "after-hours" pattern of planning activity, there is less formality, increased access to community members, and sufficient time to examine agendas. Larger communities, with more trained professionals on the city staff, generally have abandoned a nocturnal style in favor of a pattern where most official planning activities occur during the day.

A second characteristic distinguishing between the small rural town and the large town is the amount of flexibility each has in allocating resources. The difference is striking. Small rural communities have almost no capability to shift the community's human and financial resources from one sector to another, or to withstand unexpected costs. Thus, rural communities may be forced to let certain services deteriorate (e.g., roads or municipal buildings), because they lack short-range capability, or even long-range capability, to shift resources from the operating budget to the capital improvements budget. Larger towns and urban communities, on the other hand, exhibit a much greater ability to shift human and financial resources from one sector to another, to stave off a crisis in capital outlay by long-term debt structuring, and to locate state and federal funds.

Third, small and large communities differ in their cultural and economic diversity and in the quality and quantity of public services. Larger towns often have one of everything, and urban communities several of everything, but it does not follow that small rural communities will have one of everything. In fact, the lack of diversity sets small towns apart. They must continually import goods, as well as nearly all professional services, and must export their population to obtain common cultural and recreational benefits. Likewise, communities are distinguished by their public service levels. Town services typically include fire and police protection; sewer, water, and electrical distribution; trash removal; and library, recreation, and ambulance services. In very small rural communities of 150–300 people, almost none of these services can be offered, and the community must depend upon its township, county, or neighboring community for services. In slightly large rural towns (those in the range of 750–1,000 persons), partial service—usually fire, police, recreation, and shared library—is common. Full-service communities are generally found in the 2,000–2,800 population range. These communities offer many, if not most, of the array of services that commonly appear in towns and small cities of 5,000 to 50,000 people.

Small rural communities, like their rural counties and townships, have an interesting diversity of governmental organizations. In the smallest rural communities (those of 300 and under), there is not much to run. Usually a town laborer, who may double as a town night security officer or a part-time clerk, will serve as the only local government employee. Typically these communities are governed by an at-large commission or council, which meets once or twice a month. Voluntary efforts are often relied upon for basic services, such as garbage collection or snow plowing (Schenker, 1985). Slightly larger communities of up to 1,200 inhabitants use a commission with an employee clerk, or a "weak mayor" form of government where the mayor is a figurehead or paid to perform minor duties. Communities above this range are partial- and full-service rural towns, and nearly always fall into one of two types of rural government: "manager" or "clerk–manager." The town manager is in charge of the day-to-day activities of government, budget preparation, personnel management (which may or may not include the power to hire and fire), and liaison with other governments. The clerk–manager form is unique to small rural governments. The clerk–manager is normally appointed by the governing body; in some areas, however, the clerk's position is an elective office, as on the county level. In many rural communities, the clerk appears to run the entire administration of local government. The clerk serves *de facto* as the "strong mayor," personnel director, financial manager, and official spokesperson for the town council. Unlike town managers, who change jobs frequently, town clerks often retain their jobs until retirement. Office tenures of 30 years and more are not uncommon for town clerks.

Special Districts

A final type of government is the special district. These districts comprise, if school districts are counted, over half of the 80,000 governments in the United States. The main feature of the special district is the provision of a single particular service. Thus, special districts can be described as "unifunctional," whereas towns, townships, and counties are frequently called "multifunctional" in terms of the public services they provide (Hallman, 1977, p. 23).

States have granted special districts unique powers, financial arrangements, and administrative structures. Their jurisdiction may overlap small towns, urban areas, unincorporated areas, entire counties, and even multistate regions. Special districts are usually small operations governed by elected, nonpartisan boards, and have few employees. On the other hand, special districts may be large organizations that require in-depth planning. In several states, notably California, Texas, Pennsylvania, and

Washington, special districts have the authority to provide a number of municipal-type services and are called "community service districts."

Special districts play a vital role in rural areas by providing infrastructure (e.g., water, sewers, roads, and drainage) to small towns and unincorporated areas. Many small towns have become full-service communities only because special districts were first created to deliver needed services. Typically, a special district that provides an infrastructure service has jurisdiction over a large area consisting of several small towns, unincorporated settlements, rural subdivisions, and farms.

Other Sources of Public and Private Employment

Local governments and special districts are not the only sources of employment for planners in rural areas. Although the number of agencies and companies that employ rural planners are too extensive to cover exhaustively, these organizations mainly include regional agencies, utilities, agri-industry, tourism, and health organizations.

Regional agencies are generally extensions of local government or agencies of the state or federal government. Some regional agencies are basically single-purpose planning organizations, such as the Missouri River Basin Planning Commission. Others, commonly called "Councils of Governments" (COGs), are councils of elected officials and appointed citizens. The powers and duties of COGs vary widely from region to region in the United States. Generally, a COG's staff applies for and administers federal grants; provides technical assistance to local governments; and prepares regional plans for health care, energy conservation, or economic development.

Utilities that deliver electricity, telephone, natural gas, and sewer and water service frequently employ small planning staffs to prepare expansion and service plans for rural areas. Utilities make comprehensive projections and estimates in order to determine future service demands and capital needs. Utilities also provide economic development planning assistance to rural areas and small towns. Several utilities—Northern Natural Gas in Nebraska, for example—not only assist small communities in community development activities, but also promote "betterment competition" among small communities by awarding yearly honors for significant accomplishments in planning and development.

Large agribusiness firms often establish rural planning and economic development capabilities to assist rural regions, small communities, and farms. The rationale for doing this is simple. Like the utilities, agri-industries have an interest in developing and strengthening markets in rural areas.

Tourism and recreation continue to be significant economic activities in several rural areas. To date, most rural recreation and tourism planning takes place on a regional or state level. Recently, however, a trend to employ rural planners is developing among large private resort operations, such as ski areas. These private enterprises are becoming more aware of the role they must play in planning and economic development at the local level.

Regional health planning is among the latest planning activities in nonmetropolitan areas. Community health services, which are vital to rural viability, have experienced numerous problems over the past 30 years. Rural hospitals generally have not been able to offer the overall quality of care, range of services, or specialty staffing of their urban counterparts. Even more serious is the fact that rural areas have experienced increasing difficulty in attracting medical personnel. Recently, problems of rural health care have been brought to a crisis level with soaring insurance costs. Health services often financed by communities, private investors, and special districts can no longer afford the high level of medical liability coverage required to run their operations. Large regional hospitals and private health management companies, either through management contracts or through direct purchase, now direct many rural health care services from urban centers. Rural planners on the staffs of regional hospitals and health agencies now play an important role in the coordination of area-wide health management.

Organizing for Rural Planning and Development

Although many private firms and levels of government have planning staffs in rural areas, the counties and small towns are the predominant employers of planners. Often complex organizational relationships exist among counties and small towns. A "circuit-riding" planner serving several small towns is a popular arrangement, or many towns and a county may share the services of one planner. Shared services are a fine idea for cost-sensitive rural areas. On the other hand, service sharing should be approached with some caution and only after written agreements are made. Intergovernmental feuds have often arisen over who shares with whom, how much each shares, and who pays the bills for planning tasks that are often poorly delineated. The planner is often the loser in shared-service situations, especially when conflicts of interest over land use or regulatory matters occur between governments. Planners expecting to succeed under shared-service agreements should take special care to make agreements, financial arrangements, and work expectations clear, and to provide for alternate or neutral planning services when conflicts of interest arise.

There are two important components in organizing or reorganizing a rural county planning office: structure and function. The structural relationship determines where the planning office is situated in county government. The possibilities are numerous. Some county planners, in addition to their normal responsibilities, serve as support staffers to other county departments and thus report to many bosses. Other planning offices are located within the highway department or in the county clerk's office, and some are even located in the noxious weed department! Lines of authority are not direct, and immediate supervisors may take a highly skeptical view of the planning process. Generally, the most satisfactory structure for the rural county planning office is direct responsibility to the county commission or county manager. This relationship allows flexibility for tasks, clearly puts planning on the same staff level as other county departments, and uses the planning office to serve other county departments by minimizing interdepartmental conflicts.

Good working relationships in rural counties can lead to successful planning and development. The key to success is usefulness in a multitude of tasks. Too often, rural planning operations concentrate on land use planning and regulation, with regulation leading the way. Planners in a rural government are too valuable to spend a majority of their efforts on regulation. Balancing the regulation and development efforts of rural planning should be the major planning goal.

Good working relationships are also crucial for creating a successful planning operation in the rural small town. Establishing a satisfactory management relationship with the town council or manager is as necessary as developing a meaningful and useful program for public services. Equally important for the small-town planner is an active communication network. The rural county planner often sits at the center of a communication network, since all area organizations and local governments must come to the county for information, support services, and financial matters. The small-town planner, on the other hand, must build an information and communication network that extends from local decision makers to special districts, other communities, the county, and state and federal governments (Toner, 1979). Networking is one of the most important tasks of planning, but most small towns suffer from severe communication deficiencies.

Task Management

Task management in rural areas and small towns has four important components: planning, regulation, coordination/communication, and service. Although many rural planning programs have strength in performing one or two of these tasks, it is clear that the only way to assure long-term stability and effectiveness is to balance the amount of time and effort spent on each task.

Planning

Rural planning has evolved from "facelift" planning to comprehensive planning. Prior to 1950, only a few rural areas and small towns prepared plans for any purpose other than "civic betterment." The purpose of these plans was twofold: greater attractiveness of the community, and modernization. In other words, rural areas focused upon planning for storefronts, streetscapes, and parks. With the establishment of the U.S. Department of Housing and Urban Development (HUD) in the early 1950s (it became a cabinet-level agency in 1966), rural areas became eligible for monetary assistance (under the Section 701 matching grant provisions of the Housing Act of 1954; 40 U.S.C. § 461 *et seq.*). Plans prepared under HUD auspices and state supervision were more forward-looking for small towns than the older "betterment" plans, but they also began to appear mass-produced and offered little more than a source of data on population and housing. Rural county plans prepared at this time were more oriented toward land use than their small-town counterparts; still, the vast majority of county plans suffered from generalizations and an emphasis on land use regulation.

Much of the negative rural reaction to planning can be traced to this era. Rural areas were treated as if they were urban. They were told to expect growth and to prepare for it. Urban zoning and subdivision techniques were urged upon rural areas to avoid the coming development problems. Plans bore little relationship to rural resources, local realities, or community needs. By the 1960s, environmental and resource planning techniques were used to draft broad-range regional plans. These plans were often not specific enough to aid local decision makers, but did help states, soil conservation districts, and water planners. In the 1970s, more planners and policy makers began to display a greater understanding of the needs and capabilities of rural areas. County and small-town plans gradually began to reflect the appropriate small scale unique to rural settings. But rural planners directed most of their attention toward specific rural attributes, such as the preservation of prime agricultural land.

Rural planning today is comprehensive: It is concerned with natural resource use, economic development, specific and generalized land use, and infrastructure. Rural planners are drafting specific, medium-range plans of about three to eight years, which are designed to facilitate both growth and conservation. Plans now focus upon rural needs and the rural community, and not on rural areas as a holding basin for future urban infill growth; increasingly, planners are becoming aware of the difference between the small-scale development that sustains the rural community and the large developments that fuel urban/suburban growth.

Regulation

The regulation of land use in small towns and rural areas has most often involved three instruments: private property agreements, zoning ordinances, and subdivision control. Private property restrictions usually take the form of easements or covenants (agreements) between adjacent property owners, or owners of lots within a housing development. These restrictions, made between buyers and sellers, are designed to limit the use of property. A current owner, for instance, may restrict future owners from using a manufactured home or establishing a business in a residence. Lot buyers can be compelled to build residences that are in architectural harmony with those of surrounding homes. Often, agreements are made to prohibit the keeping of farm animals.

Property agreements frequently result in bitter conflict in rural areas. One of the major reasons why people move to rural areas is their desire for less interference with the use of their property. Many of these same people arrive in rural areas, only to find that their dream home is too small under minimum size requirements and that their horse is not allowed on their large lot (see Reeves, 1974).

Local zoning and subdivision regulations are the primary means of implementing a town or county plan. Local zoning and subdivision regulations are enacted under the legitimate "police power" of government, for the protection of public health, safety, and welfare.

The function of zoning is to divide land into various use categories, such as residential, commercial, and industrial. Within each zone are various districts that indicate the maximum intensity of particular types of activities. For example, within a residential zone "R," the districts are likely to be listed as "R-1" (single-family residential), "R-2" (two-family residential), and "R-3" (multifamily residential). In manufacturing zones, the districts are typically "I-1" for industrial park, "I-2" for general industrial, and "I-3" for heavy industrial. Within each district, certain requirements or restrictions establish the rules of property use. In residential districts, common examples of district requirements would be height limitations, minimum lot sizes, parking requirements, and yard sizes. In industrial districts, limitations would be placed on noise and odor, the outside storage of materials, and the separation of different industrial uses.

Subdivision regulations are a comprehensive set of guides for physical development. These regulations set standards for subdividing larger tracts of land into smaller lots. They also bear directly on community infrastructure, in that they make sure that all community improvements conform to the same set of standards and the same level of quality. Streets, sidewalks, drainage facilities, and sewers are examples of the types of infrastructure governed by subdivision regulations.

Subdivision regulations also perform another vital function in planning. One of the main goals of these types of regulations is to avoid haphazard and inefficient layout in development patterns. In housing developments, road design is carefully matched to the size and configuration of lots to ensure an efficient traffic system. At the boundaries of different zoning districts, subdivision controls are used to assure that there is an adequate transition between uses by employing green areas or screening barriers.

Local zoning and subdivision regulations are enacted under specific state enabling legislation that permits local government to adopt laws exercising a "police power" over the private use of land. All regulations enacted under the police power must be reasonable and related to a specific and justifiable end having to do with the public health, safety, and welfare. For example, a community may use the police power to prevent the overcrowding of dwelling units. If a community adopts a specific zoning criterion, such as requiring all dwelling units to have a minimum of 900 square feet of habitable area, the regulation will be viewed as reasonable if in fact 900 square feet is a verifiable and realistic standard that will help to prevent overcrowding. The exact square footage requirement per unit will vary by family size. Depending upon the number of persons per dwelling unit, 900 square feet may be too small or too large (the standard 14 by 60 foot manufactured mobile home contains 840 square feet). A minimum requirement of 900 square feet under most circumstances is clearly arbitrary and does not bear a reasonable relationship to the goals of the police power. A more rational approach is to determine the minimum requirements of space and facilities for an individual.

Planning, zoning, and subdivision are supervised by a "planning board" or "planning commission." This body, usually composed of 5–13 members, is appointed by the elected local government officials, hereafter referred to as the "governing body." A planning commission does not give final approval to proposals or make official changes to adopted regulations or ordinances.[2] Nevertheless, it is charged with preparing a

[2]The planning commission is an appointed board and does not make laws. However, in several states, the planning commission is charged with granting final approval to subdivision plats. It should be pointed out that nearly all final subdivision plats convey dedications

comprehensive plan and initiating zoning and subdivision recommendations that will eventually be adopted in ordinance form by the governing body. The planning commission also serves as a public hearing body for all requests for zonings, rezonings, and subdivision plats. All recommendations from the planning commission are passed directly to the governing body for its consideration. If the governing body agrees with the reasoning of the planning commission, an ordinance is passed amending the zoning map or text (generally, the governing bodies of counties adopt resolutions and not ordinances). When the governing body does not accept the recommendations of the planning commission, the matter is returned to the commission for a rehearing with instructions to reconsider certain aspects of the request, or it dies for lack of a majority vote without being returned.

The planning commission functions as the first arena for most local property conflicts. The planning commission affords the opportunity for debate and consideration by conducting a public hearing. In rural areas this debate is often grassroots and free-form, encompassing a wide range of activities and temperaments. Although some issues that are concerned with property rights are mundane and routine, the general rule at rural public hearings is often spirited conflict.

"Boards of adjustment," also called "boards of zoning appeals," serve as another arena in local areas. These boards, also appointed by the governing body, are usually composed of from three to seven members. They are subject to more limitations and have less range of discretion than planning commissions, and are empowered to hear only specific questions. The most widely known types of hearings before the board of adjustment are for variances, conditional uses, and special exceptions/permits.[3]

A variance permits a departure from the strict terms of the zoning district requirements. It is not an instrument used to grant a change in land use, but rather varies space, height, parking, or yard requirements. Conditional uses allow the board of adjustment to use "site review" to examine a specific location to determine whether a particular land use will be compatible with that of neighboring sites. Conditional reviews are used for a particular group of uses that are potentially compatible with

---

of land for utilities or right-of-way easements, and only the governing body has the power to accept dedications of land in the name of the public. Without the final signature of the governing body indicating acceptance of dedications, the plat cannot be filed. In effect, the governing body, not the planning commission, has the final say in the matter.

[3]In several states, boards of adjustment have considerably more power and authority than those described in the text. These boards frequently have an appeals function, permitting them to review actions of the zoning administrator.

their surroundings but may require additional site planning. For instance, a day care center is potentially compatible in a single-family district when this activity is conducted within a dwelling. This facility may be allowed in the "R-1" district under the conditions that a screening fence is erected, adequate off-street parking is provided, a license is obtained from the proper regulatory agency, and operation is permitted only until 8:00 P.M. If the day care center operators are willing to meet these rules of operation, the board of adjustment will issue the conditional use permit.

Special exceptions/permits are issued when unique circumstances accompany the use of property and make it socially, economically, or environmentally desirable to depart from commonly accepted practice. A good example of a special exception would be to allow an additional housekeeping unit on a lot zoned for only a single-family dwelling. "Granny flats," as they are commonly known, afford a social and economic benefit to the community without violating the basic spirit of the zoning ordinance by allowing elderly persons to live close to their families in their own dwelling units. This type of arrangement would not be possible under a strict definition of a "single-family" districting.

Boards of adjustment are "quasi-judicial" hearing bodies with the power to make legally binding decisions. Since they do not amend or change ordinances (or resolutions), which would require the official sanction of the governing body, their decisions are final and only subject to review by the courts.

The final arena, the one of last resort, is that of the courts. These vary from state district or county circuit courts (where property disputes are first heard), or federal district courts (if there is a constitutional question), to state or federal appeals courts (including the U.S. Supreme Court), where decisions of the lower courts are reviewed on appeal. As an arena to settle property disputes, the courts are used much less frequently than is commonly supposed. This avenue is expensive; actions are time-consuming, especially at the appeals level. The process of legal action for zoning and subdivision controversies is a physical and emotional strain on all involved in these conflicts. All too often, the relief being sought has little meaning because of the time delays (amounting almost always to years) involved in a court action and/or appeal.

Regulation often overshadows all other planning tasks and has greatly expanded in recent years. In many rural counties and small towns, adjacent urban growth has resulted in the need for sophisticated regulatory and management devices. Even so, many rural counties in the South and Midwest have no planning or zoning regulations.

The challenge for rural planners is to reach a point of greater maturity and adaptability in their regulatory devices than have urban planners. Simple modification of traditional urban regulations has frequently caused substantial discontent in rural areas. Rural regulatory codes and programs require a considerable amount of flexibility to meet specific needs and values. For instance, traditional zoning, which seeks to separate conflicting land uses and control density, is often too rigid for either small-town or county use. Very substantial changes and extremely creative implementation are needed to compensate for the lack of suitably prepared commercial and industrial sites, the wide diversity in lot sizes, and the mixing of land uses in small towns. Often, the small-town planner as regulator must play the role of facilitator/negotiator in code enforcement, rather than merely upholding property standards.

For example, small-town planners typically find themselves in the position of enforcing codes on the small home-grown industries and start-up businesses in a community. These types of enterprises commonly operate in noncompatible zoning districts, have inadequate parking, and create working environments that are disruptive to neighboring residents. Regulations must be modeled on a *quid pro quo* basis, with considerably more discretion given to the enforcement officer than is normally found in the urban area. Most rural counties have especially complex regulatory duties. Codes must be designed to cover the broad range of lands and uses that are found in the rural community. At one end of the spectrum are working farms and ranches, natural resource and recreation areas, or even heavy industry; at the opposite end are nonincorporated settlements, rural housing, hobby farms, strip commercial developments, and institutions. Bringing this diversity together in areas that may encompass up to thousands of square miles and creating some semblance of order are tasks far beyond the capability of traditional urban land use codes. Rural county planners have been forced to assemble an assortment of planned-unit rural development districts, overlay zones, bonus districts, large-lot agricultural zones, and environmental protection programs. Moreover, each of these regulatory programs must be coordinated with sanitary codes, utility regulations, road improvement programs, and county services.

Code enforcement, as a general rule, is separate from the staff and operational structure of an urban planning office. In the smallest rural planning operations, the planner often plays a dual role as planner and enforcement officer. Even in rural offices where enforcement and planning are performed by two or more persons, it is important for the staff to coordinate the planning and enforcement functions. A popular organizational model is one where the planner directly supervises all code enforcement activities and has the written regulatory power to make reasonable enforcement modifications and adjustments. The model, to be practicable, must be based upon flexible but not arbitrary written regulations and a tightly structured working relationship between the planner and any official code enforcement administrator.

Streamlining is an important task for rural planners dealing with regulatory codes and programs. Part-time code enforcement and planning, infrequent publication of official notices in local newspapers, and bimonthly governing body meetings add significantly to delays for those seeking development permits. Zoning changes, subdivisions plats, and site reviews can easily take 90 days if normal procedures are followed. Streamlining rural regulations to a response time of 15–30 days should be a top priority for rural planners.

Coordination/Communication

Being useful is perhaps the key ingredient to success in rural planning, especially in coordinating government services to the public. Planning operations in general have not notably succeeded in integrating their skills and programs throughout local government. A planning staff, no matter how small, can be useful to other departments within the same government through capital improvements planning, purchasing, grantsmanship, and data analysis.

Capital improvements—the scheduling of major construction projects and the acquisition of costly equipment—is a function of management, budgeting, and planning in urban areas. Rural areas, probably because of their lack of leadership, seldom pull together their middle-range capital improvements projections and acquisitions in any systematic way. Roads and bridges, because they require timing and multiyear expenditures, may in fact be on a capital improvements schedule, but other major equipment acquisitions are seldom scheduled except on a year-to-year basis. Planning can play an essential role in rural government management simply by bringing together anticipated projects and acquisitions on a three- to five-year basis and assigning preliminary cost estimates and project priorities.

The capital planning exercise is useful because it helps county and small-town officials to set an agenda of tasks and projects. Typically, requests for capital improvements for purposes other than roads and bridges (senior citizens' centers, area transportation for the elderly, or recreation facilities) suddenly surface without warning. Officials must then establish new agendas for discussion, recalculate budget demands, and often spend large amounts of time tracing expenditures that have already been committed to existing projects. Establishing and monitoring a capital improvements program can help to keep the agenda on track, to allow adequate time for evaluation and discussion, and, above all, to provide better management for rural government.

Purchasing—a traditional management function in urban government—is generally not a coordinated program in rural areas. Thousands of dollars can be saved annually by the establishment of a cooperative purchasing program, especially on a township, county-wide, or regional basis. A county and its small communities, for instance, can cumulatively develop a great deal of purchasing power for selective items, especially if school and special districts are asked to participate. This will be particularly true for items such as fuel and food. With the aid of no more than a personal computer and spreadsheet software, planners can have a substantial impact on overall expenditures, including frequent purchases, expensive items such as road materials or automobiles, items that must be stockpiled, and some services that can be centrally analyzed and purchased.

Grantsmanship—the process of locating and applying for grant funds—is a common weakness in rural government. Although external sources of funds are not as plentiful as they were during the 1960s and 1970s, sufficient state and federal monies still exist to justify the establishment of an active and aggressive grants program. Rural areas have traditionally bypassed many small but important sources of funds. Among these are low-interest bond mortgage money for first-time home purchasers; watershed planning funds; sanitary sewer and water facilities planning grants; weatherization programs; economic development assistance incentives; and funds for rural education, historic research, and health demonstration projects.

Grantsmanship (i.e., bringing in money) can be a highly visible activity in rural areas and an excellent way for planners to gain favor (Toner, 1979). Very few employees of rural governments, including city and county managers, have had experience in the preparation of grant proposals. The planner can develop useful relationships by providing the necessary research, data collection, and grant proposal writing. A key element of the grants process is the establishment of communication lines with other local governments and state and federal agencies. Rural government officials will often come to view the planner as an important asset because of these liaisons with other governments. A grant-funded program, unlike many programs developed by the rural planning staff, has a livelihood of its own. As politics and politicians change in rural areas, programs that were once popular may fall into disfavor. A successful track record in obtaining grants, however, tends to endure and gives a favorable example of planning's usefulness during periods of political change and budgetary austerity.

The availability of current data and the analysis of data have always been important aids to informed decision making. Rural government has been historically hampered by its inability to collect and analyze data. Planners have made and continue to make important contributions to the decision-making process in rural governments by detecting trends, issuing warnings, and supplying answers to the political questions of "when, how much, and what if." Obtaining data and performing careful analyses of the data are among the most important aspects of being useful in rural government.

In many cases, planners have not exploited opportunities in data collection and analysis in rural government. Usually, readily available census data and local demographic data receive the most scrutiny. However, there are many other important sources of data, especially given the use of computers for data storage and analysis. School systems, for instance, could benefit substantially from a special analysis of population, yearly estimates of property tax rate tables, or even periodic information on economic growth or decline within their boundaries. Much uncertainty could have been avoided in the 1980s if rural planners had collected accurate data on farm foreclosures, or even profiled the characteristics that were likely to lead to family farm failure. Rural planners could have issued warnings about the danger signs in local farm economies.

Interdepartmental cooperation is another important area of data analysis. Rural planners need to cooperate with the head of the road and bridge department, the assessor, the clerk, and the registrar of deeds. Although these officials receive data on a daily basis, seldom are the data analyzed on a systematic yearly basis. Simple spreadsheets and graphics software packages on a personal computer or minicomputer—now available to even the smallest rural governments—can be used to help analyze and plot a wide variety of information. Planners should be ready to assist in issuing annual reports on the business of government and the workload of the staff.

Service

The final item of task management in rural planning is general service. Most planners limit themselves to land use planning and regulation that neither exposes their true capabilities nor impresses the community with their versatility. More than any other task, service can balance the negative images that planning often conveys to the community. Service programs are very innovative and cannot necessarily be copied from one community to the next. They must be carefully selected according to local needs and interests in order to stand out and have beneficial results.

A much desired and appreciated service is the preparation of small publications related to such topics as property acquisition, zoning, subdivision, and appeals of local land use decisions. One approach, taken by a number of counties in Missouri, Kansas, and Oregon, is a pamphlet entitled *Ask First: What You Need to Know Before Buying Property*. This type of publication, written directly for the consumer (and also an excellent guide for the real estate profession), details the process of purchasing property. The potential buyer is taken step by step through items such as purchasing on contract, purchasing contingent to rezoning, how to interpret restrictive covenants, and how to obtain the necessary permits.

Another highly useful publication, which should always be available in the rural planning office, is a layperson's guide to the zoning process. This should explain the various steps, hearings, obligations, and fees that are involved in the zoning process. Information on the zoning process is vital to public communication and participation. The public *must* know in advance what types of materials should be brought to public hearings, what types of questions are normal and appropriate during public hearings, and what the rights of parties in a zoning amendment or rezoning application are; typical examples of land uses in different zoning districts should also be provided, so that the public can view them before a hearing. In a similar publication, the subdivision process should be detailed in a clear, concise, and easy-to-read format.

Another publication should be prepared to assist the public in matters that are brought before the board of zoning appeals (also called the board of adjustment), including applications for variances, special exceptions, conditional uses, and appeals of rulings by the zoning administrator. No other aspect of planning regulation is as confusing to the general public as the appeals or adjustment process. Planners seldom assist clients in filing an appeal on error from the zoning administrator, and often a client is left confused and angry when told to seek the services of an attorney for what is essentially a lay hearing. Variances, part of the adjustment process, are widely misunderstood; it is the planner's duty to explain that this type of adjustment was devised to equalize property rights, rather than to grant added benefits or bonuses to selected property owners.

Consumer information is an important aspect of public participation. The public must have information to participate; a lack of sound information results in reaction and not participation. It is often easy for planners to criticize professions such as engineering and law for their jargon and poor communication with the public. Planning and regulation, however, are confusing processes that touch the everyday life of the community. Technical planning terminology can baffle both the public and other professionals. To promote participation and service, planners in rural areas have a great opportunity to open the rules, work agendas, and procedures of their profession to public understanding.

Historic projects can be important in the local community, as well as a basis for economic development. Most small towns, townships, or counties have historical societies that are willing recipients of the planner's research and communications skills. Places of historic interest can and should be researched and evaluated, both for greater local awareness and for their potential as tourist attractions. Several planning programs have established slide and narrative presentations to present to numerous service and professional organizations. Not only are these presentations educational and meaningful; they also serve as excellent public relations programs that demonstrate the positive benefits of having a *professional* planning staff in a rural area. There is another benefit that goes directly to the heart of good planning practice: No professional can plan for an area unless the planner understands the historic process of settlement and development, cycles of change, and importance of *place* to local inhabitants. Developing an awareness of local history is essential for the planner to become an "insider."

Another service project that combines both historical research and vernacular architecture is a survey of buildings (especially residences), and, ultimately, the publication of a guide to buildings with important links to the past. At one time or another, a significant segment of the local population shows an interest in exploring their links to the past—links that are much more important to rural residents than to suburban or urban residents. Many small-town historical societies have satisfied this interest by providing self-guided tours of their communities. Such a guide is a positive incentive for indigenous tourism and weekend recreation.

Information surveying, or pulse taking, is another useful service in that a planner can provide. Community surveys are especially important for determining local needs. Needs surveys attempt to ascertain the public's opinion of what goods and services, both private and public, are valued more highly than others. Planners, for instance, use a needs survey to gain solid evidence before recommending what types of recreational services should be developed to serve local residents and where these services should be provided. The survey can also help determine what businesses and services local residents desire. The specifics will vary according to public values, the local economy, and regional settings. Activities surveys can provide the planner with information on what certain segments of the population are doing, where they are shopping, and how much they are willing to spend when they shop. Such surveys can be used to target the young, the elderly, families, or any combination of group, age, and gender.

Surveys can be a major benefit to the community. Primarily, they are an aide to decision making or policy making by governmental bodies and advisory commissions. In terms of economic development, they are essential for developing a coherent process of local revitalization. They are also an effective public relations tool. For example, one small town and the county planning staff surveyed movie preferences and habits of a rural population, and presented this to a chain of theaters specializing in automated movie houses. After reviewing the data, the theater chain decided to experiment by developing four small movie houses in selected rural areas. Movies that had proved to be popular in nearby urban theaters were moved out to the rural theaters.

Maps are as vital to a planner as lawbooks are to the legal profession. Urban staffs generally have excellent support in developing maps for their comprehensive plans, land use plans, or zoning ordinances. Rural planners, on the other hand, have discovered that detailed maps are in short supply. Planners are seldom skilled in cartography and are often forced to make use of limited data for rural area maps. Probably the best way to develop maps is by adapting and redrawing maps supplied by the Federal Emergency Management Agency (FEMA) or the U.S. Geological Survey in combination with maps supplied by state and county highway departments. Companies that supply resident directories or abstract maps (sometimes copyrighted) can be a secondary source of information.

The planner's goal should be to develop an atlas of maps that can be purchased by the public and yet is useful to government agencies and private firms. These maps should at least contain historic information, subdivision locations, road conditions, low-water crossings, and important public facilities. Many counties are now finding that they have growing needs for comprehensive maps. Rural fire districts must be coordinated, and emergency services must have explicit locational information. One Kansas county, for example, found a high public and government demand for an atlas of maps that was developed by a student intern over several summers. FEMA maps for preliminary flood analysis were adapted to develop maps on a township basis. Each township was sequentially numbered, and all roads were named using historical settlement titles or the surnames of early settlers. Important architectural resources were noted, and historic sites were detailed on each sheet. Road conditions, graded "good," "fair," and "poor," were indicated by symbols on each road. The public has been the largest consumer of this atlas, which may indicate a greater desire to know "place" than has previously been suspected by planners.

Conclusion

There are no "rural lawyers" or "rural engineers," nor are there "rural architects," so why should there be "rural planners"? One must experience the difference between rural and urban/suburban settlements and values before the answer to this question can be understood. People from small towns and rural areas have different sets of social rituals and types of local governments than do urban dwellers. The scope and process of development have been sufficiently different between urban and rural areas to produce discernibly different ways of planning, times for planning, and reasons for planning. Urban residents must learn to deal with tighter budgets, whereas rural residents have grown up accustomed to fiscal austerity; in fact they may become uneasy when someone suggests that a program is not austere. In urban areas, there are often many interest groups concerned with a planning decision. In rural areas, there are diverse groups but not many. Dealings are conducted on a face-to-face basis and do not lend themselves to secrecy. The "other people" are not unknown residents from distant parts of the city—they are neighbors, and it is easier by far to tell "other people" how to run their affairs than it is to tell neighbors.

Expectations about planners differ from rural to urban areas. Urban planners are expected to be efficient aids to decision making and specialists in one or a few planning subjects within a bureaucratic agency. Rural planners must be useful, available, flexible, and generalist planners. Sometimes they are the only professionals in rural government. Frequently, they are the only persons with graduate degrees within the entire framework of rural government. They must adapt quickly to different roles because there is no generally accepted permanent niche for planning in rural areas, as there is in urban government.

In rural areas, the need for "generalist" planners is clear. Urban planning demands traditional skills of plan making, policy analysis, analytical investigation, and regulation. Rural planning, on the other hand, demands the same type and level of skills plus a high degree of interpersonal relations and inventiveness. Frequently, one person must be capable of performing all, or nearly all, of the tasks traditionally undertaken by an entire planning staff. One planner must in effect be the plan maker, the zoning administrator, the economic development specialist, the advisor to many groups, and the catalyst for public organization and participation. Planners for rural areas must develop activities instead of being assigned tasks. For example, area transportation must be coordinated and emergency preparedness plans must be undertaken without guidance and frequently without sufficient support services from other governmental employees.

In an urban setting, planning is done in the daytime with weekly or bimonthly planning commission meetings. Tasks are oriented within the framework of a normal bureaucratic structure. In rural planning, activities must frequently match working schedules that demand nocturnal patterns. Tasks are oriented within bureaucratic systems, but planners must spend a significant part of their time in mobilizing public and fiscal resources that government may not or cannot readily supply.

Successful rural planning is not the adaptation of traditional planning skills and task management to a smaller scale; the "countrification" of planning is not a workable model. Successful rural planning demands a generalist planner who can combine inventiveness, adaptation, and personality into useful practice.

CHAPTER FOUR

Planning Law and the Regulation of Rural Land and Water

Although land use conflicts, regulation, and litigation are often associated with urban areas, a substantial amount of activity occurs in rural areas. Urban land use disputes often involve the application of a particular control or arguments over the economic use of a certain piece of property. Land use conflicts in rural areas, on the other hand, feature choices of values, lifestyles, and environmental expectations.

This chapter aims to provide rural planners with the legal principles and relevant case law that continue to influence rural planning. Unfortunately, rural land use law cannot be neatly categorized. As Chapter 1 points out, there are different types of rural areas, decisions, and programs, and a wide range of social, economic, and environmental disputes in America's hinterlands. In addition, rural land use laws often differ considerably from state to state. Conflicts over land reflect a diversity of interests and lifestyles. They include the small-towners fighting to retain their certain brand of Americana, as well as second-home owners in rural recreation areas fiercely defending their five-acre fiefdoms. They also include exurbanites, who have found their much-dreamed-of hobby farms beyond the reach of urban and suburban neighbors (see Spectorsky, 1955). And there are the farmers, foresters, and natural resource processors who wish to protect their land from encroachment by residential and commercial developments. Finally, there are many property owners who resent any regulations of the use and potential development of their property.

The Constitution and Land Use Regulation

The legal system, more than any other institution in recent times, has shaped America's policy on land use regulation. The courts hear disputes between property owners and government to determine whether or not the restrictions on the use of private property are reasonable. The courts do this mainly by balancing the economic or social injury borne by the individual against the gain or benefit to the welfare of the entire community. This balancing is accomplished by applying legal standards to individual cases of governmental regulation. Legal standards are expressed as limitations placed upon government's use of the police power. Those limitations are of several types.

First Amendment Communications

The salient First Amendment limitation on government's power to control land deals with the control of communication and the freedom of speech/expression. In rural areas, most regulations that are in contention with First Amendment issues are those pertaining to the regulation of signs, billboards, and advertising devices. Most signs in rural areas are classified as "off-site," which simply means that these signs give directions to a place or business, or advertise a product. "On-site" signs are located on the same property as the place of business. Sign control ordinances regulate the size, placement, light intensity, and quantity of advertising devices. The legality of these aesthetic controls has been firmly established. What is at issue is the different treatment given to on-site as opposed to off-site signs. In other words, why do on-site signs enjoy a greater degree of protected communication than off-site signs? Rural areas traditionally are home to the vast majority of off-site signs, billboards, and other advertising devices. Accompanying these signs has been a small but steady income for landowners who lease the space for these devices.

Fifth Amendment Limitations

The Fifth Amendment to the Constitution prohibits government from taking private property, unless it is done for a public purpose and the owner has been "justly compensated." On its face, the Fifth Amendment appears only to command proper procedure and a fair system of payment when the government takes land under eminent domain. But the Fifth

Amendment also restrains police power regulations. When regulations are applied to private property, sometimes the effect will be so harsh that the owner will be left with property that has very little or no economic value. Although the government has not actually taken title to the property, the use and value are diminished to a point where courts have interpreted the Fifth Amendment to require actual compensation.

Fourteenth Amendment Provisions

The Fourteenth Amendment has two major provisions: "due process" and "equal protection." Within the due process clause is the assumption that property owners have a right to be treated reasonably and fairly when government applies regulations. For instance, in zoning hearings the due process clause means that all hearings are open, that all wishing to speak have an opportunity to do so, and that the decision of the planning commission is based upon a factual record created at the hearing. Equal protection, the second provision of the Fourteenth Amendment, simply means that all property owners should be treated in the same way unless a valid and compelling reason exists for a difference in treatment. For instance, a land use regulation should not be applied to a person or group on the basis of race, color, or national origin.

Other Measures

A number of conflicts arise in land use regulation because they run contrary to specific statutes or are prohibited by state constitutions. In rural settings, these can include anticompetitiveness provisions, failure to follow guidelines for the regulation of agriculture, or simply failing to adhere to established procedures in drafting regulations.

Rural Land Use Litigation

What we term "the battle for the countryside" is not particularly well documented in American land use law. Until recently, most discussions on rural land use law automatically centered around feedlots or the question of whether hobby farms really qualified for agricultural exemptions. In the 1950s, "rural" was seldom mentioned in any meaningful way in land use controversies. But by the 1960s a nationwide movement, albeit generally limited to fast-growing rural counties surrounding metropolitan areas and some resource extraction and recreation areas, began to bring about improvement in rural planning and regulation.

Then, in the 1970s, two important events occurred. First, urban and rural regulatory programs merged in the countryside in an attempt to integrate growth and resource preservation through the use of comprehensive "growth management." The second event was the establishment of comprehensive state and regional programs to regulate the use of rural lands. A number of these programs, such as the growth management systems established in Oregon and Hawaii, in a sense share power with local government. Other programs, such as those established in Vermont and Maine, assumed state or regional control over specific areas of interest.

Like their urban counterparts, rural regulatory programs are not without legal controversy. In fact, the main vehicle involved in the battle for the countryside was and continues to be legal challenges to established and emerging regulations. There are four general categories of rural land use legal cases:

Type 1. This category, for the most part, originates with the exurban commuter and is centered around conflict in the growing small-town and rural recreation areas. These cases are characterized by an attempt to justify exclusionary regulations by invoking the "separateness of the small town" or idyllic images.

Type 2. This group includes cases from rural natural resource areas. For the most part, these cases involve environmental conflicts arising from differing values and land usage.

Type 3. This category is composed of cases dealing with agricultural land preservation.

Type 4. The final category is a collection of cases grouped around the theme of conflict at the "zone of opportunity" at the urban–rural fringe. Many of these cases deal with the intensive nature of farm operations or the commercial nature of many agricultural uses.

Type 1: The Benefits of Rural Life

Preserving the benefits of a way of life in the face of rural in-migration and recreation or second-home area residents is a classic, if not melodramatic, area of litigation in the courts. Techniques applied by rural planners, which came into being shortly after World War II, included large-lot zoning, minimum dwelling size requirements, the exclusion of mobile homes and the establishment of strict subdivision reviews. All of these techniques, basically designed to slow down the rate of in-migration by making it financially more difficult to move to rural areas, and are still very much in evidence today.[1]

The case of *Fischer v. Bedminster* (93 A.2d 378, 1952) became a landmark example of a rural area experiencing exurban migration. Bedminster, a New Jersey township with an area of 26 square miles, prepared a zoning ordinance restricting development to three zoning districts: An "A" residence district required a minimum lot size of one-half acre for residential construction; a "B" residence district required a five-acre minimum lot; and a third district for business required all commercial uses to locate within the "A" districts. Fischer, who owned five-eighths of an acre in the "B" residence district, was denied a building permit because of the inadequate lot size. He challenged the portion of the ordinance that governed lot sizes, contending that this provision was unreasonable and arbitrary and deprived him of his property without just compensation.

The New Jersey Supreme Court, when hearing Fischer's appeal (the lower court ruled in favor of Bedminster), went to great lengths to uphold the lower court's decision and characterized Bedminster as a rural area:

> [I]t is essential to get a clear picture of the defendant township which was founded by Royal charter in 1749. While it has an area of 26 square miles and is therefore more extensive than Newark, the largest city in the State, with its 23.57 square miles, it has a population of only 1,613 in comparison with Newark's 437,857. It is distinctly a rural community with no industry, light or heavy, and with little commercial activity. (93 A.2d at 379)

In 1952 Bedminster was a rolling countryside broken into wooded areas, farms, and country establishments. There were 457 families, four churches, three elementary schools, limited public water, and no public sewerage system. The court noted that "although only 40 miles away from New York, Bedminster is as essentially rural as if it were 400 miles away, as its population of 62 persons per square miles demonstrates" (93 A.2d at 379).

The court was clearly sympathetic with what it termed "rapidly developing urban problems" in rural areas. It noted that foresight would be required to preserve the countryside. Land use planners testified for both Bedminster Township and for the plaintiff, Fischer, but this only

[1]The conclusion should not be drawn that all regulatory controls in rural areas were initially established to slow down in-migrations or to "screen" new residents. Small towns and rural areas are not capable of providing, at least in the short run, infrastructure and services necessary to accommodate new growth. In addition, exurban residents and recreation development have harmed environmental and resources systems in rural areas, and some attempts were made early on to prevent such damages.

succeeded in convincing the court that higher minimum lot requirements could have been reasonably imposed. Speaking to the heart of the question, the court found that "there would appear to be ample justification for the ordinance in preserving the character of the community, maintaining the value of the property therein and devoting the land throughout the township for its most appropriate use (93 A.2d at 381).

The decision in *Fischer v. Bedminster* has been a clear signal to many exurban settlements and rural recreation communities throughout the country that large-lot controls would be condoned by the courts. *Bedminster* was at that time and continues today to be widely quoted as a rural preservation case.[2] In reality, however, this was far from the truth. One of the major underpinnings of the decision was the land throughout the township would be devoted to its most appropriate use by applying a five-acre minimum lot. Five-acre lots, or even 10- or 20-acre lots, are not in themselves techniques for preserving the resources of the countryside. Five-acre lots simply carve up rural areas into larger pieces than the traditional half- or full-acre rural subdivisions. It seems that no one, other than a few commentators who analyzed the case, took note of who owned or lived on the five-acre lots or who in the future would come to live on the five-acre lots. Certainly it was not and would not be the small farm community; its members lived and worked on parcels that averaged 104 acres. It is interesting to ask this: Who was saving what for whom?

Legal analysts and land use planners have roundly criticized the large number of cases flowing from *Bedminster* that tested the reasonableness of using excessive lot size requirements to protect the "benefits of rural living." Critics point to faulty analysis on the part of the courts. Most rural planners have little disagreement with this criticism. There is, however, room for criticism on both sides. The courts do not write zoning ordinances; planners do. A large number of these cases clearly demonstrate a failure to understand the complexity of "rural" areas and small towns. Both the courts and their critics basically believe that "rural atmosphere" or a "pleasant village" automatically applies to any nonurban area.

Certainly one of the best examples of misreading the "rural" concept can be seen in the case of *Belle Terre v. Borass* (416 U.S. 1, 1974). The town of Belle Terre, on Long Island, New York, adopted a zoning

[2]It may be of interest to note that Bedminster Township's zoning ordinance was held invalid because of its "exclusionary" character in 1974. In *Allen–Deane Corp. v. Township of Bedminster* (Nos. L-28061-71 P.W., New Jersey Super. ct., Somerset County, 1979), the new Bedminster ordinance, prepared to correct the exclusionary deficiencies of 1974, was also invalidated. However, Bedminster was allowed to retain its three-acre zoning in a large part of the township on "environmental grounds."

provision that prohibited "three or more unrelated persons from living together" in districts zoned for single family. Justice Douglas, who wrote the majority opinion for the U.S. Supreme Court, idealized the small-town and rural atmosphere of Belle Terre:

> A quiet place where yards are wide, people few, and motor vehicles restricted are legitimate guidelines in a land use project addressed to family needs. . . . It is ample to lay out zones where family values, youth values, and the blessings of quiet seclusion and clean air make the area a sanctuary for people. (416 U.S. at 9)

A fenced beach community with no industry and no commercial district (save a craft shop or two), surrounded by the nation's largest metropolitan region, bears very little relationship to rural America. Similarly, a rolling and wooded area with a "rural atmosphere" in Bedminster is related to the real rural America only insofar as it has a low population density. Rural America is a community of interests—admittedly diverse, but nevertheless a community. Bedminster, Belle Terre, and other similar places are "idyllic" communities. They house urban–suburban commuters, offer many benefits in terms of aesthetics and quietness, and are very expensive places to live. Their resources are pleasantness and charm, not timber, crops, minerals, and livestock. The real rural areas and small towns are working places and often are not necessarily pleasant places. In sum, when the courts describe the rural nature of many areas and communities, they are not looking beyond the idealized surface into how the residents of those areas make a living. Even critics of the courts often mistake exurbia for the true rural hinterland.

A good illustration of this point can be seen in the case of *National Land Investment v. Easttown* (215 A.2d 597, 1965). Easttown Township, located 20 miles east of Philadelphia, has an area of 8.2 square miles devoted almost exclusively to residential use. In 1965, roughly 60 percent of the population resided in an area of about 20% of the township (the northeast quadrant of the township). At issue was the requirement for a four-acre minimum zoning lot for residential in certain areas of the township.[3] Easttown Township justified its use of four-acre minimum lot sizes because of five pressing needs: inadequate sanitary sewer facilities, inadequate roads, preservation of the "open character" of the area, protection of the setting for historic monuments, and the preservation of the "rural character" of the area.

[3]Approximately 1,565 acres, composing about 30 percent of the township's land area, were restricted by the four-acre lot requirement. Another 17 percent of the land area was restricted by a two-acre minimum. About 5 percent of the total population of the township lived in the area zoned for two- and four-acre minimums.

The Pennsylvania Supreme Court did not buy into any of the arguments advanced by Easttown Township. In dealing with the "open character" of the area, an argument urged upon the court most assiduously, they were of the opinion that "[t]he photographic exhibits placed in the records by appellants attest to the fact that this is an area of great beauty containing old homes surrounded by beautiful pasture, farm and woodland. It is a very desirable and attractive place to live" (215 A.2d at 600). On this argument, the court concluded that if the preservation of open spaces was the objective of the township, there are "means by which this can be accomplished" (215 A.2d. at 601). The township could have utilized "cluster" provisions or the purchase of development rights by compensation. On the final argument, "protecting the rural character" of the area, the court found that "[i]f the township were developed on the basis of this [four-acre] zoning, however, it could not be seriously contended that the land would retain its rural character—it would simply be dotted with larger homes (215 A.2d at 601).

Many courts have faced the issue of using the guise of preserving rural benefits to close off idyllic areas to potential residents, but few have acted as effectively as the Pennsylvania Supreme Court in *National Land.* Local governments are required to meet the challenge of population growth without closing their doors to it. The working rural areas have met these challenges in a fashion that is creative and preservational, resulting in programs that retain both rural and residential benefits.

Type 2: Rural Natural Resource Areas

The protection of rural natural resource areas has proceeded in a direction that would clearly have been considered unreasonable and confiscatory in idealized rural communities. Minimum lot sizes and similar controls are only a small part of a resource protection plan. A great many rural areas are also regulated by environmental controls and exclusive use districts, which permit only very limited activity. Lot sizes of 20 or 40 acres are not uncommon today, but they are generally used sparingly, such as to permit widely scattered development on certain designated areas that have a very low carrying capacity for community development. Very large lot sizes, basically those above five acres, are usually allocated on a "sliding-scale" basis rather than fixed as a minimum in all cases. The information in Table 4.1 shows the application of a sliding-scale area allocation for lots in Carroll County, Maryland. The required lot size for the dwelling is one acre, but it is surrounded by a considerably larger tract that does not permit additional dwellings. Another term for this is

TABLE 4.1. Sliding-Scale Area-Based Formula for Carroll County, Maryland

Size of tract (acres)	New lots allowed	Remainder	Total number
Under 6	0	1	1
6–20	1	1	2
20–40	2	1	3
Over 40	1 lot for each 20 acres or part thereof		

New lots created from a tract pursuant to this section may not be further divided for residential purposes.

Note. From *National Agricultural Lands Study: An Inventory of State and Local Programs to Protect Farmlands* (section 13) by the U.S. Department of Agriculture (USDA), Economic Statistics and Cooperative Service, 1981, Washington, DC: U.S. Government Printing Office.

"density zoning," which has the connotation of permitting a fixed number of dwelling units for a specified amount of land.[4]

Environmental controls, which are the underpinning of effective rural land use regulation, help to prevent the destruction of sensitive lands and the misuse of natural resources. Typically, these controls limit alterations to the land or maintain density levels that are consistent with the ability of an area to absorb environmental change (the "carrying capacity"). Environmental controls frequently require infrastructure levels or adequate capital facilities, such as water and sewerage systems, to minimize resource damage.

The question of "reasonableness" or of the confiscatory nature of these rural environmental programs has been examined by the courts. One of the true milestones in case law was *Just v. Marinette County* (210 N.W.2d 761, 1972). This was a Wisconsin case that dealt with a charge of taking property without compensation when an ordinance prohibited the filling of marshy areas on lake shore lots. The natural resources in this case, lakes and wetlands, were deemed critical to the overall well-being of the state's inland water.

The shoreland zoning ordinance of Marinette County, Wisconsin, prohibited a change in the natural character of land within 1,000 feet of a navigable lake without a conditional use permit. One family, the Justs, purchased 36.4 acres of land on a lake for personal and resale purposes.

[4]In Carroll County, density restrictions apply through the areas zoned "agriculture," which constitute 87 percent of the county. Agriculture is the preferred use, and, as such, agricultural operations are permitted at any time and have preference over all other uses. Single- and two-family dwellings are permitted in the agricultural zone. The minimum building lot for each residence is one acre.

Contrary to the shoreland zoning ordinance, the Justs filled in a marshy area to prepare a lot for sale. The Justs challenged the shoreland zoning ordinance on the grounds that all or nearly all beneficial use of their lot had been prohibited by the impact of the regulation. Property values, the Justs said, had been destroyed. The court, however, found as follows:

> [T]his depreciation of value is not based upon the use of the land in its natural state but on what the land would be worth if it could be filled and used for the location of a dwelling. While loss of value is to be considered in determining whether a restriction is a constructive taking, value based upon changing the character of the land at the expense of harm to the public rights is not an essential factor or controlling. (210 N.W.2d at 771)

Just is important because of two guiding principles. First, essential resource lands may be restricted to their "natural uses." Second, sensitive lands, such as the Justs owned, are critical not only because they form a natural aesthetic fabric in rural areas, but also because of the environmental benefits provided to the entire public. On these principles, the court said:

> An owner of land has no absolute and unlimited right to change the essential natural character of his land so as to use it for a purpose for which it was unsuited in its natural state and which injures the rights of others. The exercise of the police power in zoning must be reasonable and we think it is not an unreasonable exercise of that power to prevent harm to the public rights by limiting the use of private property to its natural use. (210 N.W.2d at 771)

That environmental services are provided by certain lands is an important concept in rural land use law. *Sibson v. State* (336 A.2d 239, 1975), a New Hampshire case, advanced this concept even more clearly than the *Just* case. The court, in upholding the validity of a statewide restriction prohibiting fill and development on certain wetlands, found that limiting the use of wetlands to rice growing, study areas, or the gathering of natural products still offered the owner a reasonably beneficial use from the property. The court said, in essence, that if one pays swamp prices, one gets swamp uses:

> Moreover, the rights of the plaintiff in this case do not have the substantial character of a current use. The denial of the permit by the board did not depreciate the value of the marshlands or cause it to become "of practically no pecuniary value." Its value was the same after the denial of the permit as before and it remained as it had been for millenniums. . . . The board has not denied plaintiff's current use of the marsh but prevented a major change in the marsh that plaintiffs seek to make for speculative profit. (336 A.2d at 243)

Type 3: Farmlands and Farm Uses

Many of the natural resource cases focus upon protecting sensitive lands from harmful use or overuse. Outstanding successes such as the *Just* case must be compared with a number of failures in this type of litigation. Farmland cases, on the other hand, have scored some spectacular successes with only minor setbacks. These cases focus upon the preservation of large areas of highly productive farmland for exclusive agricultural use.

The density requirement method, mentioned in connection with Type 1 litigation, has proven to be very successful in the preservation of farmland. In *Wilson & Voss v. McHenry County* (416 N.E.2d 426, 1981), an Illinois appeals court upheld a 160-acre lot requirement. McHenry County had revised its comprehensive plan and land preservation ordinances to prevent encroachment by rural subdivisions and other nonagricultural uses. Prime farmland located in close proximity to existing municipalities was exempted from the lot requirements, to allow for growth and municipal expansion.

In the *McHenry County* case, two subdivisions, one on a 76-acre parcel and the other on 176 acres, sought a zoning change from agricultural to residential for one-acre lot construction. The court upheld the 160-acre lot requirement, even though it recognized that the land was worth substantially more if it were zoned for residential purposes. The court reasoned that the lost development opportunity had to be balanced against the obvious public interest in preserving good farmland.

The *McHenry County* case is not an isolated occurrence. The Idaho Supreme Court, for instance, sanctioned a density protection scheme on lot sizes of 80 acres (*Ada v. Henry*, 668 P.2d 994, 1983). These cases have further strengthened the resolve of rural counties to use density control techniques as the mainstay of the agricultural retention programs. Serious attempts in this area require from 20 to 640 acres.

Another popular method of protecting farmlands is the "exclusive agricultural use district." Implicit in this technique is the notion that productive farming areas should be separate (segregated) from all or nearly all nonfarm uses. Agricultural uses and farms are permitted to operate without any form of control; nonfarm uses are permitted on a maximum density basis, such as two nonfarm uses per 80 acres, or not at all.[5]

[5]Exclusive farm use zoning typically permits all agricultural, horticultural, livestock, and forest product uses and incidental uses (e.g., sawmills, veterinary offices, welding shops). Other limited uses, such as rural residences, are permitted within these districts *only if* the character of the land is such that it cannot yield an economical return in agriculture. See USDA, Economic Statistics and Cooperative Service (1981), Chapter 15.

The courts have upheld exclusive agricultural use districts. In *Chokecherry Hill Estates v. Devel County* (294 N.W.2d 654, 1980), the state appeals court upheld a protection scheme that limited certain designated lands to horticulture and grazing as the most intensive private uses. The developer in this case charged that a taking had occurred since he was not allowed to develop a lakefront farmland site that he had purchased specifically to build a housing subdivision. In the *Chokecherry* case, which is reminiscent of *Sibson v. State*, the court concluded that the plaintiff had paid farmland prices for the land, was actually farming it at the time, and therefore was supporting the county's contention that the best use of the land was for agriculture.

Another type of farmland preservation control was examined by the Utah courts in *Thurston v. Coche County* (626 P.2d 440, 1981). In Coche County, a zoning provision allowed farm operators to establish second homes on their land for their own farm use, or for use by their families or employees. In contrast, persons living on farmland who were not farm operators were required to obtain a conditional use permit if they wished to develop second homes on their land. These permits were, in turn, issued only to achieve certain agricultural goals. The Utah court found that this protection device was rationally related to the legitimate goal of preserving the maximum agricultural use of land.

Type 4: Living in the Urban–Rural Fringe

Type 4 litigation can usually be generalized into two different areas: agricultural definitions and nuisance contests.

Defining Agricultural Uses of Land

In regard to definitions, it is far from clear what constitutes an agricultural use of land. These uses often are not tightly defined by specific acreage, gross income, production, or commercial levels of output. This lack of clarity is not helped by the fact that many zoning ordinances fail to define the key words "agriculture" or "farm." This need for interpretation has required the courts to supply their own explanations. If the term "agriculture" is interpreted broadly by the courts (as it is likely to be in the heavily farm-dependent states), then activities such as animal husbandry, the running of commercial feedlots and hatcheries, feed scales, and the operation of dehydrators are likely to be included. Strictly construed, the general term "agriculture" is typically understood to permit agrifarming or crop raising and probably dairying, but not necessarily other forms of animal husbandry, forestry, or the many forms of industrial and commercial agricultural applications.

In most states, agricultural pursuits enjoy a "favored" status within the zoning enabling legislation; that is, agriculture is exempt from all zoning requirements. Kansas, for instance, exempts all land and buildings used for agricultural purposes: "No determination nor rule nor regulation shall be held to apply to the use of land for agricultural purposes, nor for the erection or maintenance of buildings thereon" (Kan. Stat. Ann. § 19-2921, 1982). This exemption was interpreted to mean the "main dwelling" or farm house by a trial court in Leavenworth County, Kansas. The court found that a definition of agricultural purpose did extend to the farm residence. Leavenworth County appealed the judgement of the trial court. On appeal in *Blauvelt v. County Commissioners* (227 Kan. 110, 1980), the Kansas Supreme Court flatly rejected the notion that a residential building on the farm was "not used for agricultural purposes."

In 1977 the Blauvelts purchased 40 acres of property zoned for "R" rural district. They intended to construct a farmhouse and accessory buildings to conduct farming operations on the property. The Blauvelts were informed that they would not be issued a building permit because they did not meet certain frontage requirements on their "lot." The Blauvelts contended that the Kansas statute cited above exempted their property and all buildings from all zoning rules and building regulations. The county argued that the use of a house is purely a residential purpose.

The Kansas Supreme Court said that a residence used on the farm was an important part of the agricultural tradition. Interesting enough, the court held that agricultural exemptions applied only to what it termed "the traditional family farm" and not to such notions as "hobby farms":

> The tract here involved was obviously an agricultural unit and we are not faced with, and do not here decide, questions involving rental units on suburban properties, small tracts with a vegetable garden and a few chickens and the myriad of other situations that might occur. Our decision is limited to the factual situation before us wherein the 40 acres of admitted agricultural land is occupied by the farmer–owner who intends to live and carry on "agricultural purposes" upon the property.

Defining agriculture by reference to a particular lot size can also be hazardous. A number of counties, in order to circumvent the agricultural definition problem, simply assert that tracts of 20 to 40 acres in size are considered to be used for agricultural purposes. Conversely, any tract of land, regardless of the use, cannot be granted an agricultural exemption if it is smaller than the minimum lot size.

A good example occurred in *County of Lake v. Cushman* (40 Ill. App. 3d 1045, 1976). At issue was whether the county could prohibit a landowner from building a poultry barn on his 3.09-acre lot. The county zoning ordinance in question provided that lots of less than 200,000 square feet in an area zoned for "A" agriculture did not conform to the ordinance. Cushman asserted that this provision was contrary to the Illinois statute prohibiting any country zoning regulation (except for setback) of structures used for agriculture purposes. The court explained the effect of the ordinance, and reason for its validity:

> The only principal permitted use on a nonconforming lot in an "Agriculture" zone is a single family dwelling; in other words despite the clear language of the statute the ordinance attempts to bar any and all agricultural uses of a lot less than 5 acres in size. This cannot be done. . . . The statute clearly provides that there can be no regulation of any land or buildings used for agricultural purposes except that setback lines may be regulated. (40 Ill. App. 3d at 1049)

The second common type of litigation is derived from nuisance suits, especially those that occur in the interaction zone of exurbia and the working farmlands.

The Common Law of Nuisances

The common law of nuisances is an ancient body of law governing the use of private property.[6] The "*nuisance doctrine*" entitles landowners to bring injunctive suits against neighbors who use their property in injurious ways. Offensive noise, dust, odor, and dangerous activities (e.g., blasting), which substantially and continuously impair the use and enjoyment of property, are prime targets for nuisance suits. The adage "First in time, first in right" has not been a successful defense in nuisance suits. Simply because a very intense activity—a large feedlot, for example—was the first to locate in an area does not legally grant it "grandfather status."[7] When nearby property begins to be developed, traditionally one cannot control adjacent land use by being a nuisance. In short, the nuisance doctrine is a sociolegal rule that limits the absolute right to use property as the owner sees fit.

[6]The law of nuisances rests upon the concept embodied in the ancient legal maxim *Sic utere tuo ut alienum non laedus*, meaning, in essence, that every person should "so use his own property as not to injure others."

[7]A "grandfather clause" grants a nonconforming use right; that is, the use does not conform to the present requirements, but it will be allowed to continue, unchanged, throughout its useful life.

Policies and Protections. Working farms and nearby nonfarm residents often do not make good neighbors. Nonfarm residents view farms as part of the "bucolic landscape that so many wish to be part of, but take exception to the very practices that make farming part of the business of agriculture." (Lapping, Penfold, & Macpherson, 1983, p. 485; see also Lapping & Leutwiler, 1987). Contemporary farming practices involve the use of very heavy equipment seasonally running on a 24-hour basis, dust, noxious odor, hazardous chemicals, and very-early-morning activities. Anyone driving past a large cattle or hog operation on a hot July afternoon quickly gives up bucolic rural images—everyone, that is, except the farmer, and to him or her it smells like money.

Rural nonfarm residents, especially subdivision lot owners, can be equally offensive to farm operators. The proliferation of water wells and lower water tables, substantially increased or altered water runoff, domesticated animals that are not properly supervised, and damage to crops and pastures by off-road vehicles are only some of the more common complaints of the farmer.

Right-to-farm laws, in part, represent an effort by the state legislatures to protect the business of agriculture from legal responsibility in nuisance suits. Specifically, these are statutes enacted by state legislatures to give agricultural activities some degree of protection against abatement by judicial enforcement of common-law public or private nuisance principles, or by administrative enforcement of local antinuisance ordinances. An important step in developing right-to-farm laws was the formulation of state policy granting agricultural pursuits a "favored" status. Other than Kansas (Kan. Stat. Ann. § 19-2991, 1982), two pioneering states have made important contributions to this policy development. New York's Agricultural District Law (N.Y. Agric. & Mkts. Law §§ 300–309, McKinney 1971) established the critical importance of agricultural protection and forbade local governments to enact laws or ordinances that would unreasonably restrict or regulate farm structures or farming practices (see Hexem, 1980; see also Collins, 1982).

A North Carolina statute (N.C. Gen. Stat. § 106-700 to 701, Supp. 1983), which is a modification of a 1915 Alabama law considered to be a model for 32 other right-to-farm laws, states: "It is the declared policy of the State to protect and encourage the development and improvement of its agricultural land for the production of food and other agricultural products (§ 106-700).

The Scope of the Right-to-Farm Statutes: Clarifications. An important aspect of the right-to-farm laws is the range of coverage they afford. Illinois, for instance, affords protection to "farms" (Ill. Rev. Stat. Ch. 5-1101, 1981), while in Kentucky "agricultural operations" are protected

(Ky. Rev. Stat. Ann. § 413.072, Bobbs-Merrill Supp. 1982). A Kansas statute covers "agricultural activities" (Kan. Stat. Ann. § 2-3201, 1982), but there is also a separate law affording protection to "commercial feedlot operations (Kan. Stat. § 45-1505, 1982). Table 4.2 lists the type of coverage for selected right-to-farm statutes.

The degree of protection afforded to various agricultural activities is obviously important. Operating farms have a vital dependency upon commercial and industrial agricultural supports. This indivisible chain of processing, storage, supply, sales, and transportation is often the target of nuisance suits. Generally speaking, those states that grant protection to agricultural operations have gone a step beyond the farming district protection by extending statutory coverage to the vitally important "middle man" and the commercial operator involved in the business of agriculture. Several states have singled out special commercial operations. The Iowa livestock feedlot law gives protection to feedlot operators from nonfarm neighbors. The Kansas agricultural protection statute, in addition to protecting feedlot operators, specifically includes commercial processing, poultry, and hatchery operations.

Conditions and Exclusions. Although many of the right-to-farm laws grant extensive protection to agricultural operations, they do not give such operations a *carte blanche* to act as they please, nor were they intended to do so. Four major factors are not protected under the right-

TABLE 4.2. Type of Agricultural Protection Granted by Right-to-Farm Legislation in Selected States

Farms	Farms–Farm Operations	Commercial Farms	AG/Operations	AG/Activities
Illinois	Michigan	New Jersey	Idaho	Kansas
Oregon			Indiana	Oklahoma
Michigan			Kentucky	Wisconsin
			Maryland	
			Mississippi	
			Missouri	
			New Hampshire	
			New Mexico	
			North Dakota	
			Rhode Island	
			Virginia	

Note. From "Right to Farm Laws: Do They Resolve Land Use Conflicts?" by M. B. Lapping, G. Penfold, and S. Macpherson, 1983, *Journal Soil and Water Conservation, 38*, 466. AG, agriculture. Reprinted by permission.

to-farm laws. The first is the matter of trespass. Under current legal theory, a "nuisance" involves interference with one's use and enjoyment of land, whereas "trespass" involves invasion of one's interest in the exclusive possession of land (see Bradbury, 1983). Since some jurisdictions have rendered judgments that accord dust, noise, and odors (traditional nuisances) trespass status, it would seem that a number of the right-to-farm laws may be rendered ineffective.

Second, the majority of the right-to-farm statutes only extend protective coverage to *pre-existing* agricultural operations—that is, agricultural activities operating prior to the arrival of nonfarm neighbors. Typically, the "first-in-time" requirements of the statutes is one year. New operations are forced to start up at their own risk. Likewise, substantial changes or applications of new technologies to existing operations are not afforded protection.

Third, nearly all of the statutes make nuisance immunity conditional upon safe and healthful operation and upon avoidance of negligent or improper practices. There are many instances where normal operations of a farm or commercial process involve safety and health considerations. Water well pollution from feedlot runoff is one example; an infestation of flies is another common health-related complaint. Negligent operation or improper practices may result from a failure to take reasonable precautions to protect others against an interfering harm.

Finally, several states grant a presumption of reasonableness if "good farming practices" are followed. "Good farming practices" are those that conform to federal, state, and local laws. An example of failing to follow a law governing farm usage would be noncompliance with sanitary regulations for feedlot operations (Kan. Stat. Ann. § 45-1505, 1982).

Legal Issues of Right-to-Farm Laws. Right-to-farm laws were for the most part enacted in a brief period of time between 1978 and 1984. Most follow the North Carolina statute so closely that there is little doubt that state legislative research staffs gave only minimal thought to the eventual legal consequences. Nevertheless, the scant litigation that has taken place has been generally favorable to agriculture.

In a Connecticut case (*DeCapu v. Cella*, No. 19-85-59, slip op., New Haven Super. Ct., 1982), the district court denied a request for a temporary injunction against a neighboring dairy operation. The 150-cow dairy farm spread manure during the summer on a small field adjoining a housing development; the manure could not be spread on more remote fields during the growing season. The plaintiff, a resident of the adjoining development, claimed that he was affected by the odor of the manure and the flies it attracted. The court rejected the nuisance suit on two levels.

First, the New Haven Superior Court did not find a nuisance in the actions of the dairy farmer. The loss or damage to the resident of the housing development was not substantial, and the odor was not continuous—both important ingredients in establishing a nuisance. Second, when the court addressed the Connecticut right-to-farm provision, it found that the "first-in-time" provision applied to the dairy operation, since it had been in operation prior to the housing development. There was no evidence of neglect, negligence, or mismanagement. In fact, the court found that Cella, the dairy farmer, was using and "promoting" generally accepted agricultural practices and could not be faulted for any inconvenience to his residential neighbors.

A Michigan court failed to sustain a constitutional attack against that state's right-to-farm law (*Rowe v. Walker*, No. 81-228769, slip op., Mich. App., 1982). A neighbor of a grain-drying facility filed suit and contended that the word "farm" was overly broad and did not apply to commercial processing of farm products. The suit complained that Walker, the owner of the processing facility, ran gas heaters and unloaded large trucks during all days of the week and late into the evenings. The court found that the Michigan right-to-farm statute (Mich. Comp. L. §§286.471, 1980) specifically included the words "lands, buildings and machinery used in commercial production of farm products." Since the dehydrator was located in an agricultural area, was first in time, and operated according to accepted agricultural and management practices, the court found that it could not be considered a nuisance.

In a landmark case (*Shatto v. McNulty*) the Indiana Court of Appeals upheld that state's right-to-farm law (509 NE 2nd 897 (IN 1987).[8] In this case, a hog production operation was located in a rural area, zoned for agriculture, where similar hog farms existed. McNulty had been raising hogs on his 114-acre farm since 1956. In 1970, the Shattos constructed a home on 15 acres directly across the road from the hog pens. The court ruled in favor of McNulty's operation and cited the Indiana right-to-farm law (Ind. Code Ann. Stat. 34.1–52.4, Burns, 1982), which protects pre-existing agriculture uses employing proper farm management practices. The appeals court noted: "People may move to an established agricultural area and then maintain an action for nuisance against farmers, because their senses are offended by ordinary smells and activities which accompany agricultural pursuits" (509 NE 2nd 897, IN 1987).

One interesting aspect of the case is that the Shattos' purchase of

[8]The *Shatto* case is very important because it is on the appellant court level. Currently, 47 states have some type of right-to-farm laws, but they remain unchallenged beyond the local courts.

their land occurred before the passage of the Indiana right-to-farm law (1981). The Shattos maintained that the law could not possibly apply to them, because their home was built and the nuisance occurred prior to 1981. The Indiana Court of Appeals did not accept this argument, and in fact ruled that the statute applied retroactively "to any premises with a history of agricultural activities" (509 NE 2nd 897, In 1987).

A number of the right-to-farm laws may raise as many legal issues as the state legislatures originally sought to solve. By far the most serious drawback of these laws is the arguable conflict with the Fifth and Fourteenth Amendments. All of the statutes can be loosely categorized as legislative authorization to interfere with the property rights of others. This is especially true of some of the more strict statutes that actually prohibit the bringing of nuisance suits, or those that forbid recovery of money for damages (e.g., see N.H. Rev. Stat. Ann. §430c, Supp. 1981). Under property theories developed over many centuries in the United States and Great Britain, an implied covenant is attached to property: a promise of quiet enjoyment, and a right to be free from harmful effects that would diminish that enjoyment. The Fifth Amendment *may* serve as a guarantee that when the government sanctions a substantial interference with this enjoyment, there must be compensation. The courts, regardless of the *Shatto* decision, may not have had the last say on the matter of legislative authorization for this partial taking.

The right-to-farm laws serve a vicarious need of rural and farm advocates to tell nonfarm residents: "You paid rural prices, you wanted rural taxes and privacy, and now you are going to have to make do with farm neighbors. Welcome to the countryside!" There is much justice in this. After all, there are few complaints about farmers invading the towns and establishing themselves on vacant community land. However, there is the matter of the Fourteenth Amendment and its requirements of due process and equal protection. There is an overriding demand in the Fourteenth Amendment for *fair play*. Most observers feel that the effects of the right-to-farm laws will be found to be "fair," but that they will also be adjudged to be very "rough."

Probably the most equitable method of dealing with the farm and nonfarm residents would be to avoid allowing the nuisance situation to arise in the first place. Who invited the "country gentlefolk" into the rural areas? It is not fair to lay the entire blame on the farm community: Many land fragments and smaller parcels have been traditionally owned by nonfarm residents who trade or sell them for speculative purposes. Even in light of the fact that many would claim that there is a coequal "right not to farm," this does not belie the feeling that much of the blame must rest squarely on the shoulders of the farm community. It was, after all, the farm community who initiated the gentrification of the countryside, and eventually it must be—right-to-farm laws aside—the farm community who must come to grips with the fundamental problems. Currently, only the state of Washington seeks to prevent the establishment of rural subdivisions that may initiate nuisance disputes. The Washington statute requires that farm operators forfeit the right to qualify for protection when land contiguous to the farm has been sold for residential purposes (Wash. Rev. Code Ann. §§ 814.04, 823.08, West, 1982–1983).

Water as a Rural Resource

Rural land development issues often depend less upon land use concerns than they do upon matters of water quality and quantity. This is especially true in the western United States, where water insufficiency—whether in supply or quality—is often a key variable in community or regional decision making. In both the eastern and western sections of the United States, many public and private development projects are simply not possible because of the lack of dependable water supplies. For instance, the intensive development of agricultural systems in many Western states has been seriously threatened by the lack of sufficient and quality water for crop irrigation.

Variables other than supply and quality may also affect water resources. The removal of shoreline vegetation, for example, may create an erosion problem that will result in the siltation of rivers, lakes, and streams. A large manufacturing concern, such as a coal-fired electrical generation plant, may necessitate the damming of a river, causing a disruption of fish and wildlife habitats and the loss of prime agricultural land. Even alternatives to water impoundment, such as diversion, can have far-reaching consequences, as would be the case in thermal pollution of the water from nuclear power generation. Lakes and underground water supplies, even those considered to be adequate for most purposes, can be slowly polluted by the overuse or misuse of septic tanks and leach fields, or by excessive or careless pesticide applications.

The relationship between water law and land use planning has not been clearly defined. One of the foremost scholars of land use law, Donald Hagman, pointed out the importance of this problem when he wrote that the term "land" is not meant to suggest that other real property is not included within land use planning. Most activities take place on "land," but planning for water activities is as closely related to land use planning as water is related to land (Hagman, 1980). Clifford Davis, a noted water law expert, is even more explicit:

> It is clear that decisions about water resources development including water quality, will affect land use and *vice versa*. . . . If a dam is built it will inundate the land, which cannot be used for other purposes like grazing or homesites, but it may provide for boating opportunities and this will create a need for land development like access roads, motels and gas stations. On the other hand, if a zoning policy is changed, high density developments in a floodplain will require expensive flood prevention constructions and maybe an expansion of the water delivery system including new dams or wells. Some people argue that decisions about land development should precede decisions about water development. However[,] the reciprocal effect of one on the other should provide for its coordination regardless of which decision comes first. (Davis, 1978, p. 657.5)

Just as land use law both informs and guides decisions, so water law aides in determining the fair allocation of water resources among competing uses. Water is usually classified into either "surface" or "subsurface" types. Both types can be further categorized into five classes:

1. *Natural surface water*: The navigable waters of lakes, rivers, streams and other natural bodies of water.
2. *Diffused surface water*: The water that is dispersed over the ground as a result of rain or melting snow (runoff) and which flows in other than natural water courses.
3. *Underground streams*: Water flowing in well-defined, traceable underground channels.
4. *Groundwater*: Water that seeps or filters through underground porous beds of earth or rock without definite channels.
5. *Springs*: Natural discharge points of groundwater or underground streams.

Riparian and Appropriation Doctrines

Two major legal doctrines have emerged to deal with surface waters in the United States: the "riparian doctrine" and "appropriation rule." The riparian doctrine was adopted in the humid Eastern states, whereas the appropriation rule, in differing forms, applies in the more arid Western states. The key distinctions between these two doctrines are noted in Table 4.3.

Riparian principles grant an owner of land adjacent to a stream or river a "usufructuary" property right in the water course. The owner, in essence, can enjoy benefits without actually owning the water—the "usufruct." This concept has its origin in English common law and allows

TABLE 4.3. Comparison of Riparian and Appropriation Doctrine

	Riparian	Appropriation
Source of the water right	The water right is tied to ownership contiguous to the watercourse. The water is not owned; the landowner has a "usufructuary" right only.	Contiguity of land to the watercourse is not a factor. Water rights are acquired by actual use. The first user acquires the best right, the second user the second best, etc.
Effect of nonuse	Rights to use water are not lost by abandonment or nonuse. A riparian who has not been using water may at any time commence a use even though this may require previous users to reduce their withdrawals. There is, however, the chance that established users may get rights by prescription.	Nonuse of an appropriation right may result in its loss by abandonment.
Place of use	Many riparian states indicate that the water must be used on the riparian land itself; states permit use on nonriparian land as long as other riparians are not measurably harmed.	The appropriator may transport to, and use the water on, nonriparian land; in fact, use in another watershed is permitted.
General nature of the water right	Riparians are thought of as correlative cosharers in a usufructuary right to make reasonable use of the water; there is accordingly no fixed quantity of water assured to any riparian.	Once the appropriator has established his right by proof of earlier use, he is entitled to a specified quantity of water as against later appropriators.
Natural flow	Earlier case law emphasized the natural flow requirement of a water-wheel economy; namely, that after using water the riparian was to return it to the watercourse as it was "wont" to flow. Today concepts of public rights of public trust are more effective in preserving minimum stream flows or levels in lakes.	There is no natural flow notion. The appropriators can take as much water as they are entitled to take even though it exhausts the watercourse. It is this aspect of assumed appropriation law which aroused conservationists. Some Western states, however, permit the state to file for and ultimately acquire a right to the unappropriated flow and thus preserve such flow.

Note. From "Appropriation Water Law Elements in Riparian Doctrine States" (p. 449) by J. Beuscher, 1960–1961, *Buffalo Law Review*, *10*, 448–458. Copyright 1961 by University of Buffalo. Reprinted by permission.

those who live and work along the banks of the river to divert water for use on the land. The right of the owner of riparian land is not unlimited, and the use formula depends on whether a particular state recognizes the "English rule" or the "American rule."

According to the "English rule," each riparian owner receives a part of the flow of a stream course largely unchanged except for the amount used by upper riparians. Riparian owners may withdraw water for domestic use, but are not permitted to alter or divert the stream course, nor may they diminish the quality of the water, except that for which would naturally occur through domestic purposes. Persons who do not own land that is adjacent to the stream (nonriparians) may not divert water to the land.

The courts in the United States have modified the restrictive nature of the English rule to form the "American" or "reasonable use" doctrine. Because the concept of "reasonableness" has undergone extensive review and reinterpretation in the American courts, the law has changed considerably, though the underlying principle—to protect the rights of downstream users from excesses of upstream users—still remains. A riparian's reasonable use of water is subject to the equal rights of other riparian owners to make a reasonable use of that water (Davis, Coblentz, & Titelbaum, 1976).

The riparian system of water allocation is well-suited to areas with abundant water and few claimants. With the rise of irrigated agriculture in the Western states, however, it became clear to policy makers that there were uses of the water other than for simple domestic consumption and that the "highest beneficial uses" would require unequal shares of water. Under the appropriation doctrine, a user must satisfy certain criteria before a water right is granted: (1) There must be the intention to use the water (not to sell or hoard); (2) water must be diverted from its channel; (3) the water must be used for a beneficial purpose; and (4) where applicable, the user must be issued a permit from the state, which usually fixes the amount of water that can be appropriated or used and how it will be used. In summary, the appropriation signifies to all other users that the permittee has a "reserved right" to the specified amount of water, which cannot be lost (except through misuse) or infringed upon by others.

The appropriation doctrine establishes a legal shorthand: "First in time, first in right." Generally speaking, those filing first for water rights will be granted their permitted water prior to those filing claims at a later time. This has often created certain inefficiencies in the system, and current permit programs aim at correcting these. Likewise, the notion of "beneficial use" has undergone substantial changes in the courts, with the result that societal rather than individual claims to limited water resources are gradually coming forward.

Legal doctrine pertaining to groundwater is even more abstract and complicated than that dealing with surface water. Hydrologists have typically divided groundwater into zones of "aeration" and "saturation." The aeration (or "vadose") zone extends from the surface down to the water table, and consists of pockets and voids in the soil and rock occupied partially by water and air. (For a comprehensive discussion, see Page, 1987, pp. 1–24.) Groundwater occurs in quantity in the saturation zone, where all voids are filled by water. Groundwater does not normally flow in defined channels, but will move by gravity from a place of recharge to a place of discharge, where the water table intersects the land surface. Groundwater, unlike surface water, is constantly moving. However, the real rate of advance is seldom observed due to the extremely slow movement (from about five feet per day to five feet per year).

Over the past 150 years legislation and judicial rulings have evolved over the allocation and quality of groundwater. Under the English rule, the owner of land has an absolute right to all water below the surface of that land. The user may pump as much as he or she chooses, regardless of the intended use or the effect such use may have on neighboring groundwater levels. Nearly all jurisdictions have abandoned this rule, and it has limited application in only a few Eastern states.

The harshness of the English rule led the U.S. courts to adopt a modified rule of reasonable use, or the American rule, soon after the turn of the century. Under this rule, a landowner can use groundwater only for beneficial purposes with a "reasonable relationship to the use on overlying land" (see Ausness, 1983, p. 547). Thus, unlike the English rule, the American rule prohibits the use of percolating water on nonoverlying lands or for wasteful and harmful purposes. As long as the purpose for taking the groundwater is "reasonable," the owner is not liable to adjacent landowners for injury caused by the use. On the other hand, no protection is granted to those who sell groundwater for use off the premises. This system has been criticized because it does not take into account the relative value of the water resource to the public. A landowner who uses water on his or her property is immune from liability, even if this results in little or no benefit to the community. Conversely, the landowner who takes groundwater and removes it from the land faces liability for harm from other overlying landowners, even if the purpose is highly beneficial. In consequence, several states have adopted a modified American rule. In essence, this rule states that an appropriator who withdraws groundwater and uses it for a beneficial purpose is not subject to liability for interference with the use of water by another, unless:

1. The withdrawal unreasonably causes harm to a neighboring appropriator through the lowering of the water table or a reduction in artesian pressure.
2. The withdrawal exceeds the appropriator's reasonable share of the annual supply or total store of water.
3. The withdrawal of groundwater has a direct and substantial effect upon a watercourse or lake and unreasonably causes harm to a person entitled to the use of its water.

This restated rule offers an open-ended view of water as a resource and community benefit. The following example should serve to illustrate the operating principles of the two rules. Owners A and B have adjacent properties with a common source of underlying water. Owner A operates a truck garden requiring substantial water each evening, while owner B operates a nursery. The water withdrawal by A substantially diminishes the pressure, which results in a drop in nursery production on B's land. A makes a reasonable and beneficial use of the water on overlying land, which leaves B in a precarious situation with no viable options. Under the restated rule, B would be entitled to have the courts or a water resource authority determine a more reasonable use formula.

An emerging system of rules, often called the "conjunctive use doctrine," is likely to gain more favor in the western United States. This doctrine, unlike other riparian or groundwater doctrines, seeks to target the interrelationship between groundwater and surface water and to develop common appropriation formulas. The benefits of the conjunctive use cannot be overstated. Integrated supplies of water can be balanced over time to sustain both quality and supply. Conjunctive use has the advantage of allowing for alternating supplies during high use or stress periods.

Although management of and protection over groundwater have traditionally been the province of the states, the federal government has recently shown a greater sensitivity in areas operating on a multistate basis. This is a direct result of a U.S. Supreme Court case, *Sporhase v. Nebraska* (455 U.S. 935, 1982). The Sporhases operated a farm straddling the Nebraska–Colorado border. A decision was made to irrigate the Colorado portion of their farm from groundwater 55 feet across the border in Nebraska. Nebraska officials, however, refused to issue the necessary permit because the two states did not provide reciprocal rights to withdraw and transport groundwater.

The Supreme Court ruled that groundwater is a commodity in interstate commerce subject to the restrictions and protections of the commerce clause of the Constitution, and therefore to the legislative authority of Congress. *Sporhase* does not preclude the states from reasonable regulations to conserve water resources, but, in adopting such rules, they may not unduly restrict interstate commerce. In short, multistate users of groundwater must be subject to a uniform set of standards; additional legislative considerations may not be applied to out-of-state users.

Water Quality Concerns

Mounting concern over the quality and supply of underground water is rapidly emerging as a major environmental and economic issue. This is the result of rapid depletion of groundwater resources in the Western states, significant quality changes that have occurred in aquifers beneath heavily populated regions throughout the nation, and soil erosion and chemical water pollution from farm production. The case of the Ogallala Aquifer, in the Great Plains, is a leading example of concern for groundwater resources. This large aquifer, which spreads throughout portions of eight states, has experienced water demands that far exceed its minimal recharge capacity.

What may prove to be the most serious resource issue in rural America is not water supply, but overall water quality. Throughout the United States, landfills, petroleum products and uncapped oil wells, sewage and farm runoff, insecticides and fertilizers, and even the massive volumes of salt used for melting snow are known sources of water pollution. Many agricultural processing industries, such as dairy products, grain mills, meat products, and canned fruits and vegetables, have the potential to reduce water quality substantially. To met this challenge, the Environmental Protection Agency (EPA) is empowered by the Clean Water Act of 1977 to create a national monitoring and permit system.

National Pollution Discharge Elimination System Permits

Section 402 of the 1972 Federal Water Pollution Control Amendments to the Clean Water Act established the National Pollution Discharge Elimination System (NPDES). In rural areas, agricultural operations that produce point-source discharges and meet certain criteria established by the EPA, such as large cattle feedlots, must obtain NPDES permits setting conditions under which they can operate (Keene, 1984). EPA has the overall responsibility for the NPDES program, but a number of states has assumed full or partial responsibility for issuing permits and administration.

Section 208 Requirements

NPDES requirements do not cover a wide range of agricultural activities, such as runoff from farmland, irrigation return flows, and discharges from small feedlots. These types of activities are classified as "non-point-source" pollution. Section 208 of the 1972 amendments to the Clean Water Act requires that states establish area-wide wastewater treatment and management programs. The eventual goal of these programs is to design a comprehensive 20-year plan of action to bring non-point-source pollution under control. Although Section 208 has the potential of becoming a powerful tool for the rural planner to manage rural growth, at the same time it will continue to bring unrest and controversy to rural areas and small towns. EPA guidelines now stress erosion control as the most effective means of controlling agricultural non-point-source pollution.[9] Very few local or county governments have attempted to integrate erosion control policies into their regulatory programs, and they are unlikely to do so without state intervention.

Section 404 Requirements

Section 404 of the 1972 amendments to the Clean Water Act, unlike other provisions that are aimed at controlling discharges and runoff of liquid effluent, regulates the addition of solid materials into navigable waters. The Section 404 permit program, administered by the Army Corps of Engineers, seeks to regulate discharges of dredged or fill material into lakes, rivers, and streambeds. The Section 404 program was originally the source of considerable controversy in the rural community; however, it has now been redesigned so that farm activities, such as land cultivation and the construction of farm ponds, have been exempted.[10]

The ultimate goal in groundwater management is that all levels of government in the United States will someday be able to operate from a common reference to enable governmental "institutions to work toward the shared goal of preserving, for current and future generations, clean water for drinking and other use purposes, while protecting the public health of citizens who may be exposed to the effects of past contamination" (EPA, 1984, p. 2).

[9]Along with erosion control, the EPA also attempts to integrate pest management, conservation tilling, conservation cropping, and animal management systems.

[10]Farm activities under the Section 404 program are exempted if they are a part of the established farming operation and do not aim to bring new areas into farming or ranching, or have little effect on navigable waters.

The EPA's strategy is to develop a three-tiered classification system for planning purposes. Class I areas are designated "special groundwater" and include those aquifers that are *highly* vulnerable because of their hydrological characteristics and strategic location. Class II is defined as the majority of all usable groundwater in the United States:

> Class II areas encompass [a]ll other groundwater currently used or potentially available for drinking water and other beneficial use. . . . They will receive levels of protection consistent with those now provided groundwater under the agency's existing programs. This means that prevention of contamination will generally be provided through application of design and operating requirements based on technology, rather than through restrictions on siting. (EPA, 1984, p. 5)

Class III areas are not potential sources of drinking water and are of limited beneficial use. This type of groundwater must be saline, or otherwise contaminated beyond usable levels, but also must not be connected to a Class I or II aquifer or to surface waters.

Hazardous Waste

Closely related to water law is the legislation governing the manufacture, transportation, use, and disposal of hazardous waste. The problem of hazardous waste has been a public concern for only a few years—mostly since 1978 when the Love Canal toxic dump site near Niagara Falls, New York, which had been built over with homes, caused illnesses among inhabitants. Recently, the awareness of hazardous waste has grown dramatically, and the disposal of hazardous waste is rapidly emerging as one of the most urgent environmental problems in rural America.

Today there are an estimated 10,000 hazardous waste sites that need to be decontaminated. Dumping in rural areas has been widespread, because (1) the areas are sparsely populated; (2) land is relatively cheap; (3) there is little local regulation or monitoring of dumping; and (4) there is a long tradition in rural America of haphazard dumping. Hazardous waste poses a particularly severe threat to groundwater supplies because of the serious difficulties involved in effective groundwater decontamination. Moreover, rural Americans depend on groundwater for about 95% of their drinking water. The pollution of groundwater by hazardous wastes has resulted in the closing of wells and bans on certain water supplies. Ultimately, residents may be forced to move to other areas. A rural planner should have a solid working knowledge of the dangers of hazardous wastes, regulations governing the use and disposal of hazardous substances, and state and federal agencies to contact should a problem arise.

About 80 billion pounds of hazardous waste are dumped in the United States each year. Most dumping occurs in a manner that is not environmentally safe: in mines, wells, sewers, open pits, or nondurable containers. Hazardous waste presents two dilemmas: First, hazardous wastes do not quickly break down into nontoxic substances; and, second, hazardous wastes are extremely harmful to human health and the environment. The major harmful effects of hazardous waste are as follows:

1. The poisoning of groundwater and drinking water supplies.
2. The destruction of wildlife habitat and fish kills.
3. Soil contamination, rendering land unfit for human habitation.
4. Deterioration of human health: cancer, birth defects, and death.

Currently, there are over 70,000 chemicals in commerical use, with an additional 1,000 new chemicals being introduced into the marketplace each year. It is estimated that as many as 60,000 of these chemicals are potentially, if not definitely, hazardous to human health (Sherry & Puring, 1983). These substances can be categorized as follows:

1. Radioactive waste, especially from stations generating nuclear power.
2. Heavy metals: lead, arsenic, cadmium, mercury, copper, and zinc. Prolonged exposure is harmful to the nervous system and may cause cancer.
3. Synthetic organic chemicals, including pesticides, solvents, and degreasers. Most of these substances were developed after World War II.[11]

Among our most pressing environmental needs today are the disposal, monitoring, and measurement of concentrations of toxic wastes. Several thousand toxic waste sites have been abandoned, and even licensed disposal sites have not been made fully secure. In fact, in late 1985, the EPA closed 1,100 of the nation's 1,600 licensed toxic waste disposal sites because the operators had not installed wells to monitor nearby groundwater for contamination, or the operators lacked insurance or adequate assets to cover the cost of paying for future problems that might be caused by dumping.

To regulate the disposal of hazardous waste, Congress in 1976 passed the Resource Conservation and Recovery Act (42 U.S.C. §§ 6901–6991, 1982; Supp. III, 1985) and the Hazardous and Solid Waste Amend-

[11]For a complete list, see Chapter 1, Title 10, Part 20, Appendix B of the Nuclear Regulatory Commission List of Radioactive Materials, 29 C.F.R. § 1910.2 (1986).

ments (Pub. L. No. 98-616, 98 Stat. 3221, 1984). These acts charged the EPA with the following responsibilities:

1. Identifying hazardous wastes.
2. Setting standards for record keeping, as well as for handling, packing, and transporting hazardous wastes by private firms.
3. Setting the standards for disposal facilities.
4. Requiring disposers to obtain an operator's permit from the EPA.
5. Staffing inspectors.
6. Levying penalties and fines on violators.

After the Love Canal disaster, it became apparent that thousands of abandoned waste sites needed to be cleaned up. Congress responded to this task with the Comprehensive Environmental Response, Compensation and Liability Act (CERCLA) (Pub. L. No. 96-510, 94 Stat. 2767, 1980; codified as 42 U.S.C.A. §§ 9601–9675, West 1987) and the Superfund Amendments and Reauthorization Act (Pub. L. No. 99-499, 100 Stat. 1613, 1986). The Superfund was initially allocated $1.6 billion over its first five years of operation to clean up abandoned toxic waste sites. Another $5–$10 billion has been proposed for the Superfund over the period 1986–1990.

The two major goals of CERCLA are prompt cleanup of hazardous waste sites, and fixing the cost of cleanups on the parties that were responsible for the harm. The primary methods of accomplishing these goals are the Superfund and a liability scheme ensuring that the government can recover the cost of cleanup from the parties that created the hazardous conditions. Thus far, EPA administration, implementation, and enforcement of hazardous waste controls have generally been weak. The implementation of the Superfund cleanup effort has been slow and clouded in controversy.

Some federal legislation has been created to control the use of hazardous substances. The 1972 amendments (33 U.S.C. §§ 1311(a), 1342(a), 1362, 1972) to the Clean Water Act prohibit the discharge of oil and some 300 other substances in harmful quantities. The primary goal of the Clean Water Act Amendments of 1972 is to restore and maintain the chemical, physical and biological integrity of the nation's waterways. The Federal Insecticide, Fungicide and Rodenticide Act (7 U.S.C. § 136 *et seq.*, 1971) requires that these chemicals be registered with the EPA before distribution; it also requires proper labeling and use in an approved manner. The Toxic Substances Control Act (15 U.S.C. § 2601–2629) allows the EPA to obtain information on new and existing chemicals and to control the manufacture, distribution, and use of these chemicals.

At the local or county level, planners can take steps to help protect public health and safety. A major problem has been that well water in areas with heavy pesticide use or agricultural chemical manufacturing plants is not carefully monitored. A planner can contact the state water resources department or the state department of health and environment to initiate regular monitoring of suspect water supplies.

Another problem is the location of dump sites in wetlands, in floodplains, or over aquifers. Local zoning ordinances can require that new waste sites be located away from water supplies and that existing sites be closed or relocated. Siting, however, is only one aspect of the problem. Dump sites must be secured and monitored to prevent unauthorized entrance.

No one really knows the extent of the hazardous waste threat in rural America. Undoubtedly it is considerable and perhaps monumental, and it will cost many billions of dollars to clean up. New waste disposal techniques need to be developed, and the pace of cleanup efforts must be quickened. Rural planners must be wary of any hazardous waste dump sites or heavy use chemical areas. The harmful effects of hazardous wastes will not disappear soon and can have a devastating impact on local communities (see Epstein, Brown, & Pope, 1982).

Summary

The rural countryside and small towns are no longer neglected areas for planning and the regulation of land, water, and resources. Originally, rural areas were conceived of as little more than holding zones for future metropolitan growth. They were, in effect, accorded the same status as marginal land being held for potential resale. It is now widely recognized that the social, environmental, and economic services performed by rural areas and small towns are absolutely crucial to the entire developmental framework of the nation. Rural areas once taken for granted, and thought of as conservative and somewhat backward—are now coming to be viewed as leaders in innovation, social stability, and regulatory protection. Work in the rural areas is, by all measures, at the cutting edge of planning and regulatory management innovation in the United States. Nearly all the important legal cases dealing with the taking issue—from *Pennsylvania Coal Co. v. Mahon* (260 U.S. 393, 1922) to *First English Evangelical Lutheran Church of Glendale v. County of Los Angeles* (96 L.Ed.2d 250, 1987)—are specifically rural cases dealing with matters such as the ownership of resources and floodplain regulation. Many, if not the majority, of cases involving Fourteenth Amendment (equal protection) issues are set in rural areas, small towns, and idyllic communities at the exurban edge. Environmental law (except for law having to do with air pollution) was essentially written in the rural resource areas. Water law and management were developed, for the most part, in response to conflicting demands made by urban development on rural lands. The single most important resource of America is its rural land. If this critical source of food, fuel, fiber, minerals, and recreation is mismanaged, the entire nation will feel the shock effect.

CHAPTER FIVE

Planning and the Political Economy of Rural Development

Before beginning work as a planner a person must have a sense of what constitutes the "good" community, region, or society. Such a philosophy is crucial in giving the planner direction and moral values. Simply put, what kind of town, region, or society is the planner working to help to create? Given that planners attempt to serve the public interest by helping jurisdictions create controlled change, which aspects of the jurisdiction should be preserved and which should be altered? In responding to these questions, the planner must be aware of the political, economic, and social framework of the society at large, as well as of how that framework of laws, property, markets, and status influences the local political, economic, and social relationships.

Political Economy

Politics establishes the rules by which the economic system operates. This fact is the fundamental feature of "political economy," the study of how wealth and power are created and exercised in a society. Wealth involves the ownership of land (including natural resources), labor, and capital (buildings, machines, and equipment). Owners use these factors of production in various combinations to produce goods and services. Power comes from the control of legal and political institutions that decide rights to property and methods of exchanging property. In America, private property and the marketplace are the major means of owning and trading factors of production and the resulting goods and services. Together, wealth and power determine who controls the factors of production, who works for whom, and the degree to which owners of property can influence the markets for goods and services.

Government ownership of property and intervention in the marketplace through subsidies and regulations may occur to promote "the common good" and to correct for imperfections in the market. But government programs do not necessarily produce a more efficient or equitable allocation of goods and services than the marketplace. Some government programs may actually promote the interests of the rich and powerful.

Three philosophies of political economy, each with its own set of development goals and prescription for achieving those goals, have led the debate about how rural areas develop and the role of government.

The "Neo-Classical" framework defines development as growth in the output of goods and services. Development is seen as a "gradual, continuous, harmonious, and cumulative process, with significant [geographic] spread and trickling down effects" (Yotopolous & Nugent, 1976, pp. 429–430). Neo-Classical theorists believe in the self-regulating role of prices in competitive markets; that is, markets consist of many individual buyers and sellers who respond to prices to allocate scarce resources in the most efficient way. Trade acts as a major engine of growth, and a region should produce and export goods in which it has a comparative cost advantage. With free trade, there will be gains from trade to the benefit of specialized labor in each region. The Neo-Classical model, however, ignores the prevailing distribution of income and wealth in a region or nation. The belief is that incomes in different regions will converge over time; however, no goal of equitable income and wealth distribution is espoused. Government has a limited role, and government interference in the market is discouraged. In sum, the Neo-Classical philosophy of development is one of *laissez-faire* capitalism.

The Neo-Classical school in many ways underlines the importance of the market in affecting the spatial allocation of economic activity. "Central-Place" Theory, consistent with the Neo-Classical view, predicts that a hierarchy of settlements will develop according to population size and the diversity of public and private services. A large central place will have a varied labor pool and offer a wide variety of goods and services over a substantial market area. This gives the large central place the advantage of agglomeration economies in attracting economic activity and greater economies of scale in the provision of public services. Small central places feature only day-to-day goods and private services, limited job opportunities, and limited public services. They are much less able to compete with large central places for economic growth.

The "Radical/Marxist" school defines development as a transition to a socialist state, with a redistribution of wealth, worker ownership of the means of production, and reduced dependency on other regions. The important elements of a region are the structure of the ownership of resources, the economic history of the region, origins of class conflict, and the dominance of cities (central places) over the countryside (peripheral places). The Radical/Marxists view the distribution of income and wealth as grossly favoring the capitalist elite. They believe in a diminished role for the market as an efficient allocator of resources, because capitalists control production and have superior buying power compared to workers. The result is an unequal distribution of the gains of economic growth.

Rural areas, historically, have grown because of the development of natural resources. But these areas have had to rely on financial capital and technology imported from cities and urban-based corporations. These capitalist corporations have extracted raw materials from the hinterlands and profits from poorly organized rural workers. Capitalists are able to exploit rural workers through low pay, movement of capital (e.g., closing a mill in one town and opening one in another), temporary layoffs, and refusal to allow workers a voice in how a company is operated. Radical/Marxists see value as coming from labor, and believe that capitalist exploitation of workers results in worker alienation, class struggle, and eventual revolution.

The Radical/Marxist solution is economic democracy—that is, worker control of the means of production, so that operating decisions will be made by and for workers and not the owners of capital. Workers must also gain control of the political machinery to eliminate the technological and financial domination of the central cities over the hinterlands. (For a discussion of Radical/Marxist analysis in a rural setting, see Young & Newton, 1980.)

The Radical/Marxist approach, however, suffers from at least three flaws. First, worker control means that workers then become the owners of capital and may indeed exploit themselves. Second, Marx himself scoffed at "the idiocy of rural life," not seeming to care whether or not rural people were exploited. Third, Radical/Marxists have concentrated on the problems of peasants in underdeveloped countries, almost to the exclusion of the rural workers in mature economies.

The "Structuralist" school defines development as a change in the structure of the economy over time and emphasizes that development is likely to continue to be uneven between regions and social classes. The Structuralists are interested in which sectors of an economy grow and which do not, what are the linkages between sectors, and why "dual economies" of prosperity and poverty persist within a region or nation. Structuralists answer these questions in terms of (1) a failure of the price system to produce steady growth or more equal income distribution;

(2) uneven investment in different sectors; (3) limited benefits from trade; and (4) structural rigidities, such as immobile labor, lack of training, or a lack of capital to develop local resources or raise worker productivity. The Structuralists believe in active government intervention in markets to create a more even distribution of income and wealth, both spatially and among social classes.

The Structuralists and the Radical/Marxists would argue that government is needed to stimulate economic development in small, disadvantaged central places and rural areas. There are four types of programs that a national government can use to try to bring the standard of living in rural areas up to urban levels: (1) macro-level policies; (2) sector-specific policies; (3) human services; and (4) territory-specific programs. Macro-level policies involve the use of expansionary monetary and fiscal instruments to stimulate the national economy. The belief here is that "a rising tide lifts all ships," meaning that rural areas would experience economic growth along with the rest of the nation. But it remains uncertain whether the income and wealth gap between rural and urban areas would indeed become smaller.

Sector-specific programs serve to promote the growth of new industries or to support industries with little growth or those in decline. In America, for example, billions of federal government dollars are spent each year to support the farm sector, which has shown little growth in employment since 1970 and which directly employs a small fraction of the rural labor force. By contrast, other federal subsidies helped create the infrastructure (roads and sewer and water lines) necessary for manufacturing, which increased its proportion of rural employment in the 1960s and 1970s.

Human resource programs feature investment in training, education, housing, and health care. These programs are aimed at increasing the productivity of rural workers and improving living conditions. Territorial programs are targeted toward certain "backward" regions in the hope of raising regional incomes, employment levels, and living standards closer to the national average. The Appalachian Regional Commission (ARC) is one such example in the United States.

One fundamental question is whether government grants and subsidies are well spent in these rural regions and communities that cannot otherwise attract capital. Another question is whether government regulation of private, urban-based companies operating in rural areas will achieve desired development results. And a final question is whether government or worker ownership of rural industries or public–private partnerships will improve economic growth and bring about a more equal distribution of income and wealth. These questions are of direct importance to the planner, who is often called upon to seek government grants, to attempt to regulate land use and environmental practices (and hence the economic profitability) of urban-based companies, and to help create nonprofit corporations. As a public employee, the planner can try to influence the role of government and should have an idea of how active or passive that role should be.

Although each planner must formulate his or her own philosophy of rural development, each of the three paradigms described above is important to understand. They guide the actions of many of the actors in rural areas.

We use political economy as both a descriptive and an analytical tool to understand and explain situations of dependency of rural dwellers on urban areas. The solution we propose is for rural communities to strive to become more self-sufficient and to make careful, consensual choices about change. An active government role is needed, and planning is an integral part of trying to achieve such controlled changes.

Models of Dependence and the Modern Political Economy of Rural America

Two political economies can be found in rural America. The first involves the distribution of wealth and power among local residents, as shown by the systems of land tenure (who owns the land), and among those who own the businesses and thereby employ the local labor force. The second involves the distribution of wealth and power between a rural area and other areas and markets. For example, a local mine may be owned by a company based in a distant city, and decisions over how to operate the mine are not made locally. Urban areas provide markets for rural exports, and yet rural residents typically have little influence over the prices they receive in those markets. Federal programs and regulations also comprise a significant outside force over which local residents have limited control.

The uneven distribution of wealth and power, both within rural areas and between the city and countryside, has been a significant factor in frustrating the creation of coherent rural development policies at both the local and national levels. As we shall see, the continuation of this political economy of rural dependence has meant prosperity to some, but a hindrance to sustained economic growth for many rural inhabitants.

Few American researchers have explored the causes and consequences of the dependency of rural areas on outside markets, capital, and the absentee ownership of local resources. Four kinds of dependence exist: "direct dependence," "trade dependence," "financial dependence," and "technical dependence."

Direct Dependence

Direct dependence occurs when key sectors of a local economy are controlled from afar. The ultimate, yet not uncommon, example is the single-resource "company town," where the local coal mine or lumber mill is owned by a company with its headquarters in some distant city. The company, because of its economic clout, is able to control local politics and even to exact favorable taxation and environmental concessions. The town economy and its inhabitants are largely at the mercy of company hiring decisions, the company's financial health, and the threat of a pullout. Although direct dependency can benefit a community or region, it can also be devastating. For example, the energy sector is the largest employer in Wyoming and western Colorado; as energy companies increased the exploration and development of oil, coal, and natural gas fields in the late 1970s, these regions experienced substantial booms in population and economic growth. When energy prices leveled off and then fell in the 1980s, energy companies abandoned many projects, producing widespread unemployment and economic hardship.

Trade Dependence

Trade dependence is measured by the importance of imports and exports in a region's economy (whether its balance of payments shows a net surplus or deficit), and by the terms of trade (the buying power of a region's exports and the cost of imported goods and services). Two kinds of trade dependence exist, "relational dependence" and "cyclical dependence," and the two tend to occur simultaneously. Relational dependence occurs when a rural region trades with urban centers or with companies that are not locally based. The rural area produces what is needed by externally owned companies or urban areas. In return, the rural region is a market for the goods and services produced by those companies and urban centers. Typically, rural areas provide raw materials to metropolitan areas and import manufactured goods from them. The relative purchasing power of rural residents who are dependent upon raw materials is both volatile and vulnerable. The rising grain prices and oil prices of the 1970s made the urban manufactured goods cheaper for many rural dwellers. The plunge in grain and oil prices in the 1980s changed the terms of trade between rural and urban areas, and sharply reduced the buying power of many rural Americans. Rural economics based on raw materials are vulnerable to increases in supplies by other rural areas, the development of substitutes (e.g., optic fiber wire for copper wire), and the depletion of natural resource supplies within the region. The first two situations cause lower raw materials prices, and the third usually means greater costs of production; all of these cases reduce the ability of a rural region to compete in the export of raw materials to distant markets.

Cyclical dependence describes the effect of changes in national or international economies on a rural region. Because rural areas do not constitute sizeable markets, a substantial majority of rural economies tend to be export-based, and fluctuations in national and international markets can have far-reaching impacts. When the market price of wheat, corn, and soybeans shot upward in the 1970s, farmers responded by greatly expanding their output of these crops. But by the early 1980s, world and national crop prices fell as the world supply of food increased more rapidly than the demand. American farmers who expanded, often on credit, have experienced lower incomes, and many are in serious financial trouble. Many farm communities have felt economic contractions and population losses.

Relational and cyclical dependence tend to occur at the same time. When the national economy is running smoothly, then businesses generally flourish and prices are stable. But when the national economy enters a recessionary period, businesses cut back production and lay off workers. The example of the timber industry in Oregon is instructive. From the 1950s through the 1970s, Oregon's timber industry profited handsomely from the national demand for new homes. Mortgage interest rates were fairly low and personal incomes were increasing. Timber companies opened mills in rural Oregon to meet the demand for wood. In late 1979, however, federal tight-money policies drove mortgage interest rates well above 10 percent. The demand for new housing and Oregon lumber fell, and the timber industry reduced output and began moving operations out of Oregon to the Southeastern United States. Until the early 1980s, the forest products industry was Oregon's leading industry, largest employer, and major source of public revenue. Between 1979 and 1984, employment in Oregon's timber industry declined by about one-third, and the state actually lost population for the first time in its history. The number of lumber mills fell from 69 to 52, and one company town, Valsetz, was evacuated and the buildings burned; the town simply ceased to exist!

Between 1984 and 1987, the Oregon timber industry made a strong recovery, boosted by a strong national economy and a renewed demand for housing. The cycle of dependency had come full circle, from a thriving timber industry to a severe downturn to a prosperous segment of the rural economy. Yet these cycles are upsetting to rural dwellers who must survive layoffs, either temporary or prolonged, and the possibility of losing their jobs.

Financial Dependence

Financial dependency occurs when the banking and credit systems of rural regions are influenced by urban-based financial institutions and federal monetary and loan programs. Money is generally scarce in rural areas, causing a lack of savings to generate investment in buildings, machines, equipment, and education needed to increase labor productivity and drive economic growth. Between the 1930s and 1980, federal laws limiting interstate banking and imposing ceilings on deposit interest rates tended to protect small rural banks from the competition of large city banks. In 1980, the Depository Institutions Deregulation and Monetary Control Act lifted the federal ban on interstate banking and removed the ceiling on deposit interest rates (Pub. L. No. 96-221). These changes have resulted in trends toward (1) the consolidation of banks within each state; (2) interstate banking; (3) the expansion of bank-like financial services by retail stores and credit card suppliers; and (4) the growth of huge, nationally based money market funds (Deaton & Weber, 1985).

Bank consolidation will turn many rural banks into branches of urban-based bank holding companies. Although it has been argued that this will create greater efficiencies in the industry and expanded services to customers, the cost may be an increased reluctance to invest in smaller companies, farm operations, and local communities, and a resulting flow of capital out of rural areas. Historically, small-town bankers have displayed a sensitivity to the boom–bust cycles that often affect their borrowers and to the need to reinvest in the communities they serve; whether or not this will continue into the future is a matter of considerable importance to small towns and rural areas.

The loan policies of urban banks already determine the flow of credit to many rural areas, and decisions on which projects to support have a profound effect on rural economic growth. Bank lending practices are in turn tied to the cost of borrowing money from the Federal Reserve System. As noted earlier, tight federal money and high-interest-rate policies have a particularly harsh effect on rural areas, by drying up credit in already credit-starved regions. Moreover, when interest rates climb, debt burdens grow; over time, these can result in a net drain of financial resources from one area to another. For example, farm debt burgeoned to about $200 billion in 1985 (U.S. Bureau of the Census, 1986, p. 627), and the amortization of the farm debt will tend to draw wealth away from rural areas toward the federal government in Washington and toward urban-based commercial banks and insurance companies.

Technical Dependence

Technical dependence is a region's need to import technology, know-how, and trained personnel in order to achieve economic growth. Technical dependence may either help or hinder a region's development. When a manufacturing firm locates in a rural area, it provides jobs and offers the possibility of improving local job skills. But the area also becomes reliant upon the technology and management resources provided by the company. As long as the company finds it profitable to remain in the area, the local economy is likely to expand. As markets change, however, a company may decide to scale down or remove its operation and technical personnel, burdening the region with an economic loss. This situation took its toll on in the energy boomtowns of Wyoming and western Colorado during the mid-1980s. Alternatively, efforts by rural dwellers to induce development by training their young people for technical skills may actually subsidize metropolitan regions since young people often can find skilled jobs only in urban areas.

The Forms of Dependence at Work: The Case of Appalachia

To varying degrees, all of these forms of dependence are likely to exist in a particular region. Appalachia is the prototypical dependent rural region, as Congress implied when it passed the Appalachian Regional Act of 1965,

> [W]hile abundant in natural resources and rich in potential, the region lags behind the rest of the Nation and in its economic growth and that its people have not shared properly in the nation's prosperity. The region's uneven past development, with its historical reliance on a few basic industries and a marginal agriculture, has failed to provide the economic base that is a vital prerequisite for vigorous, self-sustaining growth. (Pub. L. No. 89-4, 79 Stat. 5)

Appalachia, as defined by this act, consists of 397 counties in 13 states extending 1,000 miles from southern New York State to northern Mississippi, and 300–400 miles in width from the border of New Jersey to within a few miles of the Mississippi River. Much of Appalachia is remote and rugged, and very few of its 20 million people live close to interstate highways that would connect them to major metropolitan areas. While much of Appalachia is now a major manufacturing region, in central Appalachia, where mining is the major source of income for its 1.8 million people, unemployment levels are still well above the national average and per capita incomes well below (Franklin, 1985).

The dominant characteristic of Appalachia is a significant intrusion of absentee capital and ownership of land, which has been most apparent in coal mining but has extended throughout its economy. This situation is associated with the location of coal seams throughout Appalachia; coal and steel companies tend to be headquartered in urban centers on the periphery of Appalachia. This outside economic domination has profoundly influenced local social and political relations by making rural residents virtually subservient to outside interests.

In Appalachia, the pervasive economic dependency has frustrated the development of the region into a more self-contained area. The Appalachian Act provided $5 billion to induce internal development, increase income levels, teach skills, and generally uplift the entire region (Franklin, 1985). The expenditures of federal money relied upon state plans, with the intention that, by providing states with an enlarged role in project implementation, a coherent rural development program would follow. Ideally, the governors' staffs would include planners to prepare plans to bring federal resources and expertise to bear on specific problems. But state planning must have support at the local level in order to be effective, and many local politicians in Appalachia resisted state and federal planning efforts. Yet they were able to maintain their control of the local political process by dispensing public works jobs and welfare. Local politicians captured substantial local support by identifying with low taxes and opposing centralized government, especially federal agencies. Both local and externally based corporations offered the local politicians moral and financial encouragement. A commentary on a political boss in Mingo County, West Virginia, illustrates a common example:

> He was typical of local leaders who served not only their own interests but those of out-of-state corporations: Floyd was paid $6000 a year as secretary of the Mingo County Taxpayers Association, financed by coal companies and other large corporations, which in effect made him a lobbyist for those interests. Local politicians routinely received financial support from corporations concerned about the amount of local taxes that would be assessed. (Clavel, 1983, p. 128)

In short, local politics in Appalachia could be characterized as "dependency politics," where "exploitation . . . is the name of the game" (Clavel, 1983, p. 128). Local officials would keep local services and taxes low, and while the local economy stifled opportunity, sufficient numbers of persons would be employed in political jobs or kept on welfare to run the machinery necessary for the re-election of the local and state politicians whom they supported (Clavel, 1983).

Appalachia provides a sobering case study of the political economy of a major rural region, as well as of the difficulties facing those who seek to create opportunities that would change the local balance of wealth and power. Although the Appalachian Act has reversed out-migration trends and upgraded income relative to national averages, the act has had the effect of making the political economy of Appalachia more, not less, dependent upon external companies, urban areas, and the federal government. Still, as of the mid-1980s, Appalachia had higher unemployment rates and a lower per capita income than the national averages. Meanwhile, the budget of the ARC was cut from $358 million in 1981 to $149 million in 1985 (Franklin, 1985).

The Changing Political Economy of Rural America

Between the late 1960s and early 1980s, rural population increased, in part from the net migration of people from urban to rural areas and the location of manufacturing plants in rural settings. These trends began to change the balance of wealth and power between urban and rural regions and within rural communities. New residents and companies brought substantial economic growth to many rural areas. Improved transportation and communication systems have made urban markets more accessible and lowered the cost of importing goods and services. Although the four kinds of dependency continue to exist, many rural and urban areas have become more interdependent in the past 20 years.

Urban dwellers have been attracted to rural areas because of the perceived better environment and community-oriented way of life. Yet there has often been friction between newcomers and long-term residents. Newcomers may be resented when they attempt to challenge established political and economic relationships. For example, newcomers tend to demand a higher level of public services, which long-term residents are often reluctant to support. Moreover, many newcomers are often wealthier than long-term residents and are not afraid of demonstrating their economic power, as in bidding up the price of land and housing.

Manufacturing now comprises about one-quarter of the economic base of nonmetropolitan America. Manufacturing firms have settled in rural regions for a number of reasons. First, labor in rural areas tends to be cheaper, less unionized, and less militant than urban labor. Manufacturing plants locating in rural areas tend to employ mostly unskilled and semiskilled labor. Positions requiring sophisticated technical skills are initially filled by transferred employees, but later by locally trained personnel as communities make public investments in vocational training. In addition, manufacturers perceive rural inhabitants as having a stronger work ethic and a greater degree of company loyalty than urbanites.

Second, the construction of interstate highways and relatively cheap and abundant energy supplies, coupled with access to inexpensive tracts of land, have induced manufacturers to locate away from large urban markets.

The third, and often crucial, reason is that rural dwellers have readily welcomed opportunities to diversify their economies. Rural communities have subsidized the construction of industrial parks (roads, sewer and water lines, and buildings); manufacturers have been granted property tax concessions; and environmental laws regulating pollution tend to be less strictly enforced. Manufacturers are especially attracted by local political support, which holds the prospect of continuing influence on a community. Once a plant moves to a rural area, the pressure is on the community to keep the plant there and protect the local economy.

Since the early 1980s, however, rural population and economic growth have lagged well behind urban levels. In a return to the trends of the 1950s, several rural communities are experiencing a net out-migration of people and a sluggish rate of economic growth. Declines in the farming, energy, and rural manufacturing sectors have revealed the dependency of many regions on outside capital, externally owned companies, and international markets. This is the case in much of the rural Midwest, the Great Plains states, the intermountain West, and the South. On the other hand, some rural areas, particularly in the Northeast, have continued to receive newcomers and achieve healthy rates of economic growth.

Political economy is a dynamic process. A depressed region may become prosperous with a sudden influx of outside capital, and a few years later the region may fall into depression as capital is removed. This syndrome has occurred in many boom areas associated with energy and mining exploration and large construction projects; for example, when the energy companies pulled out of western Colorado in the 1980s, there were few ways for local inhabitants to make a living. In a different light, the development of tourism, recreation, and manufacturing has helped many parts of the rural Northeast replace an economy built on agriculture and forestry. The rural areas are still dependent on outside markets and capital, but these new industries have created much more stable economic growth than the old ones did.

In the following sections, we examine the role of four groups of "actors" in rural wealth and power. Each of these groups represents national or international influences over which rural areas have little control. The actions of these four groups of actors will continue to make the political economy of rural America more complex than in the past. This is not to suggest that a strict dichotomy exists between rural poverty and urban wealth. The urban poor and the rural rich are involved as well.

What finally emerges is a picture of growing rural diversity and a need for planning to arbitrate the many conflicting demands on rural lands and resources. The planner needs to recognize the dependency of rural areas and the costs of dependency, and to articulate an alternative vision of greater economic and social autonomy.

The Federal Government as a Center of Influence and Power

America's national government has conducted rural development programs for over 90 years, but large-scale projects and programs generally began with the New Deal era of the 1930s. Since then, the federal government has poured billions of dollars into improving the economic, social, and environmental conditions of rural America. Farm programs, featuring commodity price supports and loan programs, have long dominated federal rural policy. Although farmers comprise a small minority of the total rural and national populations, they have achieved strong political support in Congress and the U.S. Department of Agriculture (USDA).

The federal influence in rural areas is felt in a variety of ways: regional development programs; sewer and water projects; highway construction; loans and loan guarantees for housing and businesses; federal lands management; military bases; federal regulatory agencies; and national fiscal and monetary policies. A rural planner faces a substantial task in acquiring and maintaining a working knowledge of all the federal influences on a particular rural area. A planner must also realize that the impact of a new federal program is likely to be tempered by existing programs.

The legal, financial, and political limitations of state and local governments have caused much of the reliance upon federal rural development programs. Many state constitutions restrict state spending actions (e.g., by banning budget deficits), so that states cannot provide all the services their citizens require. Local governments often have bonding limitations and lack the financial resources, the constitutional or legal authority, and the technical ability to undertake needed development projects. Legally, local governments are the creation of the state government and depend on the legislature for both taxing and law-making authority; however, such authority is often inadequate to guarantee high-quality public services or effective land use ordinances.

Federal rural development programs carry the advantage of political convenience. A new or expanded federal program usually applies throughout the nation. Thus, a single federal action is likely to be more efficient than the time and effort spent in duplicating the same program or law in 50 state legislatures or thousands of local government councils.

The tax burden of federal programs is lighter than state and local programs because the cost of a federal program is shared by all of the nation's taxpayers. Federal revenues rely primarily on income and payroll taxes, and tax rates tend to be progressive, falling more heavily on those with higher incomes. By contrast, states raise money from income taxes and regressive sales taxes, which fall more heavily on those with lower incomes. Local governments depend especially on the generally regressive property tax. In addition, there are fewer people in a state or locality to share the cost of programs, particularly in rural counties and small towns. One of the sharpest criticisms of recent attempts to shift federal education, welfare, and economic development programs to the states is that state and local taxes will have to be raised and/or services curtailed. As funding declines for federal domestic programs, and with the end of revenue sharing in 1987, many counties and their municipalities are competing for the revenues to support public services.

Capital Formation

One of the strongest federal influences on rural areas involves capital formation and hence economic growth. Rural areas tend to suffer from a shortage of capital and credit, and federal spending and loan programs have played a major role in supporting rural economic development, particularly in the case of agriculture.

The federal government affects capital formation through fiscal and monetary policies. Fiscal policies determine how the federal government raises money through taxes and borrowing, and how money is spent on the array of programs. In addition to agriculture, the federal government spends money in rural areas in two ways: transfer payments and public works projects. Transfer payments include unemployment compensation, Social Security retirement and disability support, and food stamps and welfare checks for low-income households. Transfer payments help to maintain and even expand a local economy by giving people more money to spend on goods and services. They are especially important in nonmetropolitan counties with a significant retirement population or high poverty levels. But on a per capita basis, more transfer payments go to urban than to rural dwellers, because (1) seasonal farm and recreation-related employees are generally ineligible for unemployment benefits; (2) many rural people are reluctant to bear the social stigma of "taking a handout" associated with receiving welfare payments; and (3) metropolitan areas have better social service agencies to serve low-income households.

Public works projects have three desirable features. First, the inflow of outside capital stimulates the local construction and manufacturing industries. This raises employment and income and encourages growth in related businesses. Second, public works provide socially useful goods, such as parks, schools, and sewage treatment plants, which can attract private investment to a small city or town. Third, public works can promote region-wide economic development, as in the construction of hydroelectric dams, shipping canals, and highways.

Federal monetary policies directly regulate credit and the role of banks in capital formation. Monetary policies are controlled by the Federal Reserve System, which functions as the nation's central bank and operates independently of congressional or presidential authority. Created in 1913, the Federal Reserve System was intended to end extreme swings in the nation's money supply and to contribute to economic stability. The Federal Reserve System consists of the Federal Reserve Board (often called "the Fed"), whose seven members are appointed by the president; 12 Federal Reserve Banks; and several thousand member banks. These member banks represent about one-half of America's 14,000 banks and account for some 90 percent of all commercial bank deposits.

The Fed uses three instruments to control the availability of bank credit: the discount rate, open-market operations, and bank reserve requirements. The 12 Federal Reserve Banks lend money to their member banks at interest rates below those charged by the member banks on loans to their borrowers. In essence, the Federal Reserve Banks act as wholesalers of credit, while member banks are the credit retailers. The interest rate the Federal Reserve Banks charge member banks is known as the "discount rate." A decline in the discount rate means that member banks are encouraged to borrow from the Federal Reserve Banks and that interest loan rates will tend to fall, setting the stage for economic expansion. Increases in the discount rate drive up the cost of borrowing money as member banks pass along the higher cost to their customers. In general, higher discount rates reduce the demand for credit and dampen economic growth, although inflation is likely to decline. For example, between 1979 and 1984, the Fed kept the discount rate high, and interest rates stayed above 10 percent. The effects were strongly felt throughout the U.S. economy, but perhaps most in agriculture. During most of the 1970s, farmland values grew at an average rate of 16 percent a year, fueled in part by high levels of inflation (Healy & Short, 1981, p. 43). After 1980, inflation subsided because of the high interest rates; also, between 1981 and 1985 the value of the dollar soared 40 percent against foreign currencies, which made U.S. farm exports less competitive. As a result, farmland values plummeted. Between 1980 and mid-

1985, the average value of an acre of farmland fell by 35 percent in the St. Louis and Omaha Farm Credit Bank regions (Daniels, 1985). At the same time, farm operating loans have carried double-digit interest rates, putting pressure on farmers' cash flow.

Federal Reserve Board open-market operations involve the sale or purchase of U.S. government bonds and Treasury bills, and are considered the Fed's most important tool for economic stabilization. When the Fed buys government securities in the open market, the money supply increases along with commercial bank reserves, and interest rates tend to fall. When the Fed sells government securities, the money supply and commercial bank reserves fall, and interest rates tend to rise. Short-term borrowing rates are especially influenced by the Fed's open-market activities.

By law, commercial banks are required to maintain a reserve of cash equal to a certain minimum percentage of their deposits. For example, if the reserve requirement is 10 percent, $1 million in reserves can support up to $9 million of deposits available for lending. The Fed may set the legal reserve requirement from 6 percent to more than 20 percent. The Fed commonly makes two distinctions in its reserve requirements: (1) Small rural banks tend to have lower reserve requirements than member banks of the Federal Reserve System; and (2) demand deposits (checking accounts) have higher reserve requirements than time deposits (savings accounts and certificates of deposits).

The United States has traditionally relied upon many independent, relatively small local banks; this is particularly true in rural areas. For instance, Iowa, with just under 3 million people, has 620 commercial banks. There are over 4,000 rural banks in America, and most of these have assets of only $15 million to $35 million. Although rural banks hold less than 6 percent of the nation's banking assets, they serve as a critical source of financing to rural inhabitants, especially farmers. Rural banks are also an important source of funds for local economic development efforts. These banks purchase locally issued municipal bonds and make housing, commercial, and industrial loans.

Few rural banks are members of the Federal Reserve System. Since 1973, the Federal Reserve Banks have allowed rural nonmember banks a seasonal borrowing privilege. Nonmember banks can borrow money from the Federal Reserve to make up for seasonal loan increases, such as rises in the demand for farm operating loans and construction loans. Alternatively, small rural banks often arrange to borrow money from large urban banks at a negotiated interest rate.

Since 1933, almost all commercial banks have had their deposits insured by the Federal Deposit Insurance Corporation (FDIC). As of 1988, the FDIC insured deposits up to $100,000. The FDIC has provided confidence in the banking system and softened the blow of bank failure.

National Development Programs

In fiscal 1985, the U.S. government spent just over $150 billion in nonmetropolitan counties (Bradshaw & Blakely, 1987, p. 11-17). About $98 billion went to transfer payments, such as Social Security, welfare, and Medicare. Nearly $20 billion was spent on space and defense; slightly over $30 billion was spent on rural development programs (see Table 5.1).

There are currently some 400 federal rural development programs administered by 27 agencies. Federal programs affect rural transportation, housing, infrastructure, and economic development activities, as well as federal lands and environmental regulations. The 1985 federal budget allocated $7.3 billion to be spent on community and infrastructure programs in nonmetropolitan counties, along with $1.9 billion for loans. These figures are overshadowed by the huge $26 billion 1986 appropriation for farm programs alone. Not surprisingly, there has often been a lack of coordination among the various federal rural development programs. This tendency has both frustrated orderly rural development and actually increased the dependence of rural areas on the federal government.

National programs aimed at the development of rural areas have generally met with only limited success. Two persistent problems are (1) the lack of an adequate rural data base to guide policy making and program design; and (2) fragmented goals and a shortage of resources, which have resulted in piecemeal actions rather than a comprehensive approach. A common deficiency in many rural development programs is

TABLE 5.1. Proposed Fiscal Year 1985 Budgets for Rural Development Programs (Nonmetropolitan Counties)

Program type	Spending programs (in billions of dollars)	Credit programs (in billions of dollars)
Community and infrastructure development	7.371	1.91
Business and government economic assistance	1.236	1.09
Housing and credit assistance	1.504	7.267
Other selected programs (mostly agriculture-related)	3.872	7.596
Total	13.983	17.863
Total all rural development programs	31.846	

Note. From U.S. Bureau of the Census, 1986, p. 297.

that they seem to "throw money at problems," rather than allocate funds in a systematic way to a particular area to solve an interrelated set of problems (e.g., poverty, illiteracy, isolation, and the despoliation of natural resources). Often rural development programs seek a "quick fix" to boost employment and income. Although these goals are laudable, economic development is often a slow and gradual process whose goal should be a long-run increase in income and employment. Often after federal money is withdrawn, a rural area suffers because there are not sufficient local funds to sustain economic growth.

In addition, some areas, such as the Tennessee Valley and the Columbia River Basin, have become "hooked" on federal aid. The federal government has pumped $16 billion into the Tennessee Valley Authority (TVA) since 1933, and $5 billion into the ARC since 1965 (Franklin, 1985); similarly, the Bonneville Power Administration is $8 billion in debt to the U.S. Treasury. The TVA and the Bonneville Power Administration were able to attract new industries with cheap electric power. These federal agencies have indeed increased employment and income in their respective regions, but these areas are still far from being self-sustaining. Without federal subsidies, the local economics would be sharply curtailed. Under the Reagan administration, there has been intense pressure to reduce the activities of these regional federal agencies. The labor force employed by the TVA decreased from 51,000 in 1980 to 32,000 in 1985 (U.S. Bureau of the Census, 1986, p. 311), and the Reagan administration has advocated selling the assets of the Bonneville Power Administration to private utility companies.

In the role of a problem solver, a planner in a rural area will often be judged on how much money he or she is able to bring into the community from state and federal funding sources (Toner, 1979). Most rural people are concerned with one specific problem or project, rather than with more comprehensive issues and problems. Federal grants exist for a variety of projects, including transportation, housing, economic development, and public services (especially sewer and water lines); the first three of these are discussed below. Information on federal grants can be obtained from four sources:

1. The *Catalogue of Federal Domestic Assistance* (U.S. General Services Administration, 1985) provides a complete list of available grants, dollar amounts, and qualification criteria.
2. The *Federal Reporter* publishes information on new grants and federal regulations.
3. The Federal Assistance Program Retrieval System, operated by the U.S. Department of Agriculture's Soil Conservation Service, elicits information on population, economic, and social data

from rural regions and determines the eligibility requirements for a variety of grants.

4. *The Rural Resources Guide* (USDA, 1987b), available from the U.S. Government Printing Office, is a national catalogue of rural assistance programs—public, private, financial, and technical.

Transportation

Highways, railroads, waterways, and airports have a profound impact on the location and intensity of economic activity. Good transportation networks are necessary for linking rural areas to urban markets and for making rural areas accessible to urban goods, services, and recreationists and tourists. Federal highway programs and regulations on freight and passenger rates and routes have played a dominant role in the creation of national transportation networks.

During the past 60 years, the highway and the automobile have revolutionized the spatial location of development in both metropolitan and rural areas. Rural America contains nearly 4 million miles of highway and 3 million miles of rail lines. But over time, local governments have largely been relieved of road-building and maintenance responsibilities in rural areas. In 1921, town and county governments financed almost half of the highway construction and maintenance projects, usually through local property taxes. By the 1980s, state governments had assumed almost 70 percent of all highway funding, with the federal government contributing over 20 percent. The federal government currently spends about $5 billion a year on nonmetropolitan highways.

The federal government has sponsored the creation of the nation's two major highway networks: the interstate highways and the federal aid secondary (FAS) highways. Since 1965, 41,000 miles of interstate highways have been built, primarily outside of urban areas. The interstate highways were built with matching funds, 90 percent federal and 10 percent state; interstates, however, are expensive to maintain (up to $2 million per mile), and the states have been paying most of the maintenance costs. Meanwhile, state highways and bridges in many rural areas are receiving low priority for repairs. The burden of upkeep is again falling on local residents, but now they are having difficulty raising the necessary funds. A number of county governments have begun to abandon some county roads.

The FAS system was built in the 1920s as the first national road network. Even though FAS highways do not have the limited-access or high-capacity design of the interstates, they still provide important connections to many nonmetropolitan cities and towns.

By opening up formerly inaccessible rural areas, federally assisted highways have drawn residential, commercial, and industrial development out of urban centers and onto vacant rural land. The result has been urban sprawl around metropolitan areas and the revival of many nonmetropolitan communities. The decentralized pattern of development has obviated the use of mass transportation systems and increased dependency on the automobile.

Federally assisted highways and airports have proven so efficient at connecting the nation that railroads have lost passengers to cars, buses, and airplanes. But trains remain competitive in hauling freight. The United States ranks 15th among the world's countries in annual passenger rail miles, but second only to the Soviet Union in freight miles. The railroads are most competitive when hauling heavy loads over long distances, with the average load traveling 600 miles. But because trains usually require large assembly yards and break-bulk points at the originating and terminating stations, many businesses prefer to use trucks for shipping because they can haul freight anywhere.

Today the United States has over 264,000 miles of railroad track in use, down from 359,000 miles in 1960. Over the past 25 years, spur lines to many small towns have been abandoned, particularly in the Midwest and Great Plains states. Between 1929 and 1969, railroad freight increased from 447 billion ton-miles to 768 billion, and reached 875 billion ton-miles in 1985 (U.S. Bureau of the Census, 1986, p. 597).

Greater transportation competition has meant the demise of several railroad companies serving the Northeast and adjoining Midwestern states since the late 1960s. The federal government entered the railroad business in the 1970s by sponsoring Amtrak, a national passenger rail system, and Conrail, a more freight-oriented line in the Northwest.

The most important and recent federal action affecting railroads, buses, and trucking has been the deregulation of schedules, fares, and routes beginning in 1981. Fare deregulation allows railroads and trucking firms to set contract hauling prices that are not published; in essence, rates have become a matter of bargaining between the favorable rates. Route and schedule deregulation has already resulted in the reduction or elimination of direct rail and bus service to several rural communities. The performance of trucking service has varied; some communities have seen reductions in service, whereas others have seen increased service (Bradshaw & Blakely, 1987, p. 11-12). In order to maintain service to some rural communities, several states have allowed the creation of local port authorities to take over abandoned lines and virtually run the railroad system.

Inland waterways offer the cheapest mode of transporting freight, but they often rely on federal assistance. To regulate water flow so that freight can be hauled throughout the year, the federal government, chiefly through the Army Corps of Engineers, has spent billions of dollars to build dams, lock networks, canals, and channels across the nation. The volume of freight on U.S. waterways has increased from 118 billion ton-miles in 1940 to 351 billion in 1982. More than half of all water freight travels along the Mississippi River; most freight consists of petroleum and farm products.

The growth of air travel in the United States has been phenomenal, increasing from less than 10 billion passenger miles in 1950 to over 226 billion domestic passenger miles in 1983. Since the deregulation of air transportation in 1978, the number of airlines has almost tripled to over 100, including several regional airlines that provide service to many nonmetropolitan cities. The U.S. Department of Transportation currently spends about $250 million a year on airport facilities in nonmetropolitan counties.

Housing

Since the Great Depression, the federal government has built or subsidized the construction of hundreds of thousands of rural dwellings for low- and middle-income families, the elderly, and the handicapped. Out of the New Deal came the Federal Home Loan Bank System; the Federal Savings and Loan Insurance Program; the Federal Housing Administration (FHA); the Federal National Mortgage Association ("Fannie Mae"); and the Farm Security Administration, which became the Farmers Home Administration (FmHA) in 1947. The Fannie Mae system has assumed up to one-half of the nation's home mortgages, and hundreds of federal savings and loan associations have been established to make home loans. Numerous federally sponsored housing assistance programs are an important source of rural America's housing production. But housing problems in rural areas have continued to be worse than in metropolitan areas. In 1950, 9 million rural housing units—the homes of over half of all rural residents—were considered substandard, compared to 6 million substandard urban units comprising only one-fifth of all urban housing. By the mid-1970s, the number of rural and metropolitan substandard units had dropped to about 2 million each. However, nearly 4 million rural homes are overcrowded or occupied by low-income families who pay an excessive amount of their income for rent.

The major source of home financing for low- and moderate-income rural families is the FmHA. The FmHA serves unincorporated areas and towns of fewer than 10,000 people, or cities as large as 20,000 in nonmetropolitan counties where there is a serious lack of mortgage financing. The Housing Act of 1949 allows the FmHA to make subsidized or unsubsidized loans for the repair or purchase of single-family homes (42 U.S.C. §§ 1471–1490). "Self-help" programs offer interim financing to households who band together to construct each other's homes, and their work contribution becomes a down payment for commercial lending purposes. FmHA loans can be used to build housing for farm laborers, and the FmHA can make loans to public housing agencies or private companies for the construction of low- and moderate-income multifamily housing. The federal government spends over $1 billion a year on FmHA housing programs.

The U.S. Department of Housing and Urban Development (HUD), by law, must allocate between 20 and 25 percent of its funds to rural areas outside of metropolitan counties. HUD operates several programs that provide rent and home purchase subsidies to low-income families. In 1985, HUD spent about $5 billion on these programs in nonmetropolitan counties. Section 8 of the Housing and Community Development Act of 1974 offers rent subsidies to tenants of qualified housing projects (Pub. L. No. 93-383, 42 U.S.C. § 5304 *et seq.*). For example, a tenant pays between 10 and 30 percent of his or her income for rent, and the federal government pays the difference between the tenant's contribution and the market rent on the unit. Rent subsidies are available for new housing construction and substantial rehabilitation, as well as for existing units.

Through Section 235 of the Housing Act of 1937 (42 U.S.C. § 1401), HUD can help low-income families in buying homes. HUD funds are used to buy down mortgage interest rates. The home mortgage is then insured by the FHA, which encourages private lenders to make the loans. Other HUD programs provide low-interest loans and mortgage insurance, as well as rent subsidies for elderly and handicapped housing.

There is also a federal insurance program for mobile homes; this program insures lenders against losses on mobile homes that are of marginal quality or are improperly sited. Mobile homes are often the only viable housing option for low- and moderate-income rural inhabitants, yet many rural banks are reluctant to make mobile home loans. As of 1984, the federal insurance program covered almost 260,000 units with a value of $4 billion (Bradshaw & Blakely, 1987, p. 11-10).

Economic Development

Federal rural economic development programs have played a major role in modernizing rural living and promoting prosperity. Perhaps the most successful program has been the Rural Electrification Administration, which encouraged the creation of rural electric cooperatives through loans for power lines. In 1933, only 10 percent of the nation's farms had electricity; rural and farm life has been transformed by lights, running water, refrigeration, washing machines, and milking machines. In 1949, the Rural Telephone Loan Program began, and by 1982, 97 percent of all farms had telephone service. Rural Free Delivery (RFD) has provided mail service to rural areas for over 80 years. These three federal programs have increased the quality of life in rural America as well as reducing the isolation of rural dwellers.

Today, most federal economic development money is aimed at specific private business projects or the construction of public facilities and infrastructure. The major rural economic development programs include community development block grants, the FmHA, the Small Business Administration (SBA), the Commodity Credit Corporation (CCC), and the Economic Development Administration (EDA).

The Housing and Community Development Act of 1974 inaugurated a new concept in federal assistance to communities: the community development block grant. Prior to 1974, a rural community might have had to prepare multiple applications under several different federal categorical grant programs, with mountains of paperwork and red tape. The block grant program replaced several of the categorical grant programs, including urban renewal, model cities, public facility loans, water and sewer grants, historic preservation grants, neighborhood facility grants, and neighborhood development program grants, among others.

There is generally only one block grant program available to rural communities. To qualify, a community must have a community development program that is based upon a locally approved plan. Block grant funds can be used for a variety of public works projects, such as sewer and water lines, streets, child care, and the preparation of a community development plan or energy conservation strategy, among other projects. Projects must demonstrate that a certain percentage of new jobs created will go to low- and moderate-income people. A total of $3 billion in community development block grants was made available in 1987 (Milkove & Sullivan, 1987, p. 14-22). Federal block grant funds are competitively awarded through each state government to rural communities each year. Proficiency in proposal preparation and experience in program administration largely determine the distribution of these discretionary funds.

The FmHA features several economic development programs, including farm loans, community service loans, and business loans. The FmHA serves as a lender of last resort for farmers who are unable to obtain credit from commercial banks or the Federal Farm Credit System, consisting of the Production Credit Association and the Federal Land Bank. The FmHA makes farm loans or loan guarantees to purchase land or to meet operating expenses. FmHA emergency farm loans are offered to overcome the impact of a natural disaster on crops or property.

Between 1949 and 1978, some 15,000 rural communities and farms solved their water supply and sewage disposal problems through modern central sewer and water systems built with FmHA loans or loan guarantees. Over the same time period, over 3,000 communities received loans to construct public facilities. In addition, between 1972 and 1979, FmHA guaranteed 1,200 loans for buildings and other private business and industry in rural areas. The volume of these guaranteed loans has fallen from over $1 billion in 1980 to $96 million in 1987 (Milkove & Sullivan, 1987, p. 14-22).

The SBA is an independent agency created by Congress in 1953 to provide loans and technical assistance to small businesses. The SBA currently loans about $3 billion a year to businesses in nonmetropolitan counties (Milkove & Sullivan, 1987, p. 14-22). Through its regular business loan program, the SBA offers guaranteed, immediate participation, and direct loans. Under a guaranteed loan, a private lending institution advances the entire loan to the applicant, and the SBA guarantees up to 90 percent of the loss in case of default. Immediate-participation loans call for the SBA to provide a portion of the loan, and the lending institution to provide the balance. Direct loans are made entirely by the SBA; these loans may be used for crop or operating expenses, breeding livestock, farm machinery, and real estate construction and improvements.

SBA economic opportunity loans can be extended to economically disadvantaged individuals who have insufficient income for basic family needs, and are unable to obtain adequate business financing through traditional lending channels. Applicants must demonstrate an ability to operate a business successfully, and must have a reasonable prospect of repaying the loans from earnings of the business. The SBA also provides recipients of economic opportunity loans with management assistance. Similarly, water pollution control loans are made to assist applicants in meeting the standards of the Federal Water Pollution Control Act (1972) or other state standards (33 U.S.C. § 1311(a), 1342(a), 1362).

The SBA makes three kinds of disaster loans to those who have sustained losses in a natural disaster as declared by the president or by the administrator of the SBA. Disaster business loans are available to any farm business to repair or replace farm property. Economic injury disaster loans are available to farmers and their tenants who sustain losses to their homes and personal property. Disaster home loans are available to farmers and their tenants who sustain losses to their homes and personal property.

The federal CCC is a wholly owned government operation managed by the USDA for the purpose of (1) stabilizing, supporting, and protecting farm income and prices; (2) assisting in the maintenance of balanced and adequate supplies of agricultural commodities and their products; and (3) facilitating the orderly distribution of commodities.

The CCC operates two main programs: a nonrecourse loan program to boost farm income, and an export credit sales program to promote the sale of U.S. farm products overseas. Under the nonrecourse loan program, farmers may borrow money from the CCC for up to nine months at a set interest rate and pledge certain crops as collateral (especially wheat, corn, and soybeans). At the end of the loan period, farmers have the options of paying off their loans and retrieving their crops from government storage, or of not repaying their loans and letting the CCC take ownership of the crops. When crop prices fall below the official loan rate, farmers tend to take out nonrecourse loans and put up their crops as collateral. For example, in 1985 the loan rate for corn was $2.55 a bushel, and the market price averaged about $2.30 a bushel. Farmers were better off storing corn with the CCC, taking loans at $2.55 a bushel, and letting the CCC take possession of the stored corn. This way, corn farmers were able to earn an additional 25 cents per bushel.

If crop prices rise above the loan rate, a farmer can sell the crops from government storage and repay the loan. If crop prices fail to rise above the loan rate, the farmer can keep the loan, and the CCC keeps the crops. The CCC then absorbs any losses if the government sells crops from storage below cost on the open market. In 1985, the CCC loaned $11.8 billion to farmers; CCC loans that year accounted for 38 percent of net farm income (U.S. Bureau of the Census, 1986, p. 630). The CCC may borrow up to $25 billion from the U.S. Treasury or private sources in order to make nonrecourse loans to farmers. Between 1981 and 1985, the CCC spent $50 billion because of low crop prices and high levels of borrowing by farmers of nonrecourse loan funds (U.S. Bureau of the Census, 1986, p. 635). In 1987, the CCC reached its $25 billion loan limit and had to apply to Congress for additional funding.

Since 1965, the EDA has targeted communities with high unemployment for the creation of new jobs. EDA grants have supported public works projects, technical assistance, business loans, and planning and research programs. The largest portion of EDA funds has been spent on water and sewer lines and roads for industrial development, totaling $1.5 billion between 1965 and 1973 alone. In addition, EDA money has long been used to support regional planning commissions in many states. Another major program features the funding of multicounty economic development districts and redevelopment areas within those districts. Each district must have one or more growth centers with the potential to foster economic growth. These centers should eventually be able to generate sufficient private investment to alleviate the depressed economy of the region. The EDA has established economic development regions in five lagging sections of the nation: the upper Great Lakes; New England; the Ozarks; the coastal plains in the Southeast; and the "four corners" of Utah, Arizona, Colorado, and New Mexico. As of fiscal 1987, EDA spent $150 million a year in rural areas (Milkove & Sullivan, 1987, p. 14-22). This is a significant reduction from the 1980 level of $540 million.

The federal government has played and continues to play a pervasive yet fragmented role in the development of rural America. Though some programs have achieved noteworthy success, others have produced undesirable development patterns (e.g., sprawl caused by FmHA-financed rural sewer and water lines) or short-lived economic gains (e.g., the ARC). Many rural areas, such as the Tennessee Valley and major farming regions, remain heavily dependent upon federal funds. But the proliferation of federal rural development programs has not followed a systematic plan or strategy to improve the economic and living conditions in rural America. The funding levels of different programs rise and fall over time, and Congress seems to respond to the needs of specific interest groups by creating one specific program after another. Even the Rural Development Act of 1972 and the Rural Development Policy Act of 1980 did not target geographic regions or economic sectors (7 U.S.C. §§ 1013a, 1921–1995; Carter Administration, 1980). The result is a confusing maze of federal requirements and eligibility conditions that few rural communities have the expertise to master.

The rural planner can take three steps to help a community take advantage of federal funding. First, the planner should become familiar with what federal programs exist and how these programs might be used to meet local housing, economic development, and infrastructure needs. Second, the planner should contact federal and state agencies to pursue funding possibilities and to stay abreast of changes in programs. Finally, the rural planner should be cautious about getting the local community "hooked" on federal funding. Many rural communities face budget cutbacks, service reductions, or higher property taxes because of the end of federal revenue-sharing funds. Even though federal rural development programs are not integrated, the rural planner can help tailor the use of federal money and regulations to fit into a local community as part of a comprehensive planning process, especially in the formulation and implementation of capital improvements programs and land use plans.

Resource Industries

The United States has long enjoyed an abundance of natural resources, featuring fertile farmlands and forestlands, the world's foremost coal reserves, a host of mineral wealth, picturesque mountains, vital rivers and lakes, and thousands of miles of coastline. The use of these resources has been a key factor in America's economic growth, especially in rural areas. Local development and regional growth often depend on the availability and exploitation of natural resources. Rural industries have played a major role in extracting, harvesting, and processing natural resources for human consumption. These industries include mining, forestry, rural recreation, and fisheries. A single-resource industry acts as the dominant force in several rural communities. In general, the more remote the location and the more capital-intensive the industry, the greater the likelihood of local dependence on the resource. Mining companies, for example, are typically multimillion-dollar firms that can exert considerable control over remote communities, which are economically dependent on the local mines. By contrast, the fishing industry is highly labor-intensive, being comprised of independent fishermen scattered throughout the coastal states; moreover, there are relatively few rural communities that rely heavily on fishing for their well-being.

Timber companies and rural resort operators have more in common with mining firms than with fishermen. Timber companies own or lease vast acreages and often hold sway over rural communities through their lumber and paper mills. Similarly, rural resort operators are likely to be the largest employers in their towns. Yet there are also likely to be dozens of smaller businesses that feed off the vacationers drawn to a resort.

An important difference among these four industries is the amount of capital involved. Fishermen generally own their boats, and a few hundred thousand dollars usually provides sufficient funds to get started. A resort owner requires a few million dollars to construct and operate a ski area, marina, or hotel–recreation complex. Mining and timber companies often have assets of several hundred million dollars or more. Most of these companies are publicly held and operate in countries throughout the world. Of owners in these four resource industries, only the fishermen and perhaps the resort operators are locally based owners. Mining and timber companies are nearly always headquartered in major urban centers, far from the towns where the ore and wood products are produced.

Whereas fishermen and resort owners may be able to influence the local public decision-making process, mining and timber companies can also influence state governments, Congress, and the president. For instance, timber companies were instrumental in pushing the enactment of the Timber Relief Act of 1984 (Pub. L. No. 98-478). This legislation freed timber companies from about $1.7 billion in federal timber contracts—agreements to harvest some 7 billion board-feet on government lands. A true reflection of the power of these companies is that both Congress and President Reagan approved of the huge bailout at a time when many federal programs were facing funding cuts as part of a deficit reduction package. At the state level, timber and paper companies have both a major economic and a major political presence. Paper companies, including such giants as International Paper and Scott Paper, own or control vast portions of the state of Maine. Timber companies such as Weyerhauser, Boise Cascade, and Louisiana Pacific dominate dozens of towns in Oregon and Washington, and exert a powerful influence of those states' legislatures. In the small mill towns, the timber companies operate the local mills in an on-again, off-again fashion in response to distant market conditions. Local workers have little, if any, say about the employment policies. Currently, several timber companies are focusing their operations on the Southeast, where trees grow faster than in the Northwest and federal lands are few.

In Appalachia and the intermountain West, mining companies comprise a formidable economic and political force. In West Virginia, a handful of coal and energy companies own almost half of the surface land and a large majority of the mineral rights. Similarly, a handful of companies hold the majority of mineral rights on federal lands throughout the western United States. Mining, oil, and electric companies have developed numerous mines, facilities, and generating stations in remote areas of the West. With the coming of the energy crisis in the mid-1970s, the intermountain West witnessed the booming of many small towns as energy firms sought to exploit vast coal reserves in Colorado, Montana, and Wyoming; to mine uranium in Colorado, New Mexico, and Utah; and to construct huge electrical generating plants in Arizona, Colorado, Montana, Washington, and Wyoming. Often the decisions of the energy companies were made without prior consultation with the towns to be affected by the swarm of newcomers. As a result, boomtowns suffered from sprawl, environmental degradation (especially water pollution), and social problems (e.g., increased alcoholism, crime, child abuse, and divorce). Planning was notably lacking. Since the downturn in energy and metal prices in the 1980s, many of these boomtowns have experienced a loss in population and economic activity as the "bust" phase of the boom–bust cycle set in.

The survival of resource communities depends upon their ability to diversify the local economy, to accommodate social change, and to receive a satisfactory price for local resources. The potential for economic diversification is determined by location, the size of the local population, and the size and quality of the local resource base. For example, commercial fishing communities often have taken advantage of their attractive locations to add sport fishing and other recreation and tourist-related development. Some of this diversification has occurred from sheer necessity. In recent years, commercial fishing has come under more stringent regulations in response to declines in fish catches. Although Congress enacted a 200-mile limit in 1976 to protect U.S. fishermen, restrictions have been imposed on fishing seasons, on the number of boats in a fleet, and even on fishing technology.

Rural recreation areas have achieved some economic diversity as some owners of second homes become permanent residents and recreation facilities are used on a year-round basis. Mining and timber communities often have difficulty in promoting economic diversity. A mining community faces a limited time horizon because the nonrenewable minerals are sure to be depleted. Moreover, many mining towns are remotely situated, and local employment rises and falls with changes in energy and metal process. Timber towns also tend to be located in areas far from urban centers, and the prices of wood products are determined in national and international markets. For example, much of the Northwest timber is subject to competition from Southeast and Canadian lumber both at home and abroad. In sum, rural resource communities are extremely vulnerable to events beyond their control. This fact can make developers and entrepreneurs wary of investing in new businesses in these towns.

Resource industries frequently compete with each other over the use of rural land. The most common competition is between rural recreation operators on the one hand and forest companies, farmers, commercial fishermen, and/or mining companies on the other. A rural recreation area generally attracts well-to-do urban dwellers who may purchase a second home near the marina or ski resort. These urban visitors often outbid other resource industries for the use of the land. In many coastal areas, recreational sport fishermen compete directly with commercial fishermen over fish stocks and harbor space.

Another form of competition among resource industries occurs in the effects of particular industries on the environment. For instance, timber harvesting along coastal streams has resulted in soil erosion, stream siltation, and the destruction of spawning grounds of anadromous fish (e.g., salmon). Mining slag heaps and timber clear-cuts reduce an area's attractiveness for outdoor recreation. The requirement of managing federal lands for multiple uses has often brought recreation and commercial timber harvesting into direct conflict.

Much education about the benefits of planning remains to be done in resource communities. The planner must work openly with the local resource industry and understand the industry's needs and desires. Then the planner must devise strategies that harmonize with the industry's development potential and mitigate the negative environmental impacts and the community's economic vulnerability. The primary goals of economic diversification, social accommodation, and environmental protection should be openly discussed, and a consensus for local action should be developed.

Export Agriculture and Agribusiness

Agriculture continues to be an important, if not a dominant, component of many rural economies. As of 1984, farming and agriculture-related industries employed 6.2 million people, or almost 31 percent of the labor force in nonmetropolitan counties (Reimund & Petrulis, 1987, p. 4-23). About 2.7 million of these jobs were in farm production and services. Although farmers are romantically portrayed as self-sufficient Jeffersonian yeomen, reality reveals them to be business people who are often at the mercy of national and international markets over which they have little control. America's agricultural sector should be seen as another resource industry, except for one key difference. Resource firms are often vertically integrated: They own the land (mines and forests), processing plants, and even marketing outlets, which enables them to hold down the cost of inputs and have some control over the price of the final products. Farmers, on the other hand, typically own only the land and have little control over the prices they pay for machinery and other inputs or over the prices they receive for their products. Many of the decisions that affect the American farmer's livelihood are made by Congress and federal government agencies in Washington, farm commodity traders in Chicago, farm equipment manufacturers, and foreign farmers and trading partners.

The United States has often been called the most perfect place to grow food in the entire world. The soil is fertile, the climate is fairly mild, and rainfall is generally abundant. These features, combined with advances in farm technology and farming methods, have resulted in the production of more food than America can consume. Although food has been exported from America since colonial times, over the past 20 years American farmers have increasingly looked to foreign markets to absorb the bountiful surplus of crops. In the process, American farmers have become dependent on distant markets that fluctuate according to the policies of the importing countries and the degree of competition from other food-exporting nations such as Argentina, Australia, Brazil, Canada, and France. Most American food exports are shipped to well-to-do countries, which can afford to buy food in the world market at prevailing prices. Major importers of U.S. food include Japan, the European Economic Community (EEC), South Korea, Taiwan, mainland China, and the Soviet Union.

The foundations of modern food trade were set in 1948 with the General Agreement on Tariffs and Trade (GATT), which sought to remove tariff and quota trade barriers among non-Communist nations. In 1950, America exported only $3 billion worth of food. The expansion of U.S. farm exports was stimulated by the passage of the Food for Peace Program (Pub. L. No. 83-480, 7 U.S.C. §§ 1691–1736(e)) in 1954. Over the next 30 years, the United States donated or sold over $30 billion of food with low-interest loans to countries around the world (Morgan, 1980, p. 382). Nonetheless, in 1970 American food exports totaled only $7 billion (Batie & Healy, 1980, p. 1). It was not until the huge export sales of cereals and oilseeds to the Soviet Union between 1972 and 1974 that foreign markets were seen as providing a large and steady demand for U.S. farm products. By 1979, America exported $35 billion worth of food—a fivefold increase over 1970—and was well established as the world's leading exporter of wheat, corn, and soybeans (Batie & Healy, 1980, pp. 1–2).

The expansion of foreign markets has accelerated the specialization of crops and farmland use according to regions. Midwest farms are planted almost exclusively in corn and soybeans, and the Great Plains and the Pacific Northwest are prime wheat-growing areas. Between 1973 and 1976, the number of acres in wheat increased by 33 percent, corn by 17 percent, and soybeans by 37 percent. In 1985, of the nation's 331 million acres of harvested cropland, 65 million acres were planted in wheat, 75 million acres in corn, and 62 million acres in soybeans (U.S. Bureau of the Census, 1986, pp. 642, 644). In 1984, the United States exported over half of the wheat grown in the country, almost one-fourth of all corn, slightly over one-third of the soybeans, and nearly half the cotton; U.S. exports accounted for almost two-thirds of world corn and soybean exports and one-third of world wheat exports (U.S. Bureau of the Census, 1986, pp. 641, 642). In fact, one-quarter of U.S. farm income is earned from exports, and almost 1 of every 3 acres of cropland is planted for export crops (Reimund & Petrulis, 1987, p. 4-29).

As export agriculture has grown in importance, the vulnerability of American farmers has also increased. Despite the GATT treaty, many of the countries that import American food have continued to impose quotas, tariffs, and quality restrictions. Countries employ such trade practices to protect their own farmers from American competition. Several countries, especially Japan, South Korea, and the EEC, have food-marketing boards that control both the amount of food imported and domestic prices. For example, a nation's marketing board will buy American wheat at the world price and then sell it at a higher price to its own people. At the same time, foreign domestic farmers get more than the world price for their goods. Some countries also provide special subsidies for farm exports so that they will be able to undersell American farm products in the world market.

Export agriculture has assumed a role of great importance in the U.S. economy since 1972. Although raw agricultural commodities account for only about 2.5 percent of the gross national product (GNP), between 1974 and 1984 these commodities have earned a net surplus of $10–20 billion annually in foreign trade (see Table 5.2). Meanwhile, the cost of importing oil soared from $4 billion in 1972 to roughly $50 billion in 1984, and the overall national trade deficit for 1984 has climbed to over $100 billion (U.S. Bureau of the Census, 1986, p. 789). Without the earnings of export agriculture, America's trade deficit would worsen. The dollar would be worth less in other currencies, and foreign goods, especially oil, would be more expensive to import.

On the other hand, export agriculture is vulnerable to swings in the value of the dollar overseas. When the value of the dollar rises in relation to other currencies, American farm products become more expensive to import, and food-importing countries look to other food-exporting nations; Americans' farm export earnings are thus likely to fall, as they did in the early 1980s. Since 1984, the value of the dollar has fallen, U.S. farm exports have approached the peak of the early 1980s, though food imports have climbed.

Export agriculture has also been used as a bargaining tool in international diplomacy. For example, the Camp David treaty between Egypt and Israel was sealed when the United States pledged to supply Egypt

TABLE 5.2. Foreign Trade 1965–1984 (in Billions of Dollars)

Year	Agricultural exports	Agricultural imports	Nonagricultural trade balance	Total trade balance
1965	6.23	4.09	3.71	5.85
1970	7.26	5.77	1.35	2.84
1972	9.40	6.47	−9.34	−6.41
1974	22.00	10.25	−14.75	−3.00
1976	23.00	10.99	−20.67	−8.68
1978	29.41	14.80	−46.82	−32.21
1980	41.30	15.00	−50.10	−24.20
1981	43.30	16.00	−54.60	−27.30
1984	25.00	16.00	−131.30	−122.30
1985	29.20	20.00	−141.30	−132.10

Note. The data are from USDA (1983b, p. 6), and U.S. Bureau of the Census (1986, pp. 625–626).

with several million metric tons of wheat each year. However, the "food weapon" has also backfired. When President Carter imposed a ban on grain exports to the Soviet Union, in retaliation for the takeover of Afghanistan, other grain-exporting nations (Argentina, Australia, Brazil, Canada, and France) increased production; this both satisfied Russian demands and drove down the world prices of wheat, corn, and soybeans.

In sum, American agriculture has become fully exposed to the global economy. The price of farm inputs such as gasoline and fertilizers fluctuates, depending on the Organization of Petroleum-Exporting Countries (OPEC) and the world oil supply. The price of export crops is largely determined by world supply-and-demand relationships. American farmers currently rely on multinational agribusiness firms for machinery, fertilizer, pesticides, herbicides, and seed; likewise, farmers typically sell their produce to large global corporations. For example, 90% of the world grain trade is controlled by six privately held companies (Morgan, 1980, p. 20); 13% of the nation's agricultural service firms earn 68% of all receipts; and the four largest tractor manufacturers share over three-fourths of the market (Vogeler, 1981, pp. 107–108). This situation provides farmers little protection against rising production costs and falling commodity prices, resulting directly in a cost–price squeeze that has brought about the exit of many small and medium-sized farms.

In recent years, more farmers, particularly in the Midwest, have attempted to protect themselves from price fluctuations by selling their products on futures contracts at the Chicago Board of Trade. Such contracts allow a farmer to "hedge" against a drop in market prices by offering a chance to sell contracts at a set price far ahead of the actual sale date. If futures prices fall, then the farmer can buy a contract and pocket the difference between the sell contract and the buy contract. If futures prices rise, the farmer has still been protected against a price decline. But futures prices fluctuate, especially in relation to current market prices, so that there is no guarantee of receiving a better price in the futures market. Moreover, crop prices on the Chicago exchange are influenced by food-processing companies, international grain-trading companies, and speculators. The volume of commodities traded can be enormous, running into several billions of dollars in a single day.

Farmers are also vulnerable to changes in federal monetary and fiscal policies, as well as a variety of farm programs. Tight-money policies, featuring a slow growth in the nation's money supply and high interest rates, may be effective in curbing inflation but are especially damaging to farmers. During inflationary times, the value of land is likely to increase faster than inflation. Many farmers benefited from rising land values between 1955 and 1980. This growing wealth enabled them to borrow more money to expand their operations and attain greater efficiency in production. However, between 1980 and 1985 interest rates exceeded 10 percent and often 15 percent, and inflation declined sharply. Farmers suffered from declining land values, which hurt their ability to borrow, and the high interest rates increased debt burdens and seriously decreased their cash flows. Commodity price supports, the Payment-in-Kind (PIK) program, and easier federal lending policies have only partly offset the detrimental effects of tight-money policies. These policies also drove up the value of the dollar overseas, made U.S. farm exports more expensive, and reduced export earnings (see Table 5.2).

In recent decades, American agriculture has witnessed a growing concentration of economic and political power in Washington, Chicago, and overseas. Farmers have been increasingly subject to forces beyond their control. This greater vulnerability to outside influences has had a substantial impact on many rural communities. In response, the need to diversify local economies and to gain more local control over economic activities has become more widely recognized.

Conservationists, Environmentalists, and Historic Preservationists

In the late 1960s, there was a growing awareness that something was "wrong" with the natural and human-made environments. Smog in Los Angeles, the imminent "death" of Lake Erie, and the destruction of handsome old buildings served to create a feeling that the nation should establish and enforce standards of environmental quality. Since the "environmental crisis" was first popularized, wide-ranging actions to maintain and enhance environmental quality have been taken by government agencies, industry, and private citizens. Nor has the concern for natural and human-made surroundings diminished; rather, we have been sensitized to the need for the continued monitoring of air and water quality, the protection of wildlife and natural areas, and the preservation of our architectural and cultural heritage.

Environmental Concerns

The environmental movement has produced a shift in power away from the industrial domination of nature toward public power in regulating the use of the environment. Historically, America's abundance of natural resources gave little incentive to conserve. Even the utilitarian policies of

Theodore Roosevelt and Gifford Pinchot treated the natural environment as the provider of resources to be exploited. Today, the environment is viewed as offering a diversity of valuable services and benefits, from wildlife habitat to water recharge areas to scenic amenities, as well as the traditional food, fiber, and mineral resources. Concerns have been raised about the scarcity of resources, the rate of resource development, and the need to conserve resources for future generations. Air, water, and land have been identified as receptors of waste, and their ability to absorb waste depends on the rates and kinds of pollution. Clean air and water are necessary to sustain human health and a variety of other life forms. Natural and scenic areas have been recognized as scarce and in some cases irreplaceable; the more scarce these areas become, the more valuable they are.

All of these perceptions center around a single basic truth: "There is no such thing as a free lunch." The degradation of the environment is the price we have paid for a high standard of material well-being. The omnipresent automobile has produced air pollution; cheap mass-produced food has resulted in soil erosion and water pollution; and useful new chemicals have brought about problems of toxic waste disposal. The benefits of economic growth have also generated costs that are often difficult to measure and that people are often reluctant to bear. The costs of dirty air and dirty water to human health are hard to tabulate, because they usually become evident only over a long period of time. Moreover, the cost of cleaning up the environment and maintaining environmental quality are not easy to determine for two reasons: How much cleanup should occur, and who should pay and how much? The cleaner we want the air and water, the more we must pay. For example, to take a river from 40 to 60 percent "clean" will cost far less than to bring the river from 60 to 80 percent clean. To achieve higher levels of water quality requires more sophisticated and expensive treatment facilities and processes. Ideally, the benefits of cleanup should match the costs. But many ecological systems do not operate on such linear terms; rather, they are subject to threshold effects. That is, if the environment does not meet and maintain a certain level of air or water quality, life forms will cease to exist.

The issue of who pays and how much is complicated by two factors. First, air and water have traditionally been viewed as common property resources. No one owns the air or water, and yet everyone has access to them. Second, it is very difficult to tell how much damage a polluting activity has on air or water quality. Influences such as other pollutants, climate, water flow, and pollution dispersal patterns all play important roles. These two factors have limited the use of market solutions for pollution cleanup. No one owns the air or water, so that there can be no exchange of property rights to air or water. And the value of clean air and water is uncertain. Collective action in the form of regulation and litigation has been used to force polluters to reduce emissions and spend over $50 billion a year on controlling pollution (U.S. Bureau of the Census, 1986, p. 193). Although some of the benefits of these controls may be readily apparent, most of the benefits accrue over a long time in the form of reduced costs to human health.

Many rural areas may appear far removed from an environmental crisis. But as more people and industries are drawn to rural areas (or as industries threaten to pull out), conflicts over the use of the environment are likely to increase. Local decisions over the environment may be challenged by newcomers, distantly based corporate headquarters, environmental groups, and state and federal environmental agencies.

As more people settle in rural areas, they place greater demands on local water supplies, sewage disposal, and natural areas for recreation. At the same time, newcomers may want to preserve the scenic and environmental qualities that drew them away from urban areas. Often, they argue for "quality" development, which will both mitigate unsightly and environmentally harmful development patterns and enhance a community's livability. Rural governments, however, may be slow to respond to these environmental concerns, which are frequently perceived as thinly veiled attempts to curb economic development. Because rural people tend to experience more poverty and receive lower incomes than urban dwellers, jobs and economic growth are often given high priority. Environmental protection is sometimes regarded as a luxury or an elitist cause.

Nonetheless, many of the decisions that affect a community's environment are not made at the local level and represent conflicting values and interests over protecting the environment. Many of the problems of the rural environment originate in urban areas; the uses and abuses of the rural environment can be traced to government, corporations, and the demands of urban populations for recreation and second homes. Government policies have promoted the construction of over 250 dams, flooding thousands of acres, and have encouraged the siting of power plants (including nuclear reactors) in rural areas. Resource industries tend to be more concerned with short-term profits than with long-term environmental and aesthetic qualities. Mining wastes cause water pollution; strip mining for coal scars the landscape; refining processes contribute to air pollution; and tourists and recreationalists tend to build resorts and second homes away from established centers, thus increasing rural sprawl, reducing water quality, and infringing on natural areas and wildlife habitats.

The clash of these urban and rural interests raises the questions of whose rights to the environment should prevail, who should benefit, and who should suffer losses. For instance, the damming of a river to generate hydroelectricity may reduce the fishing opportunities that attract fishermen from urban areas, whereas the local economy is likely to benefit from cheaper electricity. Yet, these fishermen may attempt to block the construction of the dam, arguing that they have a right to enjoy fishing in the undammed river even though they do not live in the immediate vicinity. Local residents may thus be forced to bear the cost of restrictions in not being able to dam the river, which primarily benefits nonlocals. In some cases, exactly the opposite situation occurs. Urban areas are prolific water consumers. Often they promote large-scale reservoirs at remote upstream locations, which will have the effect of removing thousands of areas of land from production and disrupting rural community life. In this instance, there may be little beneficial use for the local community.

A similar clash between urban and rural interests occurs in the creation of wilderness areas. Local inhabitants are effectively denied the use of local land except for recreation, and yet the locals are forced to compete for a recreation experience with visitors from afar. The outcome of wilderness areas may ultimately be a shift in the local rural economy from one based on agriculture, forestry, or mining to one based on recreation and tourism.

Over the past two decades, environmentalists have succeeded in translating public awareness of the environment into political influence. At the local level, environmentalists are much more widely represented on local zoning boards, planning commissions, and governments. Environmentalists have also been instrumental in the push for statewide land use laws and environmental controls. Today, virtually every state has a department of environmental affairs or natural resources.

At the national level, environmentalists have successfully lobbied for legislation on a variety of issues. In 1969, Congress passed the National Environmental Policy Act, which requires federal agencies to prepare an "environmental impact statement" when federal funds are involved in a development project (Pub. L. No. 91-190, 42 U.S.C. §§ 4321–4361). The statement must address (1) the adverse environmental effects of proposed projects; (2) alternatives to the proposed action; and (3) the long-term and irreversible use of resources. The act also includes a comprehensive set of policies and goals: environmental quality, national heritage protection, population–resource balance, and the management of resources for both current and future generations. The National Environmental Policy Act turned the concept of environmental management into national policy and provided environmentalists with a strong legal framework for action. The idea of environmental impacts of development has since spread from federal projects to private projects, and state and local governments have often asked private developers to prepare impact statements. Since 1970, many companies have added special departments to monitor and improve the environmental performance of their operations.

Publicity over the "environmental crisis" culminated in Earth Day, April 22, 1970. President Nixon responded to Earth Day by establishing the Environmental Protection Agency (EPA) and the Council on Environmental Quality. Congress initiated a wave of environmental legislation, including the Clean Air Act Amendment (42 U.S.C. § 1857 *et seq.*, 1970), the Solid Wastes Disposal Act (42 U.S.C. § 3251 *et seq.*, 1970), the Occupational Safety and Health Act (Pub. L. No. 91-596, 29 U.S.C. §§ 657–678, 1970), the Federal Water Pollution Control Act (33 U.S.C. §§ 1311(a), 1342(a), 1362; 1972), the Federal Insecticide, Fungicide and Rodenticide Act (7 U.S.C. § 136 *et seq.*, 1971), the Coastal Zone Management Act (16 U.S.C. § 1451 *et seq.*, 1972), and the Endangered Species Act (16 U.S.C. § 1531 *et seq.*, 1973). EPA is responsible for implementation and enforcement of the following major statutes:

- Clean Air Act (see above)
- Federal Water Pollution Control Act (33 U.S.C. §§ 1311 *et seq.*, 1972)
- Noise Control Act (49 U.S.C. § 1341; 1972)
- Safe Drinking Water Act (42 U.S.C. § 300(F); 1976)
- Toxic Substances Control Act (15 U.S.C. § 2601–2629; 1976)
- Resource Conservation and Recovery Act (hazardous waste) (42 U.S.C. §§ 6901–6991, 1982; Supp. III, 1985)
- Comprehensive Environmental Response, Compensation and Liability Act (Superfund) (Pub. L. No. 96-510, 94 Stat. 2767, 1980; codified as 42 U.S.C.A. §§ 9601–9675, West 1987)
- Marine Protection Research and Sanctuaries Act (ocean dumping) (16 U.S.C. § 1431 *et seq.*, 1972)
- Federal Insecticide, Fungicide and Rodenticide Act (see above)

The EPA is responsible for the implementation of clean air and water legislation. This includes setting minimum air and water quality standards, monitoring air and water quality, regulating pollution emissions, imposing fines on polluters, and making grants to localities for water and sewage treatment facilities. During the mid-1970s, Congress granted the EPA new powers to deal with chemical production and waste disposal. A "Superfund" of over $1 billion was established in 1980 to pay for the cleanup of hazardous wastes. The Superfund has since been extended into the 1990s with funding at $5 billion.

EPA decisions affect a wide variety of rural activities, from the way a community disposes of its sewage to the control of industrial air pollution. Few of the EPA's standards and decisions are free of controversy. Rural communities often resent the intrusion of federal regulators. Yet the increase in urban demands on rural areas has sparked concerns that cities are exporting their environmental problems to the hinterlands. The recent migration of chemical treatment and storage plants to rural areas is an obvious case. But some rural environmental problems are basically indigenous, such as the depletion of groundwater, groundwater pollution, the pollution of water from agricultural chemicals, and feedlot runoff. Many rural communities depend on groundwater for drinking water, and people living outside of towns typically get their water from wells. Once groundwater is polluted, it is very difficult to clean up and poses a serious threat to human health.

In sum, rural communities are witnessing conflicting demands over the environment. Balancing the diversity of interests involved requires tough choices with long-term implications for who receive the benefits of environmental quality and who must pay the cost. Local decision makers must comply with state and federal environmental regulations, and yet must provide for economic development opportunities. Industry has generally acknowleged the need for environmental controls; however, industry complains at the cost of installing devices to comply with pollution standards, altering the scope or design of a development, delays in obtaining development permits, and associated legal expenses. Moreover, many industries continue to be very powerful in rural areas and may attempt to gain a relaxation of regulations to reduce operating costs. Such action may in turn lead to intervention by national environmental groups. Thus, local land use decisions may have larger than local effects, and localities may become the battleground between opposing nonlocal interests.

Historic Preservation

Historic preservation in rural areas has evolved as an offshoot of urban efforts to protect America's architectural heritage. Whereas environmental regulations may involve conflicting local and nonlocal interests, historic preservation offers a compromise between local residents and outside visitors. Local benefits from the renovation of old buildings and the rehabilitation of downtown areas can help improve a community's image, stimulate retail trade, and boost morale (Williams, Kellogg, & Gilbert, 1983). In turn, tourists and their dollars are drawn to scenic historic districts. It is now widely recognized that historic preservation can act as a catalyst for planning in a community. Concurrently, there is a growing awareness that historic preservation, environmental conservation, and rural planning share a common concern for the future survival of rural areas.

In 1949, Congress created the National Trust for Historic Preservation to encourage public participation in the preservation of sites, buildings, and objects of historic and cultural significance (36 C.F.R. § 68.2). The National Historic Preservation Act of 1966 established a National Register of Historic Places and has provided grants to states for historic site identification, planning, and acquisition (16 U.S.C. §§ 470–470n). Over the years, the National Trust has become extensively involved in preservation, planning, and economic development activities. Through its National Main Street Program, the trust has provided grants to upgrade over 100 commercial districts in cities of 5,000–50,000 inhabitants, with an emphasis on preserving historic architectural features. The trust's Rural Project, begun in 1979, has granted opportunities for historic preservationists and environmentalists to cooperate in preserving open space (including farmland), as well as protecting the historic built environment. This project has been especially active in the Northeast, where the preservation of architecture has been linked with the region's tourist industry.

There is a growing legislative connection among historic preservation, farmland protection, and the preservation of viable rural communities. Preservation and conservation easement legislation has been developed to protect historic and architecturally significant buildings and open spaces, including prime agricultural lands. Easement programs enable tax-exempt organizations or government agencies to acquire partial interest in a property through the transfer of some property rights (usually development rights) from a landowner, who then retains the title to the property. There has been a notable increase in the number of private nonprofit trusts with preservation goals. Trusts have received donations of land, buildings, or property rights, which enable them to conserve open space, recycle old buildings, and preserve agricultural land. Partly as a function of this growth, an umbrella organization known as the Land Trust Exchange was created in 1982 to help coordinate the activities of the more than 700 land trust and preservation groups that have emerged as an important new force in rural planning and community development.

The Planning Process: Fragmentation, Incrementalism, and the Planning Style

Fragmentation of Rural Development Policy and Responsibility

The political economy of rural areas is often more complex than one would imagine. The degree of local autonomy and economic diversification in a community will vary according to its degree of exposure to the political and economic power of outside influences, as well as the level and quality of local leadership and organization. Outside forces are largely beyond local control, yet it is the challenge of planning to establish and implement priorities that either harmonize with outside forces or else mitigate their impact. This is where the planner should strive to be proactive, expressing the community's development goals and objectives rather than merely reacting to outside forces.

Although federal and corporate policies greatly affect rural development, the responsibility for planning generally falls on state, county, and local governments. The majority of decisions affecting rural land use, however, are still made by private firms and individuals. In recent years, planning efforts by state and local governments have struggled to address (1) the population growth of rural communities and environmental protection; (2) wide-scale concern about the future of the family farm and agricultural land; and (3) the growing competition among industry, tourists, and environmentalists for the control of rural land resources. It has become apparent that rural communities alone are often poorly equipped to handle the conflicts of personal preferences, federal policies, and corporate interests. Not only is there a clear lack of continuity and coordination in rural policy at all levels of government; there is a fragmentation of power and responsibility as well. So far, America has failed to articulate a rural development vision of a working rural landscape that will provide an acceptable quality of life for rural society (Lapping, 1982a). Until a broad consensus emerges to deal with the problems and potential of both rural and small-town environments, the fragmentation, frustration, and failure of rural planning policies will remain the rule and not the exception.

Incrementalism and the Planning Style

A major obstacle to rural planning and community development has been the planner's uncertainty about his or her role and the planning function. Given the abundance of economic, political, and social influences in a community, a planner must ask, "How much impact can I expect to have?" The question is not easily asked, and the answer will depend on how much power a planner is given to enforce his or her professional advice. Another question the planner must ask is this: "Do I know what I am doing?" Planning is a process of power in the making, and involves controlling or influencing the plans of others. The planner may have nagging doubts about the economic consequences of decisions that confer or deny benefits to different parties. Moreover, a planner must work effectively with the elected planning commission, board of selectmen, and local government. Yet when the planner's advice is ignored by these authorities, the planner will begin to question the limits of his or her professional responsibility. The planner ideally serves the public interest, but this may clash with the interests of appointed or even elected officials.

Finally, much is demanded of a planner in a rural setting. The planner must be able to speak convincingly as an architect, geographer, economist, ecologist, sociologist, motivator, manager, public servant, and information coordinator! This is indeed a tall order. Because of the variety of subjects a planner needs to know, and the limits on a planner's power, the style of planning has become fragmented and based on small, incremental changes in policy. This planning style is designed to cope with day-to-day planning matters—a zoning change for a property, building permits, a new traffic signal. It is not proactive, comprehensive, or long-range in scope. Broad economic changes often are not small or incremental, and they have long-term effects. The local mill closes. A mine lays off 100 workers. The price of wheat plunges. A new shopping mall is built in a nearby town, causing local downtown merchants to lose business. The planner and the community are often slow to respond to these unforeseen events and, in many cases, respond too late.

It can be said that rural planners have generally failed to account for the structure of political economy that determines community well-being. The methodology of planning typically begins with an inventory of local and regional characteristics, including population projections, economic sectors, roads and transportation networks, municipal services, and the type and location of different land uses. From the inventory, the planner tries to identify problems and assess community needs. Then the planner makes recommendations for action by the community. These recommendations are likely to be in accord with fundamental rural development goals:

1. Promoting long-term economic, social, and political viability through greater self-reliance.
2. Expanding opportunities for productive work, especially for young people and women.
3. Enhancing and maintaining the quality of the natural and human-made environments.

4. Seeking balanced growth.
5. Satisfying local needs for public services.
6. Striving for social equality.

These are certainly worthy goals. However, they often tend to ignore the economic and political powers of outside influences. A community that relies purely upon the market to govern its internal affairs places certain limitations on the plans and actions formulated by its planner and government officials. A community's dependency on distantly controlled private industry, urban markets, and federal money and regulations places constraints on local options and actions. And it is precisely these constraints that tend to frustrate the success of local development projects and thwart the achievement of local goals.

Outside influences are seen as frequently creating unplanned developments. But as one writer has noted, "'unplanned' events usually turn out to be the unanticipated consequences of someone's planned activities. The 'lack' of planning is rarely, if ever, the total absence of planning. More often it is the failure of someone's plans to account for someone else's plans" (Ostrom, 1965, p. 338). Another writer adds, "Thus, [a planner] should expect that his policies will achieve only part of what he hopes and, at the same time, will produce unanticipated consequences he would have preferred to avoid" (Lindblom, 1959, p. 86).

A planner in a rural area is likely to face major limitations in time, money, and personnel in analyzing problems. Objectives and the evaluation of performance in problem solving may be unclear, so that a distinct relationship of means to ends is difficult to identify. Public opinion regarding a proposed plan or project tends to defy measurement. Therefore, the planner is forced to be cautious, and the planner adopts the style of incrementalism: a slow and frustrating process of "muddling through" to decisions, based more on value judgments than on the careful analysis of detailed technical information. The planner must use "successive limited comparisons" of past policy experiences as a guide to formulating solutions to problems. Changes in policy are thus not likely to be abrupt, and such a style is more likely to be acceptable to a variety of interests, especially the local government. For when decisions are made incrementally—closely related to known policies—it is easier for one group to anticipate the kind of moves another might take and easier too for a group to make correction for injury. Yet incremental decisions are often made as a reaction to a problem that was unanticipated or unplanned for.

Recognizing this, the planner must work to persuade others to adopt a proactive, comprehensive perspective, anticipating and preparing for change. This is the only way in which localities can control change. Local day-to-day decisions should be made with the guidance of long-term goals and objectives. Too often, local officials fail to see or anticipate the cumulative impact of many individual planning decisions. The planner must function as a cumulative impact memory and strive to be forward-looking and progressive. In this way, encompassing issues of dependency and local autonomy can be more openly and effectively addressed.

CHAPTER SIX

Productive Systems: Agriculture

Agriculture is not well understood by those not directly involved with farming or food processing. All too often, agriculture conjures up neo-Jeffersonian visions of small farmers happily working the land in a tight-knit rural society. Even among rural communities, many myths and outdated perceptions of agriculture persist. The most popular myth is that American agriculture is dominated by family farms.[1] In reality, the overwhelming majority of food is grown on larger-than-family farms that bear little resemblance to the farmsteads of 40 years ago. Modern farming is a capital-intensive process, requiring hundreds of thousands (if not millions) of dollars invested over a long term, with the likelihood of low returns. In fact, nearly 70 percent of all farm commodities sold and 90 percent of farm income are produced by the top 300,000 farms, or only 12 percent of all U.S. farms (Daniels, 1986a, p. 33). A second myth is that the farmer is an independent, self-reliant owner and operator. Nearly half of the nation's farmland is owned by nonfarmers (Daniels & Nelson, 1986, p. 26). One-quarter of U.S. farm earnings come from exports to distant markets, and export demands have fluctuated considerably in the last ten years. In addition, many family farmers depend on off-farm jobs to earn an adequate income. But perhaps more fundamentally, there has been a divorce between the ownership of land and the control over input costs and output prices. Farmers rely on large corporations for machinery, fuel, fertilizers, pesticides, herbicides, and seed;

[1]A significant problem with this myth is that governments have been reluctant to regulate the environmental effects of farming practices. A prime example is the pollution of groundwater and surface water by farm chemicals. On the other hand, the 1985 Farm Bill does require owners of highly erodible farmland to develop conservation plans by 1990 (Pub. L. No. 99-198, 16 U.S.C. § 3801 *et seq.*). This is an attempt to reduce soil erosion and runoff into water supplies.

they also typically sell their produce to large corporations. Hence, farmers have little control over the cost of farm inputs or the market price of farm products.

Since the 1920s, American farmers have produced a surplus of food and experienced wide fluctuations in commodity prices and farm income. The New Deal farm programs of the 1930s established a federal farm policy aimed at managing farm output and maintaining farm income. This policy has continued to shape federal farm programs to this very day, and recently has resulted in a surplus of many farm products, especially corn and wheat. Although some farmers may argue that the federal government should "get out of agriculture," federal programs have provided commodity price supports, a variety of loans, and payments for the idling of farmland to stabilize farm income and mitigate the effects of crop overproduction. In 1985, U.S. farms received an average of almost $24,000 per farm in government funds (Harrington, 1987, pp. 12-20), although large farms received more per farm than small and medium-sized farms. In 1985 and 1986, the federal government spent $50 billion on farm programs. In short, government programs have enabled many farmers to remain in business. However, because of the need to reduce the federal budget deficit, farm subsidies are likely to be lower in the future. In fact, the Reagan administration has proposed eliminating all farm subsidies by the year 2000. Other proposals have featured a "decoupling" of farm payments from production. In this way, farmers will not be rewarded by volume of output but will receive subsidies in return for reducing acreage planted (Meyers, Womack, Johnson, Brandt, & Young, 1986).

In recent years, questions have been raised about the loss of farmland to nonfarm uses (Crosson, 1982). On the one hand, an estimated 30 million acres were converted to nonfarm uses between 1967 and 1977; as of 1987, about 1.5 million acres, including 960,000 prime acres, of farmland are being converted to nonfarm use (U.S. Department of Agriculture [USDA], 1987). At this rate of conversion, the amount of land in urban use will double in about 50 years. On the other hand, the total acreage of cropland in production has stayed fairly constant since 1930; food remains relatively inexpensive, and overproduction has been a persistent farm problem. In 1987, farmers used about 330 million acres of cropland but idled some 69 million acres under federal farm programs. This level was second only to the record 78 million acres farmers idled in 1983.

Controversies about the future of agriculture have become most pronounced when urban and rural residents collide at the urban–rural fringe. Counties in fast-growth areas include less than 15 percent of the nation's cropland, but account for nearly 30 percent of the dollar value of total farm production (USDA, 1987). About 372 of the nation's 640 leading agricultural counties are either within or adjacent to major metropolitan areas (Anthan, 1988, p. 54) (see also Figure 6.1). Urbanites settle outside of central cities because they enjoy the open space and bucolic environment farming provides. But as more people relocate to fringe areas, they drive up the price of land, isolate tracts of farmland, and hasten the decline of local farming. Farmers require a stable supply of land available in order to expand their operations and thus take advantage of cost-reducing economies of size to stay in business. Yet, as land prices rise and the land base is fragmented, it becomes increasingly difficult for a farmer to expand and function efficiently.[2] Also, farmers in fringe areas have been discouraged by vandalism to their crops, livestock, and equipment. A further limitation has been the enactment of nuisance laws, which restrict or ban certain farming practices. Finally, when the opportunity cost of holding land in farming becomes too high, farmers realize they are better off selling their land for a nonfarm use.

Local and regional planners can expect to be called upon to explain, defend, or comment on a variety of technical agricultural issues. Because planners increasingly must address a wide array of farmland protection problems and techniques, they must understand the broader issues of the economic viability of farming, which are intricately tied to the future of the farmland base. A basic knowledge of agriculture as an industry and of the local agricultural community is essential for effective action.

Issues in Agricultural Structure: Where Does the Planner Fit?

Although the federal government is the major actor in farm policy, planners can influence local agriculture not only through land policies, but also through promoting local farm businesses. Agriculture has generally been recognized as a form of economic development, and farming, food and fiber processing, and farm service industries are important components of many rural communities. In fact, farming and the processing, transportation, and marketing of farm produce account for 17 percent of America's gross national product and 18 percent of the labor force (Reimond & Petrulis, 1987, p. 4-1). But farming alone directly employs only 3.1 million people, or just 10 percent of the rural labor

[2]By contrast, in rural areas far removed from cities, farmland values have declined significantly. This has made expansion and entry into farming easier for those who did not borrow to expand in the late 1970s and early 1980s, when interest rates and farmland prices were high.

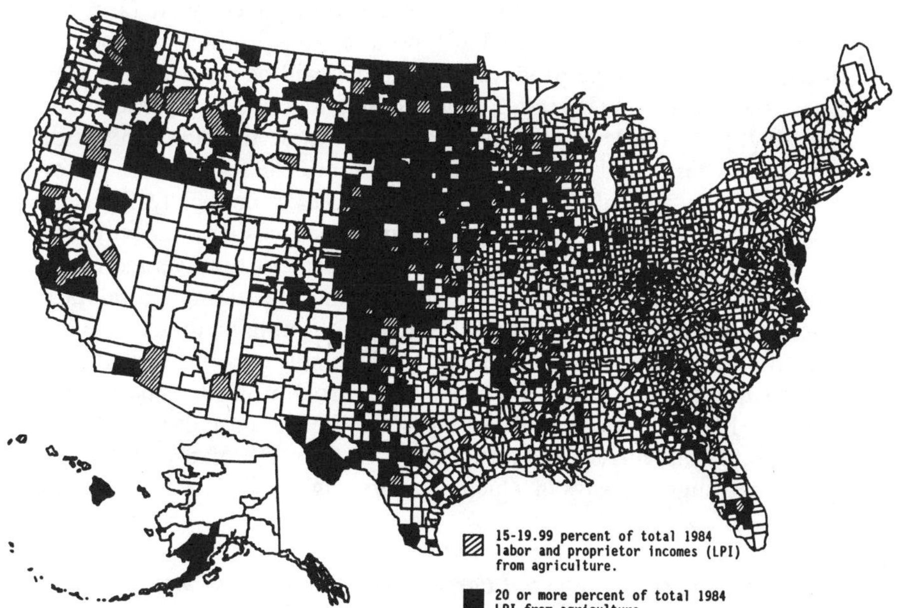

FIGURE 6.1. Agriculture counties, 1984. From *Rural Economic Development in the 1980s*, by D. L. Brown and K. L. Deavers, 1987, Washington, DC: USDA, Economic Research Service.

force (U.S. Bureau of the Census, 1986, p. 619). Still, the long-term economic viability of local agriculture must be seen as a crucial element, if planners are to be successful in helping to create healthy and diverse rural economies.

The structure of agriculture is composed of many, often interdependent, factors. These include the following:

1. The number of farms
2. Farm size
3. Forms of business: individual, partnership, or corporation
4. Freedom to make decisions and risks borne by the operator
5. The way in which inputs are procured and products marketed
6. Farm ownership: tenancy, renting, or owning
7. Ease of entry into farming
8. Transfer of assets to future generations
9. Restrictions on land use
10. Land ownership and tenure

The structure of agriculture, together with government farm and monetary policies, determine what food and fiber products will be produced, how they will be produced, how much will be produced, at what price, and for whom (USDA, Economic Statistics and Cooperative Service, 1979). At the level of individual farms, these decisions will affect land use patterns and local economies well into the future. The performance of the agricultural sector has changed dramatically in recent decades (Schertz et al., 1979). For example, food costs have declined to the point that they consume roughly 18 percent of the average family budget. Total farm output has increased sharply, though total farm acreage has decreased slightly. Still, modern farming methods have had negative effects on the environment, such as soil erosion, depletion of water supplies, and water pollution (see Crosson & Brubaker, 1982; Reisner, 1986). A close look at the structure of agriculture provides insight into both current and future trends in the performance of the farm sector, and suggests ways in which planners might influence these trends.

Fewer but Bigger Commercial Farms and More Hobby Farms

Since 1945, the most significant trends in U.S. agriculture have been the decline in the number of farms from 6 million to 2.2 million and the increase in average farm size from 200 to 455 acres (U.S. Bureau of the Census, 1986, p. 621). Several factors have contributed to these trends. New machine and chemical technologies have replaced human and animal labor and have produced a great increase in crop yields. These new technologies are particularly well-suited to large farms and have brought about reductions in the cost of producing a unit of food. The heavy use of machinery and chemicals is too expensive for small farms; equipment costs cannot be spread over many acres; and total output is quite limited. Thus, small farmers have not been able to lower their unit production costs enough to compete with large farmers. This has been especially true for row crops (corn, wheat, soybeans, cotton, and certain vegetables) and dairying. Because small farmers could not compete, they often sold their farms to neighboring farmers to create larger units or else to nonfarmers.

At the same time, hired labor has been attracted off farms by jobs providing higher rates of pay, shorter hours, and better working conditions. The resulting scarcity of farm workers has further contributed to the decline of small farms and emphasized the need for mechanization on large operations. The number of people employed in farming fell from almost 7 million in 1950 to under 3 million in 1970. As of 1987, about 3 million people were employed on farms (U.S. Bureau of the Census, 1988).

A variety of federal farm programs have also favored large farms. These farms have tended to receive the majority of government farm loans, price support payments, cropland-set-aside payments, water from diversion projects, and benefits from government-sponsored research. For example, in 1985 farms with annual sales of over $100,000 garnered almost 75 percent of all direct federal farm payments (Harrington, 1987, p. 12-5). In sum, federal programs and incentives have enabled many farmers to expand their operations. The result has been a greater concentration of both wealth and income in larger farms.

Through the 1970s, inflation increased the wealth of landowners, and the sharp rise in exports after 1972 produced a large upswing in farm earnings. These gains in income and wealth also accrued mostly to large farmers. Table 6.1 shows that the number of farms declined only slightly between 1978 and 1982. However, the average value per farm increased by more than 12 percent, suggesting a greater concentration of wealth among remaining farms.

Between 1980 and 1985, the value of American farm real estate plummeted from $988 billion to $544 billion (U.S. Bureau of the Census, 1986, p. 627). Farm debt swelled to $200 billion, and farm foreclosures became a common occurrence across rural America. In 1985–1986, the Farm Credit System lost $4.6 billion, and its loan portfolio shrank from $80 billion to $50 billion (Garcia, 1987). The wealth of rural America was draining away. Particularly hard pressed were medium-sized farms that had expanded on borrowed money in the late 1970s and early 1980s, when land prices were high and interest rates rising well above 10 percent. By contrast, in 1982, the 1.1 million small farms with annual sales of less than $10,000 had only 9 percent of the nation's farm debt (U.S. Bureau of the Census, 1986, p. 623). Many large farms had taken on considerable debt burdens, but generally they had large enough assets

TABLE 6.1. Number of Farms, United States, 1978-1986

	1986	1982	1978	% change 1978–1982	% change 1982–1986
Number of farms	2,200,000	2,241,124	2,257,775	−0.7	−1.8
Acreage	1,007,000,000	984,755,115	1,014,777,234	−2.9	+2.2
Average farm size	455	439	449	−2.2	+3.6
Average value of land and buildings per farm	$314,363	$347,974	$279,672	+24.4	−9.7

Note. The data are from USDA (1982, p. 80) and U.S. Bureau of the Census (1986, pp. 621–627).

and could take advantage of federal farm subsidies to survive. By 1987 farm debt had decreased by $50 billion to about $150 billion, and net farm income reached a record $45 billion, helped by a strong livestock sector and $23 billion in federal farm subsidies (Drabenstott & Barkema, 1987, p. 29).

An important distinction exists between large, commercial farms and smaller, part-time farms. The largest 50,000 farms in America account for more than one-third of all farm products sold, and the largest 200,000 farms account for almost two-thirds of all sales. On the other hand, two-thirds of all farms had annual sales of less than $20,000 in 1982. Large farms can be defined as having over 500 acres or annual sales of over $100,000. Medium-sized farms, typically identified as "family farms," consist of 50 to 500 acres or have annual sales of $10,000 to $100,000. Small farms, which are generally part-time operations, have less than 50 acres or sales of less than $10,000 a year.

Table 6.2 indicates that between 1978 and 1982, the number of large farms declined slightly, whereas the number of medium-sized farms decreased by 8 percent. However, small farms increased by a substantial 17 percent. This growth in small, part-time farming is consistent with the migration of urban people to rural areas and the increased location of rural people outside of population centers.

The change in the number of farms by value of sales provides a clear picture of the increasing concentration of income in large farms (see Table 6.3). The number of farms with sales of more than $100,000 jumped by 37 percent between 1978 and 1982. Meanwhile, medium-sized farms decreased by over 12 percent and small farms increased by almost 2 percent. Certainly some medium-sized farms have become large farms

TABLE 6.2. Farm Size, 1978–1982

Farms by size		1982	1978	% change 1978–1982
Small	1–9 acres	187,699	151,233	+24.1
"	10–49 acres	449,301	391,554	+14.7
Medium	50–179 acres	711,701	759,047	−6.2
"	180–499 acres	526,566	581,631	−9.5
Large	500–999 acres	203,936	213,209	−4.3
"	1,000–1,999 acres	97,396	97,800	−0.4
"	2,000 or more acres	64,525	63,301	+1.9

Note. From *1982 Census of Agriculture* (p. 80) by the USDA, 1982, Washington, DC: U.S. Government Printing Office.

TABLE 6.3. Farms by Value of Sales, 1978–1982

Annual sales		1982	1978	% change 1978–1982
Large	$250,000 or more	86,776	56,450	+37.0
"	$100,000 to $249,999	216,188	165,791	"
Medium	$40,000 to $99,999	333,047	360,423	−12.3
"	$20,000 to $39,999	249,063	299,398	"
"	$10,000 to $19,999	259,258	299,421	"
Small	$5,000 to $9,999	281,895	314,245	+1.9
"	Less than $5,000	814,897	762,047	"

Note. From *1982 Census of Agriculture* (p. 85) by the USDA, 1982, Washington, DC: U.S. Government Printing Office.

in terms of sales, due in part to inflation; however, almost an equal number appear to have gone out of business. The growth in small farms can be attributed to the surge in farms with sales of less than $5,000. In fact, almost half of all U.S. farms produced less than $10,000 in annual sales in 1982!

What we are seeing, then, is the emergence of two distinct kinds of farming systems: commercial operations and part-time farms. Farms with annual sales of greater than $10,000 make up just over half of all farms and account for 97 percent of all farm sales (Daniels, 1986a, p. 33). These farms average over 700 acres in size and have an average value of over half a million dollars. Moreover, 80 percent of the operators of these farms list farming as their principal occupation.

Among small farmers, only about one-quarter consider farming their principal occupation. On average, the operator of a farm with sales of less than $20,000 a year makes the majority of his or her income from nonfarm employment. Most small farmers and many medium-sized farmers must rely on nonfarm jobs in order to continue farming.

The debate over the number and size of farms is often emotional rather than rational. The large majority of medium-sized family farms are located in the Midwest and South. Here, the settlement pattern features hundreds of small towns that often depend significantly on surrounding farms. The loss of family farms poses a direct threat to the health, if not the actual existence, of many of these rural towns. Another potential danger in the trend toward fewer and larger farms is the greater concentration of economic power in fewer hands, which could result in the manipulation of food prices and supplies. Moreover, Walter Goldschmidt (1975), in his classic study of two California communities—one featuring corporate tenant farming, the other family farms—found that the family farm community was a more vibrant, progressive town. The fear is that with fewer and larger farms, there will be both fewer rural towns and more rural towns with less social cohesion. Thus, the concern with saving the family farm is partly one of demographics, partly a social statement, and partly a mythologizing of the self-employed owner–operator who works the land. But in terms of production methods, there is really no difference between family farms and larger-than-family farms.

Capital and Entry

The growing financial requirements of modern farming and the declining rate of entry of new farmers have had a major impact on the structure of American agriculture. Since 1950, about two people have left farming for each one entering. Meanwhile, the average age of a farm operator has increased to 52. This "graying" of farmers raises serious questions about succession on existing farms and the ability of new farmers to get started in agriculture.

Several factors have acted as barriers to entry into farming:

1. Large capital requirements for machinery and chemicals
2. Large land requirements
3. Increased land prices
4. High interest rates
5. Operating losses
6. Competition over land from nonfarm investors
7. Higher-paying nonfarm employment

The trend toward larger and fewer commercial farms underscores the huge amounts of capital necessary for a young person to get started in farming. A popular adage suggests that there are only two ways to become a farmer: "Inherit the farm or marry into one." The USDA has estimated that a minimum viable commercial operation requires an investment of about $300,000. Of this amount, roughly equal shares would be needed to purchase land and machinery and other equipment. The cost of purchased inputs and land has risen considerably since 1970. Inflation-increased commodity prices and greater competition for farmland all contributed to the 16 percent a year increase in farmland values through the 1970s (Healy & Short, 1981, p. 43).

Even with the retreat of farmland prices in the 1980s, historically the number of farmers continues to fall (U.S. Bureau of the Census, 1988).

New farmers have struggled (often unsuccessfully) to meet operating expenses and loan payments, and even established farmers who expanded in the 1970s have had serious cash flow problems or been forced to leave farming. In 1984, a USDA survey reported that 15 percent of farm households contacted had negative total income, and another 18 percent had positive income below the poverty line (Reimund & Petrulis, 1987, p. 4-7).

In 1950, total farm indebtedness stood at $12 billion. Today, the figure is about $150 billion (Nash, 1987). In 1986, over 11 percent of all farmers had debt-to-asset ratios of over 40 percent and negative cash flows, indicating financial stress. Financial stress was greatest (1) in the northern plains, the Great Lake states, the mountain states, and the Corn Belt; (2) on farms with gross sales of $40,000 to $250,000 a year (family farms); and (3) on dairy and grain farms (Reimund & Petrulis, 1987, pp. 4-8, 4-22).

Several sources exist for farm loans (see Figure 6.2), and loans are classified as longer-term loans (for real estate) and short-term loans (for short-term operating expenses). The Farm Credit Administration, an independent federal agency, consists of the federal land banks, intermediate credit banks, production credit associations, and banks for cooperatives. Each of these four lending institutions has offices in the nation's 12 Farm Credit Districts. The federal land banks are the leading holders

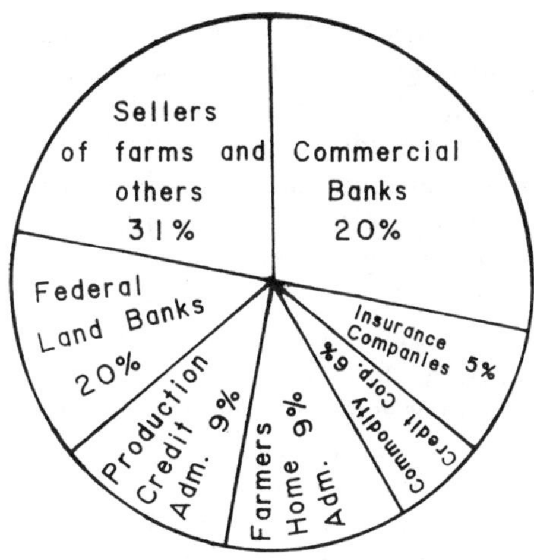

FIGURE 6.2. Sources of farm credit. From *Handbook of Agricultural Charts* (p. 8) by the USDA, 1983, Washington, DC: U.S. Government Printing Office.

of farm real estate debt (Cramer & Jensen, 1982, p. 249). The intermediate credit banks and production credit associations provide about one-quarter of the nation's short- and medium-term operating loans. In order to borrow from the Farm Credit System, farmers must also purchase stock in the local associations. Nationally, there are some 500 land bank associations and 400 production credit associations.

The Farm Credit Administration has suffered serious financial stress since 1984. The downturn in farm commodity prices, exports, and land values has reduced the value of the debt held by the administration from $80 billion to $50 billion. In 1985–1986, the Farm Credit System lost almost $5 billion, and in late 1987 Congress responded with a $4 billion package to enable the system to survive.

Farm Credit Administration lending programs, together with commercial banks and insurance companies, are geared toward large, established farmers. These farmers are viewed as good credit risks, having sufficient assets and income necessary to meet debt payments. The average net return on farm investment in land, buildings, equipment, and livestock is only 3–5 percent. Thus, the larger a farmer's assets are, the larger his or her income is likely to be.

A farmer's net income after paying all production costs is generally a small percentage of his or her gross earnings from the sale of farm products. A small drop in gross earnings could have a large impact on net income. Such a drop would be more likely to have a more severe effect on a small farmer or a new farmer with a large debt burden. Typically, established older farmers carry less debt in relation to their assets than new young farmers, and the income of a large farmer is also likely to be more stable, thanks in part to government commodity price support programs.

The Farmers Home Administration (FmHA) offers loans and loan guarantees to farmers who are unable to obtain credit elsewhere. The FmHA makes both real estate and operating loans and holds about 20 percent of all farm debt (Cramer & Jensen, 1982, p. 249). The FmHA is a major source of funding for new farmers, and the average age of a borrower is about 35 years old. As of 1987, there were about 250,000 FmHA farm borrowers nationwide. Many FmHA funds have been used to finance rural housing and development projects. Critics have charged that too much lending has gone to "farmettes" and not enough to support the entry of new commercial farmers.

The important source of nonrecourse loans is the Commodity Credit Corporation (CCC), discussed in Chapter 5. Under a nonrecourse loan, a farmer may borrow from the CCC at a fixed rate per bushel and put up his or her crops for collateral. When a loan comes due, the farmer then has the option of (1) keeping the loan if the market price is below the loan rate and letting the CCC take ownership of the crops (nonrecourse), or (2) if the market price is above the loan rate, repaying the loan and selling his or her crops from government storage (Cramer & Jensen, 1982). This means that in the first use the CCC absorbs the loss, and in the second case the farmer pockets the profit.

Another source of funding for new farmers is from private individuals. Some landowners may be willing to help finance the sale of their land and equipment to new farmers. Since 1979, North Dakota has allowed a person selling land to a beginning farmer to exempt half of the income he or she receives (up to a maximum of $50,000) from his or her state income tax. Also, the interest income received from the sale of land to a new farmer is tax-exempt. These measures are aimed at lowering the prices new farmers will have to pay for land. In 1976, Minnesota enacted the Family Farm Security Act (Minnesota Family Farm Security, Stat. ch. 41, 1976) to provide credit to individuals to purchase farms. Successful applicants must have farm experience, be Minnesota residents, and have less than $50,000 in assets. Between 1979 and 1984, $17 million was loaned to 109 farmers. In 1981, federal inheritance laws were changed to permit up to $600,000 of an estate to be passed on to the next generation free of federal inheritance taxes. This tax reduction and corporate forms of ownership have somewhat eased the problem of farm transfer within families.

Adding to the problem of farmer entry is the opportunity of higher income in nonfarm employment, which has attracted many farm children away from the farm. In the face of generally lower expected incomes, people enter farming out of personal preference and enjoyment of farming as a way of life. To ease the pressure on farm incomes, the federal government has, over the years, granted farming certain tax advantages. But depreciation allowances and investment tax credits for farm real estate, livestock, and equipment have also attracted millions of dollars from nonfarmers in search of tax shelters.

The rising value of farmland through the 1970s drew investors seeking a hedge against inflation. Because farmland is often the easiest land to develop in terms of slope and drainage, speculators and developers have purchased farmland for conversion to nonfarm uses. This investment in farmland by nonfarmers has helped drive up the price of agricultural land, particularly in rural–urban fringe areas. The trend of absentee ownership of farmland has also increased, furthering the separation of ownership and control of the farmland base. All farmers, both new and old, are increasingly put in the position of relying on rented and leased land. In fact, about 40 percent of all American farmland is rented or leased (Daniels, 1986a, p. 33). This situation is likely to reduce the long-run stability of farm operations.

Technology and Cost-Efficiency

Technology has a direct impact on the structure of agriculture and the future adequacy of the farmland base. Advances in farm technology have made possible the substitution of machinery, chemicals, and water for land and labor, while greatly expanding total output and reducing the cost of producing a unit of food. Between 1950 and 1980, the annual use of farm fertilizers increased by 500 percent, the number of tractors increased by 30 percent, and 50 million acres were brought under irrigation. Crop yields per acre soared, as total crop production increased by 97 percent but the amount of land in crops was only 3 percent greater; meanwhile, the number of farm workers fell by 63 percent.

These land- and labor-saving technologies have facilitated the movement toward the creation of larger and fewer farms. As more chemicals and expensive machinery are employed, it is difficult to spread these costs over a small land base. Output is quite limited by small farm size, and gains in output are often not sufficient to reduce the cost of producing a unit of food. But as farms become more mechanized and expand, they are able to take advantage of "pecuniary economies of scale." That is, even though total farm costs rise (in terms of both land and capital equipment), the *average cost* of producing a unit of food declines, up to a point.

Figure 6.3 illustrates this point. In the top panel, returns to scale are increasing from point A to point C. This means that the total output of

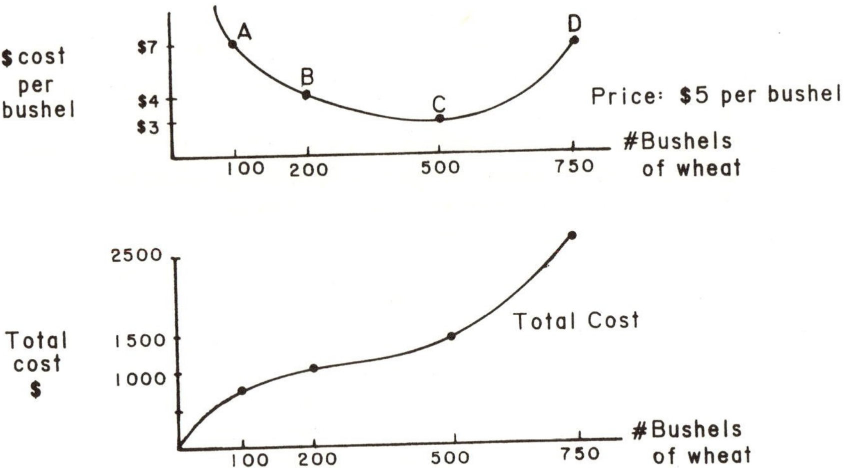

FIGURE 6.3. Economics of size in farm production. From *Handbook of Agricultural Charts* (p. 10) by the USDA, 1983, Washington, DC: U.S. Government Printing Office.

wheat is increasing while the average cost of producing a bushel is falling. The farmer at point C pays a greater total production cost than the farmer at A, as shown in the bottom panel. However, the point C farmer produces wheat at a lower average cost per bushel and can thus afford to receive a lower price. For example, assume that the market price of wheat is $5 a bushel, and farmer A grows 100 bushels of wheat at a cost of $7 a bushel; A loses $200 and will soon go out of business. Farmer C produces 500 bushels of wheat at a cost of $3 a bushel, and thus makes a profit of $1,000.

Because farmer C is producing wheat at a minimum average cost, he or she will be more profitable than farmer B. Farmer B produces 200 bushels at an average cost of $4 a bushel, but makes a net profit of only $200. Thus, farmer B grows 40 percent as much food as farmer C, but earns only 20 percent as much! Farmer B would be better off expanding his or her operation to point C.

From point C to point D in the top panel, returns to scale are decreasing. More bushels are produced, but the average cost of growing a bushel is also increasing. Thus, profit will decline. Farmer D, in fact, loses $1,500 and must either reduce the size of his or her operation or look for another line of work.

Exactly where points A through D exist will vary among farmers and different crops; however, this simplified example of achieving cost-efficiency through economies of size accurately reflects actual farm production choices. Farmers who attempt to grow wheat on a few acres with heavy machinery will not be able to reduce their average costs enough to make a profit (point A). These farmers must "get big or get out" of farming. Farmers who have achieved economies of size (point C) will be better able to absorb declines in crop prices and to profit more from price increases than less efficient farmers (point B). Some farmers may have overexpanded and need to cut back the size of their operations to earn a profit (point D).

Despite the past gains in farm productivity, concerns have been raised about the extent of future crop and livestock yields and the impact of heavy machinery, irrigation, soil erosion, and chemical pollution on the environment. As noted above, the new technologies have achieved much greater output without an increase in the cropland base. Currently, research in biotechnology holds the promise of yielding major breakthroughs in crop genetics and livestock breeding, which could greatly reduce the need for agricultural land and reduce the number of farms (U.S. Congress, Office of Technology Assessment, 1986). However, if new advances in technology do not occur and if the quality of agricultural land deteriorates, then productivity is likely to be made stagnant or even to decline. This situation would tend to (1) sharply increase production costs; (2) reduce the cost-efficiency of capital-intensive farming; and (3) necessitate the return of millions of acres of land to agricultural use to retain the level of crop output.

Since 1970 the cost of machinery, energy, and water has notably increased, but crop yields have also increased. For example, the average yield of an acre of corn land in Kansas rose from 100 bushels in 1980 to 136 bushels in 1986 ("Crop Estimates," 1987). Even so, the environmental impacts of farm machinery, chemicals, irrigation, and cropping patterns threaten to reduce future yields. According to the *1987 Resources Conservation Act Appraisal* (USDA, 1987), 40 percent of America's cropland is suffering wind and water erosion in excess of five tons per acre per year—the rate at which topsoil can still be replenished by natural processes without a significant loss in productivity. The loss of high-quality topsoil, along with soil compaction from heavy machinery, decreases the land's productivity and forces farmers to use greater quantities of costly fertilizers, pesticides, herbicides, and fuel merely to maintain crop yields. The resulting increase in production costs often leads farmers to plant more acreage to boost total output. This practice tends to bring more fragile marginal land under the plow, and erosion becomes worse. Furthermore, much of the eroded soil, together with the farm chemicals, is washed into streams, rivers, and lakes; soil siltation and chemical pollution pose serious threats to wildlife habitat and both surface and underground drinking water supplies. This problem is frequently referred to as "non-point-source" pollution, since it does not originate at one place or point, but at many places in the farming landscape.

Despite 50 years and $20 billion in federal conservation programs, soil erosion continues to be a major barrier to increasing future farm productivity. Of the nation's 406 million acres of nonfederal rangeland, 250 million acres are considered to be in less than good condition (USDA, 1987). Rangeland is an especially important component of the Western beef industry. Overgrazing, especially on publicly managed lands, has saddled the industry's future with the likelihood of much higher production costs. Many soil conservation techniques exist, such as no-tillage or minimum-tillage farming, contour cropping, terracing, and crop rotation. These methods are likely to be cost-effective in the long run; yet farmers are generally more concerned with short-term yields, costs, and profitability. Many farmers may not be able to afford to adopt conservation methods. As a result, the soil is "mined" like a nonrenewable resource. Clearly, farmers understand this problem, but they are often prevented from adopting conservation strategies because of the immediacy of mortgage and other loan repayments, which force them into short-run financial allocations rather than long-run resource solutions. Moreover, because federal farm subsidies are based on volume of output, farmers have an incentive to maximize output rather than conserve soil (see Lovejoy & Napier, 1986).

In 1985, Congress passed the Food Security Act (also known as the 1985 Farm Bill), which included the creation of a Conservation Reserve Program (CRP) aimed at removing highly erodible cropland from production, and thus reducing soil erosion and crop surpluses (Pub. L. No. 99-198, 16 U.S.C. § 3801 *et seq.*). An initial 45 million acres within five years was the target of the law when implemented in 1986. As of mid-1988, nearly 25 million acres were enrolled in the program. The conservation reserve is being amassed through ten-year contract agreements between the USDA and individual landowners. Landowners submit bids based on an annual rent per acre they are willing to accept in order to retire cropland. The USDA may accept or reject bids. When a bid is accepted, the landowner receives an annual rental payment for ten years, and up to half the cost of approved conservation measures such as grass seeding and tree planting. The landowner must submit a conservation plan for the enrolled acreage, and the plan must be approved by the local conservation district.

It is anticipated that the CRP will cost $1.1 billion a year over a ten-year period (Daniels, 1988). One clear benefit of the program is that it has established a floor under farmland values, especially in the Midwest and Great Plains states, which contain about two-thirds of the enrolled acreage. Critics point out that monitoring compliance on the 200,000 participating farms will be almost impossible. Some critics postulate that the CRP will have little effect on grain surpluses, particularly corn. Farmers have an incentive to use their remaining land more intensively. Should crop prices rise back to 1970s levels, the incentive to conserve highly erodible lands could again evaporate.

The Food Security Act also included "sodbuster" and "swampbuster" provisions to protect fragile grasslands and swamplands from being converted to cropland. Farmers who plow up these sensitive lands will not be eligible to receive any federal farm subsidies. But the most sweeping aspect of the act was the "conservation compliance" requirement that owners of some 120 million acres of highly erodible cropland develop conservation management plans by 1990. These conservation plans are an attempt to regulate farming practices—a major shift in federal policy.

Techniques to Retain Farmlands

Nearly all states and many local governments have enacted programs encouraging the retention of land in agricultural use, and about 50 million acres are now subject to state and county programs (Anthan,

1988, p. 52). Although the federal government has lately begun to act on preserving prime lands, farmland policy is essentially a state and local matter. Yet it is widely recognized that protecting farmland by itself does not necessarily guarantee a farm's financial success (Lapping, 1982b). In the following discussion, it should be kept in mind that none of the techniques described directly address questions of farm economic viability—namely, getting a price (marketing), new farmer entry, reducing costs, and access to credit. The need for integrated land and agricultural policies becomes evident. But, so far, the federal government has been reluctant to interfere with state and local jurisdiction over the use and regulation of privately owned land. In turn, the states and localities lack the funds to create an array of farm loans and subsidies.[3]

The Federal Role in Farmland Protection

The federal government has no formal farmland protection program, even though agricultural policy is largely determined by federal agencies and congressional mandates. Although the federal role has featured more rhetoric than action, some positive steps have been taken. The USDA has directed the FmHA to target its funding of rural sewer, water, and housing developments away from agricultural areas. In 1977, Congress passed the Soil and Water Resources Conservation Act, which requires the USDA to issue an appraisal of the nation's land and water resources every ten years (16 U.S.C. §§ 2001–2009). This report then serves as the basis for national land and water conservation programs. In 1978, Congress enacted the Agricultural Foreign Investment Disclosure Act, requiring foreign-based owners of more than one acre of farm or forest land to report their holdings to the Secretary of Agriculture (7 U.S.C. §§ 3501–3508). By the end of 1985, foreigners held only 12 million acres of the nation's farmland (U.S. Bureau of the Census, 1986, p. 625). But foreign investment in farmland has been reported in 1,831 of 3,041 counties (Majchrowicz & DeBraal, 1984). In some areas at certain times, foreign purchases have been significant and have driven up local land prices. However, the ownership of agricultural land by foreigners does not constitute a national problem.

Since 1978, the Soil Conservation Service (SCS) has been conducting a mapping program to identify important farmlands. To date, about

[3]In 1988, Vermont enacted a one-year subsidy for its dairy farmers, who could receive an extra 50 cents per hundredweight of milk up to a limit of $5,000 per farm. In subsequent years, farmers will be eligible for property tax abatement of up to an estimated 95 percent of their tax bills. The state will reimburse towns for taxes lost.

500 county maps have been produced to enable local governments to pinpoint their best agricultural lands. In 1979, the USDA joined with the Council on Environmental Quality in conducting the National Agricultural Lands Study. Two years later, a widely discussed report of this study was issued (USDA, Economic Statistics and Cooperative Service, 1981); it outlined the extent of farmland loss throughout the United States, the costs of such losses, and methods to preserve farmlands and enhance agriculture. Largely in response to the report, Congress passed the Farmland Protection Policy Act as part of the 1981 farm bill (Pub. L. No. 97-98). The act set out three major programs. First, the USDA has been directed to ensure that the actions of federal agencies do not contribute to the loss of agricultural land from productive use. Second, the SCS has been authorized to provide technical assistance to states, local governments, and nonprofit organizations to develop farmland preservation programs. And third, both the USDA and SCS have promulgated criteria for the use of a "land evaluation and site assessment" (LESA) system to rate the quality of land for agricultural uses and to rate sites for their economic viability. Each of these ratings is added together for a total score of overall agricultural viability. The higher the score, the more valuable the land is as farmland. Sixteen criteria have been proposed and must be addressed by federal agencies when assessing program decisions that affect the conversion of farmland to nonagricultural uses.

The LESA system was first tested in 1981. As of mid-1988, an estimated 400 counties had incorporated a LESA system into their land planning and farmland protection efforts. Currently, LESA has been adopted at the state, county, or local level in 39 states, though all states are considering its use. The LESA system holds promise for targeting poor agricultural land for future development while preserving farmland with long-term economic viability. The goal is to achieve economic growth along with maintaining a diverse economic base. But the relative weighting of farmland productivity versus off-farm factors will have a pronounced effect on the performance of a LESA system. For example, a heavy weighting of farmland productivity would probably lead decision makers to allow little nonfarm development, whereas a strong weighting of off-farm factors would lead decision makers to recommend significant conversion of farmland to nonfarm uses.

State and Local Programs

State and local governments have recently become increasingly active in regulating land uses in rural areas. Despite this growth in land use planning, the majority of decisions affecting land use are still made by private individuals. Many of these private decisions are shaped by county and municipal governments, and farmers in particular have often been exposed to a confusing array of incentives and restrictions. Local farm protection programs have often failed to coordinate zoning ordinances, property tax relief, and right-to-farm laws. If farmers are to bear restrictions on conversion, they must be allowed to farm as efficiently as possible. Also, property tax breaks to farmers represent a public investment in farmland, and mechanisms need to be created to protect that investment. The following techniques are discussed separately, but often two or more are employed together to form a farmland protection package. This is necessary because a single technique generally deals with a single problem, and, as we have seen, the problems involved with the farmland are both numerous and complex.

Table 6.4 presents a summary of the different farmland retention techniques being used in the different states. Table 6.5 indicates the tradeoffs between equity and efficiency in keeping land in farm use that

TABLE 6.4. State Programs for Farmland Preservation

State	d	c	t	az	ad	r	a	e
Alabama	X		X			X		
Alaska	X		X			X		
Arizona	X				X	X		
Arkansas	X					X		
California	X	X	X	X	X	X	X	
Colorado	X			X		X		
Connecticut	X		X			X	X	
Delaware	X					X		X
Florida	X	X				X	X	
Georgia	X					X		X
Hawaii	X	X		X	X	X	X	X
Idaho	X			X	X	X	X	X
Illinois	X			X	X	X	X	
Indiana	X			X		X	X	
Iowa	X			X	X	X	X	X
Kansas	X			X		X	X	
Kentucky	X				X	X	X	
Louisiana	X							

State	d	c	t	az	ad	r	a	e
Maine	X	X	X	X		X		X
Maryland	X		X	X	X	X		X
Massachusetts	X		X			X		
Michigan	X			X	X	X		X
Minnesota	X			X	X	X	X	X
Mississippi	X					X	X	
Missouri	X					X	X	
Montana	X					X	X	X
Nebraska	X			X		X	X	
Nevada	X						X	X
New Hampshire	X		X			X		X
New Jersey	X	X	X	X	X	X		X
New Mexico	X					X		
New York	X		X	X	X	X		X
North Carolina	X		X			X	X	X
North Dakota	X			X		X	X	
Ohio	X				X	X		X
Oklahoma	X					X	X	
Oregon	X	X		X		X	X	
Pennsylvania	X		X	X	X	X	X	X
Rhode Island	X		X			X		
South Carolina	X					X	X	X
South Dakota	X			X		X	X	X
Tennessee	X			X		X		
Texas	X					X		
Utah	X			X		X		
Vermont	X	X	X			X		
Virginia	X			X	X	X	X	X
Washington	X		X	X	X	X		
West Virginia	X		X			X	X	
Wisconsin	X	X		X		X	X	
Wyoming	X			X			X	

Note. The data are from National Association of State Departments of Agriculture Research Foundation (1985); Klein (1980, p. 12); Lapping, Penfold, & MacPherson (1983, p. 466); and Authan (1988, p. 53). d, differential tax assessment; c, centralized land use policies; t, transfer/purchase of development rights; az, agricultural zoning; ad, agricultural districting; r, right-to-farm laws; a, absentee landownership regulations/restrictions; e, erosion and sediment control legislation.

TABLE 6.5. Equity and Efficiency Tradeoffs of Farmland Retention Techniques

Technique	Equity	Efficiency
Property tax breaks	Favors large landowners at moderate taxpayer expense. More equitable than efficient.	Questionable efficiency if not tied to zoning, agricultural districts, contract not to develop, or rollback tax. Voluntary programs may not attract a critical mass of farmland.
Agricultural zoning	The higher the minimum lot size, the lower the equity. Less compensation for the restrictions on property use. Favors those who can afford to buy the minimum lot size.	The greater the minimum lot size, the greater the efficiency in preserving commercial agriculture. This may not be the case near urban areas where truck farming is prevalent. Variances and exceptions reduce efficiency. In general, greater efficiency than equity.
Agricultural districts	Fairly equitable. Voluntarily formed by landowners. Compensation incentive from property tax break.	Limited efficiency. Works better in places far from urban development pressures where land values are low.
Purchase of development rights	Greatly benefits farmland property owners at high taxpayer expense. Low equity.	Public ownership of development rights to farmland. Highly efficient at farmland retention, but voluntary.
Transfer of development rights	Fairly equitable. Farmland owners compensated for development restrictions by developers.	Difficult to implement. Needs clear preservation and development areas.
Capital gains taxation	Falls on land sellers, but sellers may shift tax burden to buyers. Fairly equitable.	The higher the tax rates, the greater the likelihood of land remaining in farming. Limited efficiency in an active land market with escalating land values.
Land banking	Fair to high. More equitable than efficient.	Expensive.
Urban growth boundary	Favors landowners within growth boundary.	Efficient. Contains urban sprawl.

are associated with different retention techniques. "Equity" relates to fairness in dividing the cost of farmland retention among farmers, taxpayers, and other groups. "Efficiency" refers to how much farmland will be kept in farm use and for how long. Generally, the greater the efficiency of a technique, the more the burden must be borne by either landowners or taxpayers, reflecting reduced fairness; the more equitable a technique is, the lower the efficiency of farmland retention is likely to be. The planner should also be aware that the success of a particular technique depends on how well the technique is implemented and administered. A further concern is whether a particular technique is feasible in a jurisdiction, given political attitudes and budgets. For example, a program for purchase of development rights may have political support, but there may be little or no money available to fund the program. Alternatively, there may be little political support for exclusive farm use zones in a remote county with little development pressure.

Property Tax Breaks for Agricultural Land

"Differential assessment" is the most common farmland retention technique in the United States. Every state has adopted some form of property tax break for farmland. Differential assessment programs fall into three main categories: (1) "preferential assessment," in which agricultural land is assessed for tax purposes on the basis of current use as farmland, rather than of the fair market value of its "highest and best" use; (2) "deferred taxation," in which preferential assessment is combined with a penalty (often called a "rollback") to recoup foregone property taxes when the land is converted to a nonfarm use; and (3) "restrictive agreements," in which a landowner and the local government agree to restrict the use of land for a certain period in exchange for preferential assessment (a penalty is levied if the land is converted before the agreement has expired).

The consensus among land use analysts is that use–value property taxation has not been very successful in retaining land in agriculture. This is especially true in urban–rural fringe areas. Differential assessment programs in general have not been able to dampen the increasing value of farmland for nonfarm uses, and they have failed to curb scattered development patterns that erode the farmland base. Preferential assessment programs impose no responsibility on the landowner to maintain a working agricultural use of the land, nor is there a penalty for converting farmland to a nonfarm use. Rollback penalties, when applied, are rarely large enough to discourage conversion. Farmers in fringe areas are often reluctant to enter restrictive agreements that limit their options to sell land. In many cases, speculators holding agricultural land have benefited from property tax breaks before converting their land to nonfarm uses. In addition, differential assessment causes the property tax burden to shift from owners of farmland to nonfarm owners. Local revenues may fall as a result, prompting a search for more intensive development to bolster local tax bases.

Agricultural Zoning

Although zoning has been the best-known program to influence urban land use, relatively little zoning has been done in rural areas. The constitutionality of zoning as a legitimate exercise of local police power was established in several cases, especially *Village of Euclid v. Ambler Realty Co.* (272 U.S. 365, 1926). But governments cannot use zoning to restrict a landowner's rights unreasonably and purely on the basis of policy. A major barrier to the use of zoning to preserve agricultural land is the lack of objective standards to determine whether or not property is being restricted in a reasonable way. Although zoning need not permit the most profitable use of the land, agricultural zoning may produce little or no benefit for farmers. When a regulation imposes burdens without any compensating benefits, the regulation might be considered an unconstitutional "taking." (For a discussion of the taking issue, see Chapter 4.)

Over 400 counties currently have agricultural zoning ordinances, and a few states—Hawaii, Oregon, and Wisconsin, at this writing—have enacted statewide zoning programs. Three kinds of zoning approaches exist: minimum lot size, exclusive agricultural zoning, and compensable zoning.

Minimum lot size restrictions require that land in an agricultural zone cannot be broken into parcels below a certain designated size. If lot sizes are sufficiently large, then they should be too expensive for residential uses, and they should retain agricultural land in big enough blocks to be farmed in a profitable way, either individually or as a collection of parcels. Thus, the intrusion of nonfarm uses into an agricultural area is discouraged, and the farmland base is not harmfully fragmented. Perhaps the most difficult aspect of the minimum lot size approach involves deciding what that minimum lot size should be. Lot sizes may be too small to support farming and yet too large for low- and middle-income families; such lot sizes could be challenged as exclusionary and discriminatory, as established in the decision of *South Burlington County NAACP v. Township of Mount Laurel* (336 A.2d 713 (NJ); 1972). Perhaps a more efficient use of minimum lot size restrictions would be to allow the clustering of development on small lots while preserving adjacent farmland in large blocks.

In an exclusive agricultural zone, only farming is allowed. Hawaii pioneered the use of exclusive agricultural zoning in its 1961 State Land Use Plan (Haw. Rev. Stat. § 205). Other jurisdictions that have tried this approach have permitted certain nonfarm uses, but only under the condition that they not conflict with existing farm operations. Even so, the mixing of farm and nonfarm uses has not been very beneficial to preserving agricultural land, because of the higher nonfarm value of the land base. In Hawaii, agricultural land may be converted to nonfarm use, but this land must be contiguous to urban or rural residential districts. In true exclusive agricultural zoning, the market value of the land and its value in agricultural use tend to converge. This facilitates farm expansion and serves to keep down farm property taxes.

Under compensable zoning, a landowner receives payment for the loss in development value to his or her property, due to the restrictions placed on the land. The owner receives compensation only at the time he or she sells the property to another farmer. Administrative problems, together with the costs of condemnation, make this program expensive. (For other compensation programs, see the discussions of purchase of development rights and community land trusts, below.)

The most obvious weakness of zoning as a device for agricultural land retention is that even the most carefully prepared zoning maps and ordinances are subject to variances, zoning amendments, rezonings, and special exceptions. As a result, zoning is notorious for its lack of permanence. Zoning decisions are normally made by politically vulnerable local governments, and ordinances are likely to change in the face of development pressures. Furthermore, zoning decisions are made on a case-by-case basis; as a consequence, the cumulative effect of zoning is not fully recognized. Also, the lack of coordination among jurisdictions can easily frustrate comprehensive regional agricultural zoning and compensable zoning programs.

Oregon: A Package of Farmland Retention Programs

The Oregon Land Use Act (1973) has offered a novel approach to farmland protection (O.R.S. 215.505 *et seq.*). Under the act, cities and counties are required to adopt comprehensive plans for approval by the state government. These plans must identify all agricultural lands and place them in "exclusive farm use" (EFU) zones. Farmland owners are granted farm tax deferral and subdivision restrictions; minimum lot sizes are also employed to retain farmland in large blocks. Some nonfarm uses are permitted in EFU zones, but farmers may not be subjected to nuisance laws that would restrict standard farming practices. Counties may designate "rural residential" zones to channel rural growth away from farming areas. Cities and counties are required to cooperate in establishing "urban growth boundaries," which mark the limit of urban service extension and thus seek to curb urban sprawl. So far, about 16 million acres have been placed in EFU zones, and the loss of farms and farmland appears to have slowed. However, there has been a sharp increase in the number of hobby farms, and it appears that local jurisdictions have been lax in enforcing the state's land use planning law (Daniels & Nelson, 1986, p. 31).

Wisconsin: A Second Package

The Wisconsin Farmland Preservation Act of 1977 has also presented a collection of measures to protect farmland (Wisc. Stat. 71.09(11); 91.11–91.79). The Wisconsin program combines agricultural preservation plans, agricultural zoning, and "circuit-breaker" tax credits against state income taxes to relieve "excessive" property taxes. The amount of the tax credit depends on the farmer's income, the level of property taxes, and whether the county has an agricultural preservation plan and zoning ordinance to implement the plan. Tax incentives increase as the method of protection becomes stronger. For example, a farmer is eligible for 70 percent of the credit in a county with only zoning and 100 percent of the credit in a county with both zoning and an agricultural preservation plan. Also, a farmer may enter into a 10- to 25-year contract requiring that the land remain in agricultural use in exchange for income tax credits. If the land is rezoned or sold for a nonfarm use before the contract expires, then the landowner is liable for up to ten years of tax credits received. Between 1977 and 1984, about 4.5 million acres from more than 20,000 farms were enrolled in the program (Johnson, 1984, p. 147).

Agricultural Districts

Agricultural districts can be seen as a compromise between differential assessment and statewide mandatory zoning. Agricultural districts were begun in New York in 1971 and have since been employed in 14 other states. Under the New York program, farmers may voluntarily form a district of at least 500 acres, thus providing a critical mass of agricultural land. Farmers are eligible for deferred property taxation and are exempt from nuisance laws restricting farm practices; in addition, controls are placed on extensions of public sewer and water lines and roads into the districts. However, there is no penalty for withdrawing land from a district. Between 1971 and 1978, over 300 districts had been created, covering approximately 5 million acres, or well over one-third of all farmland in New York (Lapping, 1980, p. 163).

Districting has been popular in those rural and semirural areas where the likelihood of selling farmland at greater than farm values in the foreseeable future is not great. In developing or suburban areas, the program has met with resistance, and few districts have been formed within a 25-mile radius of major cities. Farmland owners in the rural–urban fringe anticipate an imminent conversion of their land to more intensive uses, and hence fail to initiate farm districts or enroll their land.

Purchase of Development Rights

The purchase of farmland development rights has been used by governments as an alternative to the outright purchase of farmland. Under this method, a farmer sells the right to develop his or her land (in the form of an easement) to a state or local government. The farmer retains ownership of the land except for this one right. The development limitation runs with the land and thus binds all subsequent landowners as well.

The public purchase of development rights has both advantages and disadvantages. Buying development rights is likely to be cheaper than outright purchase of the land, and the landowner maintains control. The capital received for the development rights can be invested in the farm operation. After the development rights have been sold, the farm can be taxed only on its agricultural use–value.

The major and obvious problem in purchasing development rights is that of cost. In rural areas, where development pressures are less intense, the cost of buying development rights is lower. But in areas under conversion pressure, the cost of acquisition is often high. For example, by 1980 New York's Suffolk County, situated on Long Island and within commuting distance of New York City, had spent $21 million to buy the development rights to about 4,000 acres—an average of $5,000 an acre (Lapping, 1980, p. 157). The program is anticipated to expand to purchase the rights to a total of 15,000 acres at a total cost of $90 million. But two-thirds of the Suffolk County farmland in the development rights program was owned by nonfarmers, and in some cases the development rights cost as much as 85 percent of the fair market value. In King County, surrounding Seattle, Washington, $50 million has been raised through the sale of bonds to buy the development rights to about 10,000 acres. Initial purchases have been made, and the tendency has been to buy the rights to parcels of less than 50 acres, which have no apparent relation to farming. Both the Suffolk County and King County programs are really aimed more at open-space protection than at farmland preservation.

On the state level, Massachusetts, New Jersey, New Hampshire, Connecticut, North Carolina, Pennsylvania, Maryland, Rhode Island, Vermont, Washington, and West Virginia have established programs for purchase of development rights. In all of these states, farmland losses since World War II have been great (approaching 60 percent in some cases). Funding has generally been limited to a few million dollars, and it remains uncertain whether enough farmland can be purchased to preserve a critical mass of farms needed to justify the continuation of agricultural support services. Massachusetts so far has been the most aggressive state, appropriating $45 million since 1977 to purchase the development rights to over 9,000 acres. In 1987, Massachusetts launched a $500 million bond issue to purchase open space; New Hampshire appropriated $20 million for the purchase of land and development rights; and Pennsylvania earmarked $100 million for the purchase of development rights to farmland and open space (American Land Resource Association, 1987, p. 1).

Transfer of Development Rights

A possible way around the expense of purchasing development rights is the transfer of development rights (TDR). Under a TDR program, a certain area is designated as a preservation zone that is to be kept in agriculture and free of any other development. Landowners in the preservation zone are given tradeable development rights in exchange for the loss of development rights on their land. Lands within development zones may be developed at greater than currently permitted densities to absorb the growth deflected from the preservation district. Development may exceed allowable densities only if transferable development rights are purchased from landowners in a preservation zone. Thus, development rights are transferred from the preservation district to the development zone according to the market price of the rights.

The main advantages of TDRs are control of the timing and pattern of development, and the low cost to the taxpayer after the program is established. However, there may not be sufficient demand among developers in the development zone for TDRs, meaning that landowners in the preservation district would receive little compensation for the loss of the right to develop their land.

TDRs have been tried in urban settings and particularly in historic preservation areas. For agricultural use, TDRs are just now being implemented, and their effects have yet to be evaluated. In 1980, Montgomery County, Maryland, a suburb of the nation's capital, enacted a TDR program aimed at preserving agricultural land. Palm Beach County, Florida, enacted a TDR program in 1982 with the goal of protecting the county's farmland base and $870 million-a-year citrus fruit industry. In general, TDRs await practical application, and it is uncertain whether these programs will become a popular alternative to other systems of farmland retention.

Capital Gains Taxation

Capital gains taxation on the profits from the sale of land has been used in Vermont since 1973 to discourage land speculation (Daniels, Daniels,

& Lapping, 1986). The tax rates vary according to the amount of gain and the length of time the land is held (up to six years). The rates generally decline over time, but many increase if greater profit is earned. The highest tax rate is 84 percent in the first year, the lowest 5 percent in the fifth year. Results so far have been mixed. In agricultural areas, the gains tax has been of help in slowing the loss of farmland; in residential and recreation areas, however, land sales remain fairly brisk.

Land Trusts

Both private land trusts and community land trusts offer some promise for preserving agricultural land. The private trust is a private, nonprofit corporation that accepts gifts and donations with the objective of holding land in an open or agricultural use. The relative success of such a trust depends upon the wealth of private donors, who are motivated by philanthropy and the deductibility of the value of the land for income tax purposes. Although private land trusts are rare, the Ottauquechee Regional Land Trust in Vermont has received title to more than 12,000 acres of agricultural land and forestland since 1974.

A local government and a farmer may voluntarily agree on a community land trust, which allows the farmer to concentrate the development potential of all his or her land on a certain portion if the remainder is dedicated to a community trust. The farmer is thus granted permission to develop part of his or her land (e.g., 20 percent) at a density that would create a value equal to that of the entire farm in farm use. The area selected for development should be the one that is least favorable for agriculture and still allows for development to be economically rewarding as well as consistent with the community master plan. The farm owner would have the option of either selling the developable land or retaining it for personal use. The farmer and his or her heirs would reserve the right to farm the land dedicated to the trust in exchange for a nominal fee. This way, a farmer could sell some land for intensive use and invest the proceeds in the farm operation. So far, experience with community trusts is very limited, and their long-term feasibility remains to be seen.

Restrictions on Land Ownership/Purchases by Foreigners

Some 27 states have imposed restrictions on farmland ownership or restrictions on alien land purchases. These range from a total ban on the ownership of agricultural land by foreigners (as found in Iowa, Minnesota, and Missouri), to allowing ownership by resident aliens (as in Mississippi), to placing a time limit on the holding of land by aliens based on their intent to become U.S. citizens (as in Indiana and Kentucky). Although it is too soon to evaluate the impact of these restrictions, foreign investors have been drawn to American land as both a secure and a sound investment.

Land Banking

Land banking involves the acquisition of farmland by governments for future sale or leaseback to farmers. Land banking has not been extensively used in the United States, but programs in Canada and Sweden have proven successful (Lapping & Forster, 1982). Land banking could be helpful to small and beginning farmers and would probably require state or federal government funding. It would probably be too expensive for local governments. In leasing arrangements, government agencies traditionally make the substantial investments in land improvement and machinery required to sustain long-term productivity.

The Canadian province of Prince Edward Island has used land banking as the cornerstone to managing the island's 1.4 million acres. About 110,000 people live on the island, and roughly 70 percent of the island is in agricultural use. Because of its scenic landscape and moderate climate, Prince Edward Island has experienced heavy tourist-related development pressures and substantial land purchases by nonresidents. The increasing ownership and management of land in foreign hands and rising land prices were seen as obstacles to improving the island's agricultural productivity. Local inhabitants first responded by limiting the amount of land a nonresident could own to ten acres, unless permission to own more acreage is granted by the Lieutenant Governor in Council. Next, a Land Development Corporation was established, with the authority to purchase land and sell or lease it to farmers for farm expansion or consolidation, with priority given to small family farms and beginning farmers. Land unsuited to agriculture may be acquired and managed for other purposes, such as forestry, wildlife habitat, recreation, or watershed protection. The Land Development Corporation does not actively seek to buy land, unless so requested by a government agency. People ordinarily must apply to sell land to the corporation, but not all applications are accepted. The corporation has a policy not to compete with the local real estate market. For example, the corporation does not usually purchase entire farm units, except where age, health, and financial hardship are crucial factors. The corporation pays the use value of land devoted to farming, forestry, or other uses, so as not to increase or decrease local land values. Also, the corporation has purchased some land from nonresidents who could get permission to own more than ten acres.

Since 1970, the Land Development Corporation has purchased nearly 150,000 acres, or 10 percent of Prince Edward Island. In turn, the corporation has entered into some 800 leases covering about 50,000 acres and has made approximately 700 sales totaling 40,000 acres (Lapping & Forster, 1984). The most common method of acquiring land from the Land Development Corporation is through a five-year renewable lease with an option to buy. In leasing, the corporation makes specific demands concerning cropping and proper soil conservation. "Mining" of the land (i.e., exhausting the land's productivity by not investing in conservation measures) has not been a problem, because most farmers buy the land at the end of the lease.

The Prince Edward Island experience in land banking illustrates the interconnection of preserving the land base through the enhancement of farm viability. Land tenure (who owns the land) has been recognized as being of the same importance as land use, resulting in the development of information systems to identify tenure and land use capability. Farming as a means of earning a living has been supported, and provision has been made for the entry of new farmers. And perhaps above all, a land ethic has evolved, indicating an awareness of the importance of land as a means of production and as a determining factor in the future of a rural economy.

Is There a Right to Farm?: Issues Between Neighbors

The right to farm affects both farm viability and farmland retention, and merits special attention. Where agricultural and nonfarm uses are contiguous, there are likely to be "spillovers" from one use to another. People settle in the countryside because of the open space and pleasant environment. But these nonfarmers may also be subject to noise from farm machinery, livestock odors, dust, traffic congestion from slow-moving machinery, and exposure to toxic chemicals. In response, some nonfarmers have pushed to enact nuisance laws that restrict or ban certain farming practices. However, as limits are placed on farm operations and vandalism to farms increases, nonfarmers hasten the exit of farmers from farming and increase the likelihood of the conversion of farmland to nonfarm uses. Thus, the bucolic surroundings become urbanized, and those nonfarmers who sought to escape the city are once again within its grasp.

The important point is that there is no such thing as farmland without farmers. If nonfarmers are to enjoy the amenities of a working rural landscape, then they must either learn to tolerate farming practices or else settle at a distance from farm operations. The friction between farmers and nonfarmers involves a clash of property rights that cannot be resolved in the marketplace. Instead, legislative bodies and the courts must act as referees.

At least 47 states have enacted "right-to-farm" laws, which favor agricultural uses above all others and supersede local nuisance ordinances. Farmers perceive these laws to be beneficial in providing the freedom to operate and to earn a living. This is particularly true where land has been zoned for agricultural use. Most of the laws require that farm operations predate competing land uses by at least one year and that a farm continue to be managed according to "good" or "standard" farming practices. Farmers must comply with environmental laws and land use controls. Right-to-farm laws serve to educate a public long separated from the process of food production. The laws emphasize the fact that modern agriculture is an industrial process that needs to be protected from conflicting uses and increasing local population.

Toward an Integrated Approach: The Swedish Model

Sweden is a predominantly urban nation, with only 7.5 million acres (10 percent of its land mass) of arable quality. Over the last 20 years, Sweden has developed comprehensive agricultural programs that protect agricultural land, assure favorable returns for farmers, and guarantee an adequate supply of food to the public at reasonable prices. From the mid-1960s to 1977, agricultural programs had two main elements: (1) the containment of urban expansion; and (2) farm rationalization—the creation of larger family farms. The limitation of urban encroachment onto rural lands has reduced the competition for farmland and helped keep land prices at or near their agricultural value. Also, the link between the aesthetic qualities of agriculture on the one hand and tourism and recreation on the other has been recognized. The assembly of land into larger farms has enabled farmers to reduce the unit costs of food production with the adoption of new technologies. An important element of the program has been the transfer of land from retiring farmers to young farmers.

Since 1977, Sweden has attempted to limit the size of commercial farms while promoting larger family-run operations. Evidence has suggested that family farms in Sweden are more efficient in production than larger-scale commercial units.

To oversee the agricultural policies, two boards were created. The National Agricultural Marketing Board is responsible for setting food price controls, regulating food imports, and ensuring that farmers earn an adequate income. The board is in charge of farm rationalization and coordinates the activities of 24 County Agricultural Boards (CABs). The CABs have four major responsibilities: (1) distribution of financial assistance for farm rationalization; (2) development and revision of county rationalization plans; (3) regulation of land acquisition and the ability to engage in land purchases; and (4) provision of counseling and planning services for both retiring farm families and those staying in production.

Grants and loans are available for expanding farms and purchasing equipment, and deferred-interest loans are offered to new farmers. All agricultural land acquisitions not between family members must be approved by the CAB. The CAB may purchase land to keep it from going to a nonfarm use or to redistribute it to other farmers. Money to finance CAB land purchases comes from a National Land Fund. About 40 percent of land for farm enlargement has come from CABs, and CABs have seldom held land for more than two years. Some farm sales have been denied as "speculative" or "contrary to the public interest." Other sales have been denied initially because high sales prices would distort the local real estate market (Lapping & Forster, 1982).

On a national level, Swedish land use planning directs industrial location, resource development, and urban expansion. These urban policies and actions serve to complement rural devices of exclusive agricultural zoning, land banking by CABs, and a land speculation tax. In this way, urban and rural policies are meshed into a coordinated whole. Such integration does not exist in the United States. Federal, state, and local governments are generally reluctant to intervene in the land market to influence the location of development. Many rural counties do not have even rudimentary planning policies or zoning ordinances. State and local farmland protection programs are divorced from federal farm income programs. Ironically, farmland protection efforts are often most visible near urban areas, while federal subsidies have most benefited large farmers located in rural areas where conversion pressures are low.

CHAPTER SEVEN

Productive Systems: Forestry

Forests are complex and highly diversified ecosystems. With nearly one of every three acres in the United States covered with trees, forestlands constitute one of the most substantial and significant rural land uses. There are 159 nonmetropolitan counties with 20 percent or more of labor and proprietor income derived from forestry and wood products (Weber, Castle, & Shriver, 1987, p. 5-23). As Figure 7.1 indicates, these counties are located in the Pacific Northwest, the Southeast, northern New England, and the upper Midwest. Figure 7.2 shows the percentage of land in each state that is forestland.

Forests differ from one another in terms of their species and age composition, origin and topography, soil, water, climatic characteristics, and other factors. Marion Clawson, a noted specialist in forest policy, writes that there are seven major forest uses: "Maintenance of an attractive forest environment, provision of opportunity for relatively intensive recreation, provision of opportunity for a wilderness experience, provision of a habitat for wildlife, watershed functions, general conservation including minimization of soil erosion, and the production of wood for various uses" (1970, pp. 30–31). In many cases, only the last group of outputs has been "priced" in the marketplace, and considerable criticism has been directed against government and private firms for a perceived bias in planning and policy toward wood production. But as forest resource planning grows in its sophistication, and recreation and wilderness interests become particularly articulate, a wider range of forest uses is becoming the goal of nearly all planning processes.

Because of their renewability and versatility, forest resources are exceedingly valuable. Current management philosophy has tended to underscore the principle of "multiple use–sustained yield," under which forests are to be managed for many compatible uses that will generate

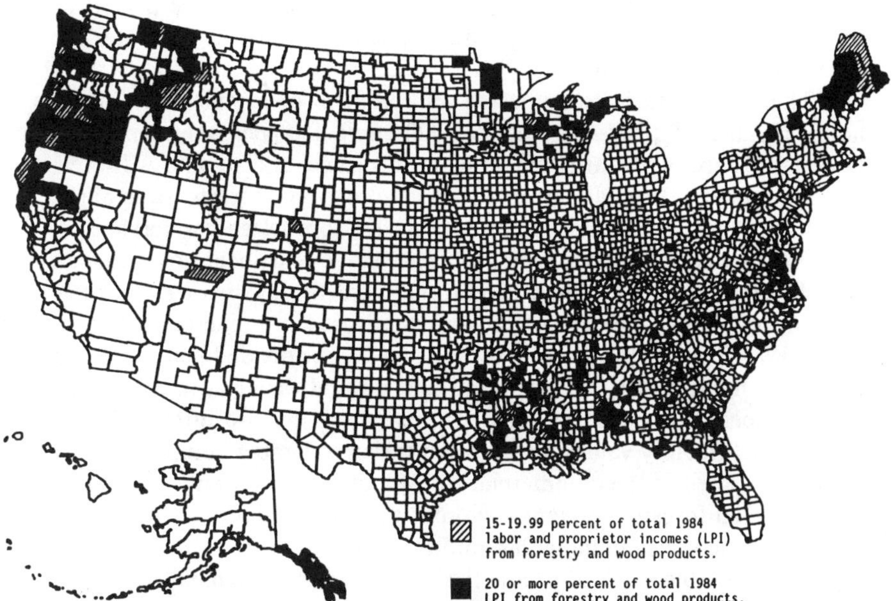

FIGURE 7.1. Forestry and wood products counties, 1984. From *Rural Economic Development in the 1980s*, by D. L. Brown and K. L. Deavers, 1987, Washington, DC: USDA, Economic Research Service.

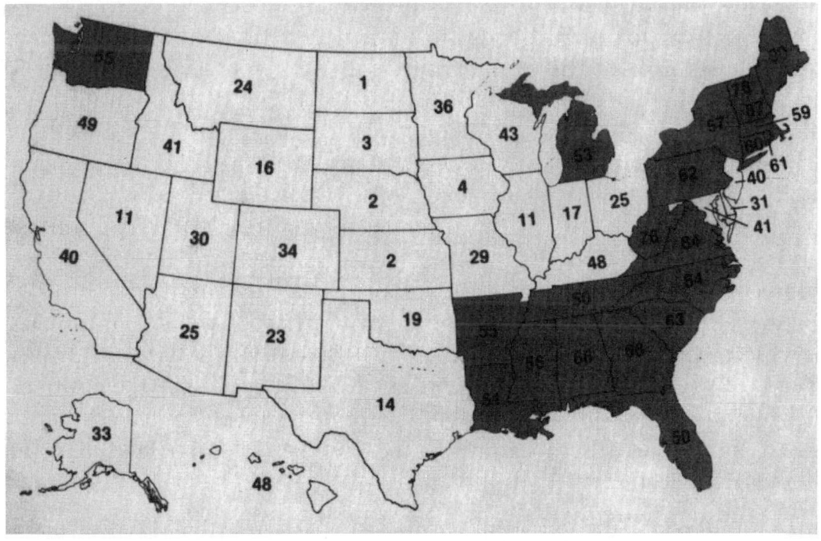

FIGURE 7.2. Forestland as a percentage of total land area. (Shaded areas have over 50 percent of their area as forestland.) From *An Assessment of the Forest and Range Land Situation in the United States*, 1980, Washington, DC: USDA, p. 30.

services, products, and incomes over a long and continuous period of time. Ultimately, what a forest produces and the incomes it generates are determined by ownership.

The United States contains 488 million acres of commercial timberland, capable of producing at least 20 cubic feet per acre per year. Ownerships are of two fundamental types, private and public. All levels of government—through national parks and national forests; state parks and forests; and local and community parks, forests, and watersheds—are forest owners and managers. Federal, state, and local governments own 136 million acres, which produce 40 percent of America's marketable timber and 60 percent of the softwood used in home construction (Cole, 1982, pp. 14, 16). Private ownerships can be divided between large industrial holdings on the one hand and nonindustrial farm and small private types of ownerships on the other; the latter are often held for investment, amenity values, or personal reasons. Timber companies own 68 million acres of the commercial forestland, and thousands of private nonindustrial forests (i.e., lands without processing plants on the premises) comprise the remaining 284 million acres.

The available stock of public timber and the management of private nonindustrial forests are issues of concern among foresters and rural planners. Elaborate planning systems are evolving for forestlands under government ownership. Though a great deal of highly sophisticated planning takes place on lands controlled by industry, the primary planning environment shaping decisions on private lands is defined by ownership objectives, tax policy, and forest practices acts. Although these tend to be less direct forms of planning, they are nonetheless of critical importance for the planner.

Over one-fifth of the privately owned commercial timber is held in parcels of 100 acres or less (Healy & Short, 1983, p. 121). At least one study has shown that timber parcels of this size are generally not managed for commercial production (Thompson & Jones, 1981). The combination of rising demand for forest products and shrinking forest resources is expected to put pressure on private nonindustrial forests. Between 1962 and 1977, 19 million acres of private commercial timberland were taken out of production (U.S. Bureau of the Census, 1985). The U.S. Forest Service has projected that by the year 2030, the national demand for softwood will rise by 93 percent, and the demand for hardwood by 193 percent; meanwhile, the commercial forestland base is expected to decline by 33 million acres (Cole, 1982, p. 11).

Forestry, more than any other private land use, demands a willingness to act for the future. Many of the decisions affecting future production must be made years in advance, given that commercial timber takes between 25 and 100 years to mature. The economic viability of a timber stand depends on (1) distance to the mill, (2) quantity of board-feet, (3) quality of the wood, and (4) market prices. Although future prices remain uncertain, production costs are likely to rise, because

> unlike most other American industries, the timber industry has suffered offsetting decreases in productivity—distance that logs must be hauled has increased, log quality has undoubtedly declined, logs are now much smaller—and freight rates for finished products have increased as the industry has moved further from population centers. (Gregory, 1972, p. 117)

Because of the lengthy time involved in timber production, forestland is vulnerable to changes in timber prices, price competition from other land uses, and annual property taxes based on the market value of the land and timber rather than the use for forestry. Even major forest products corporations have developed their forest holdings for residential uses, and vacation homes. According to one researcher, "There is considerable evidence that land speculation and the purchase of forests for purposes other than wood production have pushed the price of forestland in many areas to levels so high that a purchaser cannot hope to grow wood and from wood growing alone make a good return on his investment" (Clawson, 1979, p. 279).

As forestland is subdivided and purchased for residential and recreational uses, local land values rise and increase the opportunity cost of holding land for timber production. The creation of small forestland parcels suggests that a large amount of forestland may not be available for future wood production (Healy & Short, 1981, p. 216). Typically, most of the forestland sold in small parcels was not previously industrial forestland. But nonindustrial forests are often an important source of industry timber, as well as adjacent to lands in commercial forest use.

Small parcels suffer from diseconomies in timber production, having high fixed costs relative to the timber stock (Row, 1978). Owners have little incentive to spend time and energy becoming informed about forest management. Not surprisingly, large forest parcels tend to be well stocked, while small parcels are often poorly stocked (Clawson, 1972, p. 133). Moreover, small parcels inhibit the aggregation and retention of parcels for commercial use. Boundary restrictions of small tracts often prohibit the construction of roads needed for timber management and harvest. Fences, telephone and power lines, and buildings render movement more difficult. The likelihood of spillovers from forest practices (e.g., herbicide and pesticide sprays, slashburn) onto nearby residences increases; in addition, the more people living in forest areas, the greater the risk of forest fires. Finally, as more forestland is effectively converted to nonforestry uses, a critical mass of timber may be lost, making timber harvesting uneconomical in the long run. And when local mills close, commercial timber production may virtually disappear from an area.

The National and State Forest Planning Systems

The legal framework for planning on national forestlands is contained in three pieces of federal legislation: the Multiple Use–Sustained Yield Act of 1960 (Pub. L. No. 86-517. 16 U.S.C. §§ 528–531); the Forest and Rangeland Renewable Resource Act of 1974 (Pub. L. No. 93-378. 16 U.S.C. §§ 1601–1610); and the National Forest Management Act (NFMA) of 1976 (Pub. L. No. 94-588. 16 U.S.C. §§ 1601–1614). The Multiple Use–Sustained Yield Act established the overall goal for the federal government to manage lands for the "harmonious and coordinated management of the various resources . . . and not necessarily the combination of uses that will give the greatest dollar return or the greatest unit output." The government thus recognized the renewable nature of forest systems; the need to create even flows of products emanating from forests over the life of a forest (i.e., indefinitely); and basic compatibility of managed timber harvesting with wildlife, watershed, recreation, and soil conservation.

The Forest and Rangeland Renewable Resource Act sought to strengthen this public mandate through periodic assessments of the nation's renewable resource base and demand projections, such as the number of board-feet of pulp wood needed to meet domestic and export demands in a given year. The first such assessment was carried out in 1975 and was updated in 1980. The act also required the development of a program to establish specific goals for the national forest system.

The NFMA was a series of amendments to the Forest and Rangeland Renewable Resource Act that clarified the planning of the national forest system. The NFMA attempts to give direction to the U.S. Forest Service, the largest manager/planner of public forestlands in the United States, by emphasizing the unique attributes of a given forest in the forest planning process. (For a map of the Forest Service's holdings, see Figure 7.3.) For example, if one of the goals of the national forest system in the eastern region of the United States is to provide a certain amount of timber for pulping in a given year, the NFMA provides guidance to the Forest Service planners in the distribution of harvesting over the entire region's national forests. For example, the Ottawa National Forest in Michigan may sustain a greater level of harvesting activity than the Shawnee National Forest in Illinois, because its intrinsic resource capa-

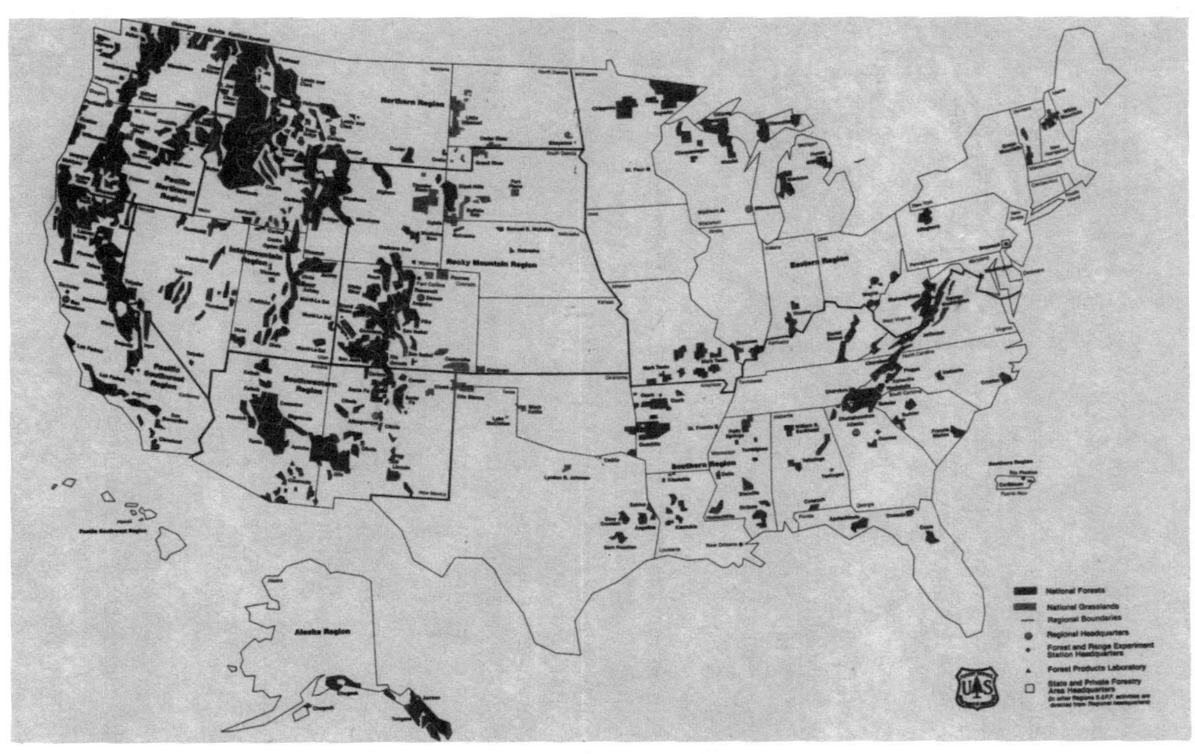

FIGURE 7.3. Holdings of the U.S. Forest Service. From *Land Areas of the National Forest System*, 1986, Washington, DC: USDA.

bility is greater for fiber production than the Shawnee's, whereas the Shawnee may be a superior habitat for certain wildlife or recreational pursuits. Under the NFMA, regional planning guides are prepared for an entire national forest region; national forest plans are then drafted for each forest in the region or several forests in the region, assuming some relationship between these forests. Once adopted, these plans become the land use planning and management guidelines for the forests. In this way, the management of an individual national forest is keyed to regional output goals, which then are aggregated at the national level to define the resource outputs to be achieved by the national forests as a total national system.

Each plan for a national forest must, according to NFMA, include the following: critical issues facing the forest and its region as identified by the public; criteria to evaluate different planning objectives; an inventory of the forest resources; analysis of the present status of forest management; drafting of alternatives to meet regional forest output goals; evaluation of the biophysical and socioeconomic impacts of alternative management options for the forest; selection of a given management option; the plan approval process; and a mechanism for monitoring the impact of the ensuing forest management policy.

Although this is a rather sketchy portrait of the planning process for national forests, it highlights several important realities. First, forests provide a multiplicity of goods and services, and each forest does so in a distinctive way. Second, the planning process must clearly address many forest functions, not merely the production of timber. Third, the process of public forest planning requires extensive citizen participation.

The control or "ownership" of the national forest planning process appears to reside with the U.S. Forest Service; however, this is simply not the case. Although responsibility for the process may rest with the Forest Service, ownership lies with the people. The rural planner can function effectively as a facilitator of the larger process of participation. This is an appropriate and necessary role for several reasons. First, many local and regional economies are defined by forest-based activities. In some areas of northern California or the upper peninsula of Michigan, for example, forestry-related employment is predominant. Thus decisions made about the management of a national or state forest can have implications for the entire structure of a local economy and the fabric of community life. Second, forest activities are of direct consequence to local tax bases. Federally owned lands do not pay property taxes. Rather, a fairly elaborate system of "payments in lieu of taxes" (PILTs) has evolved, in which the federal government pays to the local county government a percentage of the annual revenues generated by the sale or use of resources from the national forest. These funds can be very significant for the revenue base of any rural county or jurisdiction. Different management options will create different PILTs. For example, in a given year, a recreation area may generate less revenue returned in the form of PILTs when compared with a major timber harvest. But over a ten-year period, a recreation area may actually generate more revenue than one substantial harvest in the same time period.

The planner may be in the best position to evaluate this situation and to advise citizens of the options that lie before them. An understanding of the PILT system enables the planner to project local revenue sources relative to local service requirements in such a way that fiscal planning can effectively take place. And where local property taxes are an important source of government revenue, as in rural areas and small towns, this may be critical.

Decisions made about public forestlands often set the stage for decision making in the private sector. The determination to develop an intensive recreation area in a national forest, such as a ski area, will have numerous and potentially profound implications for the communities and lands peripheral to the public land base. Thus, participation by local planners in making decisions about public lands is highly relevant.

Forestland planning is not solely the domain of the federal government. Indeed, the individual states are among the largest forestland owners and managers in the country. In rural New York State, for example, the Adirondack and Catskill Forest Preserves are among the largest and most valuable assemblies of land resources in the entire state. With some support and encouragement from the federal government, notably through the State and Private Forestry Division of the U.S. Forest Service, states are increasing their involvement in forestland planning. Although some of this activity is directed to the planning of state-owned public lands, state-level forestland planning generally addresses those forest resources managed and owned by both the public and private sectors.

As one group of planners has stated, "many state planning processes are modeled to great extent on the Forest Service's RPA process" (Brooks, Hendren, & McCann, 1982, p. 585). By 1982, many states were actively involved in forest resource planning. Several of these programs have received their impetus through support generated by the U.S. Forest Service, though it is the individual state forestry/natural resource bureaucracy in each case that administers the program (see Brooks, 1980, 1981a, 1981b). Michigan's program was the first to be undertaken though other outstanding models exist in California and New York. Beyond the obvious implications for rural planning, state-level forest resource planning is important to communities, because local and regional issues or problems tend to "drive" the process. Unlike the federal approach, which seeks to respond to national resource demands and pressures, state forestry planning tends to reflect local pressures and problems more accurately.

One of the great benefits of such a locally oriented procedure is that it generates considerable data of great utility to local planners. One regional planner in northern Minnesota has used this data to analyze the potential impacts of various governmental programs that could result in the withdrawal of timber from productive use, the accessibility of the resource relative to transportation costs, and the demand for timber by regional industries and firms (Krmpotich, 1980). Once again, local planners find that their ability to understand and guide decision making has been enhanced by the forestland planning process. This planning activity can provide the public with some answers to the following problems:

> How should forestlands be allocated and managed among the different demands for timber production, recreation, wilderness, wildlife, and watershed protection? What role should private non-industrial, public, and industrial forests play in meeting these demands? How can the state best insure that forest resources are adequately managed and protected, not only from fire and insects but from land conversion, management, and neglect? How can the states get the most social and economic benefits of each tax dollar? (Cole, 1982, p. 64)

Of course, no planning process can really deliver all of the answers to these or other issues. What is imperative, however, is that the questions be asked and that a rigorous process be established to respond to these concerns. The participation of local planners and the rural community in this process is both fundamental and well understood in those places where the process has been implemented.

Public Policy and the Forest Environment

Outside of formal planning programs, a considerable array of public policy exists that influences forestland decision making. These policies must be understood by the planner, since they shape both the environment for decision making and some local planning options.

The most common policy deals with the taxation of forestland as property. Unlike agricultural lands, which often produce income on an annual basis, forestlands rarely do. The time horizon for forestry decisions is measured in decades rather than years, given the relatively slow growth of the resource, and this factor complicates taxation schemes built on annual assessments. The variety of state forestland taxation rules requires that planners understand the current policy both in their own jurisdictions and in other states, since the lack of uniformity between states may promote or inhibit forestry activity in different places. Some states (Alabama, New Jersey, and Tennessee, among others) exempt timberlands from property taxation either permanently or for specified periods of time. Other states, such as Washington, permit the taxation of the land at a "use" value rather than its market value, with a special assessment made against the timber at the time of its harvest. This is known as a "yield tax" and has the advantage of allowing the forestland owner to keep a stand growing toward maturity, rather than forcing a premature harvest because of a heavy property tax burden. Forest taxation experts have been advocating such mechanisms as an alternative to the more common differential taxation approaches, because the latter often shift the tax burden in inequitable ways, and also because speculators can be major recipients of such tax benefits. Still, the differential taxation schemes—or "tax-modified programs"—are by far the most common in the nation (Kemperer, 1976; Max, 1981).

No state has been more aggressive in moving against speculation in rural lands, including forested lands, than Vermont. With the passage of that state's capital gains tax in 1973, a system was implemented wherein variations in the tax rate depend on both the degree of gain from a land sale and the length of time the land is held prior to the sale. The tax rate rises as the percentage of gain from a transaction increases, but declines over time. Long-term owners are taxed far less than short-term, fast-turnover owners. The purpose of the law is obvious: to tax the capital gain on land transactions so heavily that speculation loses its attractiveness and profitability. The program works to the advantage of the committed forest landowner and against the subdivider of forested lands who is seeking to meet the demand for rural building lots.

Vermont's program has the indirect benefit of providing a penalty structure that discourages rural land parcelation or fragmentation (Baker, 1975; Daniels, Daniels, & Lapping, 1986). This is a growing problem in many forested regions of the country. Fragmentation or parcelation often creates many tracts of small ownerships that generate such low returns as to make timber management an unattractive option. Forest products firms that historically depended upon deriving the resource from private landowners may now find it impossible to assemble enough timber locally, and at reasonable cost, to operate efficiently. This should not be interpreted to mean that only large ownerships of forestlands are desirable. Rather, it is indicative of the problems inherent in decision making about almost any resource: A tradeoff is involved, and it is often the planner's responsibility to define both the benefits created and those foregone with any given decision. Parcelation may be something a community wishes to encourage to increase the number of building lots. But one of the costs may be the creation of increased difficulties for the efficient operation of existing local forest-dependent industries.

States directly influence forestland decisions through the regulation of specific forest practices, especially the harvesting of timber. Forest practices acts regulate the management of land rather than land use. The genesis of such state laws really had little to do with concern over the timber resource itself. The issues that motivated state involvement were the negative externalities of harvesting, particularly water pollution. Like many other land use planning laws, forest practices acts are an expansion of the nuisance law concept: It is the state's responsibility to prevent the creation of a public nuisance through the unrestricted exercise of any individual's or firm's private property rights.

A major review and analysis of forest practices acts by Ellefson and Cubbage (1980a; see also 1980b) revealed that such laws have gone through two periods of maturation. Older laws (pre-1969) tended to regulate harvesting in a minimal way, featuring replanting or restocking requirements for newly harvested sites. The goal was to establish vegetation on a site so as to increase soil stability and to prevent erosion and the siltation of nearby bodies of water. "Second-generation" forest practices acts (post-1969) are found mainly in the Pacific Northwest, where forestry is not only a major land use but also a major element of the regional economy. These programs tend to be

> comprehensive, detailed attempts to regulate private, and in some cases public, forest management practices. They are backed by strong enabling legislation, mandate complex regulations of forest practices, require detailed administrator and landowner actions, and are backed by strong penalties and enforcement tools. They regulate most timber management practices, including harvesting, regeneration, shielding, and road building. In some cases they also control application of chemicals and fertilizers, disposal of slash, and pre-commercial thinning. (Hagenstein, 1984, p. 88)

In some states, such as Oregon, rural planners have numerous opportunities to participate in decisions to be made about local forestlands; in other jurisdictions, a planner's role may be less direct. But in all cases the protection of an area's environmental quality and the enhancement of its forest economy requires something of a balancing act. The "jobs versus environment" issue, which is often raised in such situations, is really a bogus one, and planners must recognize that the stakes are so high that they cannot easily abdicate their responsibility in the process of forestland decision making.

Perhaps the most traditional way in which rural planners participate in forestland planning is through the land use planning and controls process. The reality is, however, that forested lands have too rarely been the focus of the attention of planners. Where forests have been a concern of local planning, they have been dealt with through large-lot zoning or high minimum lot sizes. The assumption has been that large lots "preclude use for strictly residential purposes and encourage food and fiber production on the land" (Miller, 1984, p. 168). But the effectiveness of such measures has not been conclusively confirmed.

California has gone an important step further and instituted "timber production zones" (TPZs), whereby counties can designate lands for timber production to the exclusion of all other uses, such as housing. Landowners in these designated zones receive substantial tax relief in exchange for a binding agreement to keep their lands in forest use. A landowner may petition to remove the land from the TPZ, but this may be granted only after an extensive public review and hearings. And even then, a landowner must keep the land in question in forest use for a decade after the TPZ designation has been removed. A student of the California approach concludes:

> Zoning land as TPZ helps control the influence of urban pressure on increasing land values. Land use is restricted, and speculative pressures are dampened because the zone runs for ten years. Available evidence shows that lands zoned at TPZ sell for less than lands not so zoned. This is some indication that with TPZ, land can be acquired at a price related to its ability to grow timber. (Cromwell, 1984, p. 158)

In Oregon, rural forestlands capable of producing 50 or more cubic feet per acre per year of Douglas fir (the main commercial species) are placed in "timber conservation zones " as part of the county comprehensive planning process. Although just over half of Oregon's 62 million acres are in state and federal ownership, about 13 million private acres are now in timber conservation zones. Landowners in forest zones may not be subjected to nuisance laws that would restrict standard forestry practices. In addition, landowners may apply for property tax deferral use–value assessment: Both the land and timber are assessed according to forestry value. Harvested forestlands must be restocked for use–value assessment to be maintained. If the forestland is sold for another use, then the seller must pay back taxes at the development value.

In the forest zones, counties impose subdivision regulations together with minimum lot size restrictions to retain forestlands in large blocks. Minimum lot sizes vary from 40 to 160 acres, but most counties employ two sizes. A large size (80–160 acres) applies to areas of prime commercial forestland that already consist of large tracts, tend to be owned by timber companies, and are located at higher elevations or in remote areas. The smaller minimum lot size (40 acres) applies to areas of mixed farm and forest uses where existing tracts are not large and are closer to developed places.

Outright permitted uses in a timber conservation zone include (1) the harvesting and processing of forest products; (2) open space; (3) some outdoor recreation on a commercial basis; and (4) the grazing of livestock. Considerable controversy has arisen over the construction of forestry-related and non-forestry-related dwellings. The partition of forestlands and the construction of forestry-related dwellings must meet a commercial forestry standard, which requires that a new forestry operation must contribute to local markets and at least be of similar size to existing local operations.

Counties may permit some nonforestry uses in forest zones, but only on marginal land and in places where development will not interfere with commercial forest operations. Counties may also conditionally allow non-forestry residential construction on parcels below the minimum lot size. This provision recognizes that high-quality resource lands are often mixed with lands of low productivity. Moreover, the potential for selling some land for nonforestry uses offers an important source of income for owners of small private forestlands and increases the political acceptability of forest zoning.

The planning approach of the state of Maine, the nation's most heavily forested state, is especially noteworthy. Although the state has a number of planning initiatives that affect forests, none is more important than the Maine Land-Use Regulation Commission (LURC). The LURC, whose members are appointed by the governor, has jurisdiction over the unorganized areas of the state (that part of the state, roughly half of Maine, without a municipal government infrastructure). These lands tend to be among the most heavily forested, and the largest landowners tend to be such forest products companies as Great Northern, International Paper, St. Regis, Georgia–Pacific, and Boise Cascade. The LURC is a permitting commission with broad powers. It requires permits for the building or placement of any structures, the subdivision of lands, the development of any roads, and all farming and logging operations in its jurisdiction. The LURC also controls logging roads (and the subsequent creation of backwoods subdivisions) and actual logging practices. Even when permits are not required for a particular operation or land use, performance standards may be attached by the LURC. Mountain zones, defined as areas above 2,700 feet in elevation, receive special attention from the LURC; all forestry uses in these zones require a permit before operation. Timber-harvesting standards applied by the LURC pay particular attention to the mitigation of soil erosion, road and trail placement, skidding techniques, and other silvicultural treatments. Areas such as the "high mountain" zones or "aquifer recharge" zones are defined by intrinsic resource factors or limitations rather than developmental criteria. Taken together, the LURC's policies take a positive approach to forestry while seeking to protect Maine's high environmental quality—undoubtedly one of the reasons for the very substantial in-migration to rural areas that the state has experienced in the past two decades (Pidot, 1982).

Forestry Resources: A Rural Development Imperative

One of the subtle themes underlying much of forest planning is the assumption that the perpetual maintenance of the forest resource is the key to the stability of those communities and regions dependent upon forestry for their economic base. Certainly, this is one of the elements of the "multiple use–sustained yield" policy. If the resource can be managed in such a way as to sustain yield over time, so the argument goes, then the economic base of an area will be maintained. Although few would argue with this view, it only partially explains matters. In writing of Canadian rural communities where forestry is the economic base, Bryon has correctly noted that "even flow regulations *per se* cannot achieve short-term stability of employment of incomes when the forest industry of a region produces primarily for volatile export markets" (quoted in Marshak, 1983, p. 306). Communities dependent upon forestry tend to be single-resource towns, in that their base—perhaps their very *raison d'être*—is linked to the well-being of that industry. If a strong market exists for forest products, some degree of stability may be evident. But the timber industry is very vulnerable to increases in interest rates that drive down the demand for new homes, for example. Moreover, as the entire world's economy grows ever-increasingly interdependent and new centers of supply emerge, a community's stability may be threatened. For example, the sharp rise in interest rates between 1980 and 1982 produced a similar downturn in the timber market, resulting in the closing of 69 lumber mills and plywood plants in Oregon. Many other plants closed temporarily and then reopened. The impact on rural Oregon was devastating. Unemployment rates of 20 percent were not uncommon, and many local economies hit a virtual standstill. In 1985, forestry-dependent counties had an average unemployment rate of 10.3 percent (Weber, Castle, & Shriver, 1987, pp. 5-28).

Of course, rural communities and small towns are not the only types of place susceptible to boom–bust economies. The cities of the heavily industrial Northeast, which are dependent on the steel and automobile industries, have also seen themselves become redundant to some extent.

But what complicates the situation for rural communities is their rather narrow economic base and the fact that those making decisions about the local industries and firms rarely live in those communities.

Two intertwined strategies have been put forward to combat the instability inherent in forestry-dependent communities (Lapping, 1982c). First, it has been argued that until these communities develop the capacity to process and manufacture goods out of the resource, they will simply continue to export raw materials to those metropolitan centers where the "value-added" product is fabricated (see Table 7.1). The profitability of the industry, in terms of both income and employment, lies in the "value-added" industries. Second, diversification of the local and regional economy must be a goal, in terms of both the spinoffs from forestry and other income-generating employment. Although this may be obvious to the planner, a more nebulous objective must first be achieved: the wide-scale recognition that forestry, as much as agriculture, is central to any effort to achieve rural development. Forestry and wood products have not been rapidly growing industries in nonmetroplitan areas over the past 15 years (see Table 7.2), but forestry and wood products wages tend to be higher than wages in service industries and agriculture.

An important objective of public policy should be the protection of a critical mass of forestlands to maintain forestry support services and economic diversity. Governments need to implement policies that direct new development onto lower-quality soils, into tighter settlement patterns, and away from commercial forest operations. The loss of produc-

TABLE 7.1. Average Employment in Forestry and Wood Products Industries in Metro and Normetro Counties, 1969–1984

	Metro		Nonmetro	
Industry	Thousands	% of total metro employment	Thousands	% of total nonmetro employment
Forestry and wood products (total)	1,123	1.39	712	3.21
Forestry	5	0.00	8	0.03
Paper and pulp	510	0.64	176	0.80
Lumber and wood	285	0.35	377	1.71
Furniture and fixtures	323	0.40	151	0.67

Note. From "The Performance of Natural Resource Industries" (pp. 5–21) by B. Weber, E. N. Castle, and A. L. Shriver, 1987, in U.S. Department of Agriculture, Economic Research Service (Ed.), *Rural Economic Development in the 1980's: Preparing for the Future.* Washington, DC: U.S. Government Printing Office.

TABLE 7.2. Average Annual Growth Rate in Employment in Forestry and Wood Products in Nonmetro Counties, 1969–1984

Industry	Annual growth rate (%)
Forestry and wood products (total)	0.78
Forestry	5.61
Paper and pulp	0.57
Lumber and wood	0.79
Furniture and fixtures	0.98

Note. From "The Performance of Natural Resource Industries" (pp. 5–24) by B. Weber, E. N. Castle, and A. L. Shriver, 1987, in U.S. Department of Argiculture, Economic Research Service (Ed.), *Rural Economic Development in the 1980's: Preparing for the Future*. Washington, DC: U.S. Government Printing Office.

tive forestland can no longer be measured solely by the number of acres converted to other uses. The degree of conversion also depends on economic parcel size, access to local mills, and the compatibility of commercial forestry with adjacent land uses. Most importantly, however, the rural planner must come to see the forest resource as a truly productive element of the rural landscape worthy of attention and concern.

CHAPTER EIGHT

Productive Systems: Recreation, Fisheries, Mining, and Energy

Local development and regional growth in rural areas are often highly dependent on a particular natural resource or recreation activity. Several towns in the Rocky Mountains and the Northeast rely on nearby ski resorts; Appalachia has long been known for coal mining; many coastal towns still earn much of their livelihoods from commercial fishing; and a number of Western towns have seen their fortunes rise and fall along with the value of local coal, oil, and natural gas deposits. Each of these communities—recreation, fisheries, mining, and energy—shares elements of dependency discussed in Chapter 5. Situations of dependency are particularly relevant to natural-resource-based communities, because their fortunes are often determined by international economic and political events. These communities are vulnerable to increases in the production of raw natural resources by other areas of the world, resulting in lower resource prices and hence lower local purchasing power. The development of substitutes, such as plastic for certain metals, can reduce the demand for a local natural resource, leading to price declines in the resource and even threatening the profitability of the local industry. Finally, a community's natural resource base may become depleted, especially in the exploitation of nonrenewable energy and mineral resources.

As noted in Chapter 5, four kinds of dependency exist. "Direct dependency" occurs when key sectors of a local economy are controlled by absentee owners. Decisions about production and employment are often made in distant company headquarters, rather than locally. Direct dependency most often exists in mining towns and in towns with large energy development projects; these towns may be "company towns" in which the local mine or energy project is the economic mainstay of the community, but local residents have little if any control over operating decisions.

"Trade dependency" describes the importance of imports and exports in a local economy, whether there is a trade surplus or deficit, and the relative buying power of local exports. Prices for locally produced commodities are generally established in distant urban centers, and fluctuations in price affect the profitability and survival of those commodity industries. For example, recreation areas rely on the importation of tourists. If the price of gasoline rises, fewer tourists are likely to visit recreation areas; and yet recreation areas have no control over gasoline prices. Similarly, in recent years, the price of copper has fallen because of the development of glass fiber optics for communication wires. Because the buying power of copper has fallen, it has become more expensive to import goods and services into the copper-mining towns of Arizona and Montana.

"Financial dependency" refers to the influence of urban banks and of federal monetary and fiscal policies on the availability of credit in rural areas. Federal tight-money policies and huge budget deficits in the 1980s have produced high interest rates and have limited the use of borrowed money for the development of recreation and energy projects. In addition, high interest rates have placed a heavy financial burden on fishermen who borrowed money to buy fishing boats. Interest rates and lending policies can be a major hindrance to local development and are generally beyond local control.

"Technical dependency" describes a region's need to import technology and trained personnel in order to achieve economic growth. Energy towns, for example, have "boomed" during the influx of construction workers and engineers. Other rural areas remain stagnant because of an inability to attract outside industry, capital, and skilled labor.

Planning in a single-resource community presents several challenges. Whether a planner approaches a single-resource community from a local or regional office, the planner should have a good understanding of the commodity the community produces and the situations of dependency involved with that commodity. Two major needs are likely to arise: economic diversification and the accommodation of social change. Part-time small-town governments have often been unable to address these needs. The local planning style has been incremental and uncoordinated, responding to crisis situations rather than anticipating needs in a comprehensive fashion. Much education about the value of planning remains to be done in single-resource communities. The planner must work to build a public consensus on strategies that either harmonize with outside forces or else mitigate their impact. In the process, the planner can help a community become more self-reliant (e.g., by forming local development corporations and obtaining economic development grants), with greater control over its well-being and future.

Planning in the Single-Resource Community: Recreation

Rural recreation areas became popular in the 1960s and 1970s as a result of national increases in personal incomes, greater amounts of vacation time, a growing concern with health and exercise, and improved transportation afforded by the interstate highway network. In 1970, there were 81 million acres of recreation land in private ownership; by 1980, the figure was an estimated 103 million acres; and by the year 2000, there will be an estimated 126 million acres of private recreation land (Healy, 1976, p. 29). Rural recreation areas feature second-home developments that are located near organized facilities, such as ski resorts and marinas, or else in more remote regions of scenic beauty. These second homes are predominantly owned by urban dwellers and are used for vacations or in conjunction with weekend outdoor recreation. Second homes comprise about 3 percent of America's total housing stock; the highest concentrations of second homes are found in California, Colorado, Maine, Michigan, Missouri, New York, North Carolina, Pennsylvania, Texas, Vermont, and Wisconsin (Wolf, 1981, p. 432).

Recreation differs from tourism in that recreation involves a sport or physical activity, whereas tourism is largely passive. Recreation can be seen as a form of land consumption in which the recreationist consumes land services through the outdoor recreation experience. Because recreation is often associated with a specific location, it is a commodity that cannot be transported.

Recreation consists of three related elements: physical, economic, and psychological. Access to a recreation area and available modes of travel affect the number of likely visitors. The more remote the area, the fewer the number of visitors who can be expected. Related to access and travel is the actual cost of the recreation experience. Economists have long used a "travel-cost" method to determine a visitor's demand or willingness to pay for a certain activity, based on the cost of getting there and returning home. The psychological aspects of recreation reflect the visitor's preferences of where to go and what kind of recreation experience to seek. The psychological aspects last far beyond the experience itself through memories, pictures, and descriptions. Some recreationists may seek distant, vigorous activities such as rafting through the Grand Canyon, where others may opt for more local pursuits such as hiking. Individual choice of recreation is likely to vary according to income, vacation periods, and time of year. People with larger incomes and more leisure time can afford to travel further to enjoy more remote and expensive forms of recreation, and a greater variety of outdoor recreation is available in the summer than in the winter.

When rural recreation areas developed rapidly in the 1960s and early 1970s, they underwent a series of growth stages similar to those of energy boomtowns. In the "preboom" phase, a town is typically remotely situated with a population of a few hundred to a few thousand people. Public services are not sophisticated, but are generally adequate. An attitude of mutual self-help and neighborliness exists among the townspeople. The local economy may be based on forestry, agriculture, or mining, as well as retail sales and some tourism.

Then an area is "discovered" for its recreation potential. Capital, labor, and technology are brought in to develop specific kinds of recreation projects, such as a ski resort or marina. This "construction boom" includes both the creation of a recreation facility and the development of second homes and hotel/motel units to house visitors. A construction boom normally lasts from two to five years.

About midway through the construction boom, a "secondary boom" occurs as local businesses expand and new businesses move in to provide more goods and services to the increased population. Unlike construction workers brought in from outside, newcomers in the secondary boom generally intend to become permanent residents of the town. The secondary boom in recreation areas often exceeds the construction boom in length of time and total dollars invested. This is the opposite of what has been observed in energy project boomtowns, where the cost of building a huge energy facility is greater than the value of dollars invested in the secondary boom.

In the "postboom" phase, a rural recreation area may (1) decline in popularity; (2) diversify its economy and continue to grow; or (3) maintain a stable economy and population. The postboom phase will also depend on how well a town is able to maintain a high quality of life in accommodating economic, social, and environmental changes. The large majority of rural recreation towns in the United States are now in the postboom phase and have experienced sharp increases in land prices, a decline in average parcel size sold (signaling more intensive development), and increases in absentee ownership of land and housing.

Economic Impacts

The most obvious effect of rural recreation facilities is that they often induce greater economic activity. The size and length of the economic impact will be directly related to the size of the recreation project. The influx of capital and labor during both the construction phase and the secondary phase tends to raise local employment and incomes, which in turn produces an increased demand for goods and services, especially housing. Rising land prices often attract speculators who may hold land off the market as they wait for land prices to climb still further. Often the withholding of prime building sites results in haphazard development patterns that are costly to service. Moreover, second homes are almost exclusively owned by outsiders who have greater wealth and income than the local residents. The locals are often priced out of the land market as the number of second homes increases and the town becomes more dependent upon recreation for its livelihood. For example, land sales data in Vermont indicate a trend of fewer purchases by local residents in ski resort towns (Daniels & Lapping, 1984, p. 507). The price of locally sold goods and services tends to increase as well, driving up the cost of living for local residents.

Along with increased population and economic activity, newcomers—both part-time resident second-home owners and year-round residents—tend to demand more services than the indigenous population. Financing the construction of new schools, roads, and sewer and water systems, and hiring additional fire, police, and other personnel, place a sudden heavy burden on local taxpayers. Long-time residents often see their property taxes escalate to pay for services they formerly were able to do without. In addition, there often is not enough public money available to address all of the public service needs. Thus, some problems are attacked before others (e.g., constructing a new school rather than paving streets).

Once recreation becomes the focal point of the local community, a major problem is the irregularity and peaking of use associated with recreation activity. Recreation seasons may be limited by law (hunting, fishing) or by climate (skiing, boating). Facilities may be idled for months at a time, causing employment layoffs and slowdowns in retail and service outlets. If average use of a recreation facility is much lower than the peak use, the cost per unit will be high if an effort is made to provide housing and public service capacity to accommodate peak demand. This situation of providing peak capacity is especially troublesome for public services—roads, sewerage, and water—because of the tax burden placed upon property owners. The underprovision of sewerage and water facilities, however, can have detrimental environmental effects.

Social Impacts

Rural recreation areas have brought about many pronounced changes in local social structure and personal interaction. Newcomers consist of well-to-do part-time residents, transient workers seeking seasonal employment, and members of the secondary boom who intend to become permanent residents. Each of these three groups causes a change in the community's image of itself.

The well-to-do may give the community a reputation as a "playground for the rich," but at the same time, they may increase the variety of social and cultural activities. Transient workers bring out a lack of permanence and commitment. New store owners often compete directly with established businesses. And in some recreation areas there is a growing retirement population. The overall result of the influx of newcomers can be a decline in social cohesion. The local population becomes more heterogeneous, made up of distinct interest groups, and often geographically segmented according to social status. Indigenous residents may resent the intrusion of newcomers or disagree with their personal values, and an increase in crime, alcoholism, and drug abuse may be brought about by both recreationists and transient workers. Changes in local leadership may occur, with greater representation of the newcomers; this may or may not be beneficial to the community.

Environmental Impacts

The influx of recreationists and new residents, as well as indigenous population growth, has increased the pressure on rural environments. Greater use of the outdoors has heightened the competition between humans and plant and animal life. Many rural recreation areas consist of sensitive environments. The ability of these environments to support development is often limited by their soils, lack of water supplies, or steep slopes. Second-home developments have had profound effects on the local environment, and often these effects are not limited to a specific property. Sewer and water problems associated with second-home developments in ski resort towns led to the passage of Vermont's Land Use and Development Law (Act 250, 1970; VSA 10 §§ 6091, 1973; as amended Supp. 1975). Construction on steep slopes and thin soils was resulting in effluent runoff that polluted wells, streams, and rivers below. Yet, even today, there are bumper stickers in Vermont that describe a particular ski resort as "where the effluent and affluent meet."

A major source of environmental problems can be attributed to a lack of well-defined subdivision regulations, in terms of both location and a minimum parcel size with an adequate individual or municipal septic system. Land ownership patterns often are not obvious to the naked eye, especially before the construction of housing commences. In the past, developers have been allowed to subdivide land into hundreds or thousands of building lots in a single project. In fact, there are curently an estimated 20 million recreational lots in the United States (Wolf, 1981, p. 424). Yet there has rarely been careful review as to whether these lots have road access or adequate carrying capacity in the form of sewer and water facilities. Indeed, many of the lots created have been unbuildable. Wolf (1981, p. 423) notes that there is a low ratio of homes constructed in new recreational subdivisions, compared to the large number of lots subdivided. In California the ratio is 1 house per 33 lots; in Arizona, 1 per 60; in Pennsylvania, 1 per 120; and in Florida, 1 per 73,268! The idling of land in recreation subdivisions precludes the consolidation of those lots into larger units that might be used for farming, forestry, or other enterprises to help diversify the local economy. Some development schemes have been underfinanced, with the second-home project collapsing and leaving hundreds to thousands of underdeveloped lots.

Recreation development may be aesthetically harmful to a town and its surroundings. High-rise condominiums and the construction of strip-type commercial development tend to detract from an area's natural setting. Moreover, the cumulative environmental impact of growing populations and development is difficult to measure. Plant and animal ecosystems will suddenly "give out" if the extent of human intrusion and pollution becomes too great. For this reason, the expansion of recreation areas may have more damaging effects than the original project.

Challenges to Planning

Planning should seek to protect those qualities that make a rural recreation area a desirable place to live, work, and vacation. The planner should realize that there are four basic factors beyond local control: (1) the price and availability of gasoline; (2) the weather; (3) the tastes and preferences of recreationists; and (4) the number of outsiders who choose to move to a recreation area. The price and availability of gasoline determine the cost of accessibility. The more remote a recreation area, the more sensitive its livelihood is to the world oil market. Weather also plays a large role in the success of many recreation areas. For example, if sufficient snow does not fall, a ski resort may suffer financial losses from a lack of skiers. The tastes and preferences of recreationists are subject to change as new forms of recreation are developed, and the choice of recreation will also depend upon the level of income of urban dwellers and economic climates. As incomes rise, a greater proportion of income will tend to be spent on recreation; in times of economic recession, however, recreation may be one of the first services sacrificed.

The diversification of the local economy can serve to increase local control. In most recreation areas, the demand for second homes will not be as high as in the past. Thus diversification may be necessary in order to sustain economic growth. More year-round employment and use of recreation facilities, such as the use of ski resorts for conferences in the off season, can promote diversity. The growth of the local permanent population through in-migration and natural increase can place additional public service demands on the community, as well as increase the competition for land and environmental pressure.

To maintain the quality of the recreation experience, planners can gather adequate information from which to devise such town management strategies as a comprehensive town plan, zoning and subdivision regulations, and a capital management plan. Through an inventory of land resources, a planner can identify fragile areas with low capacity to support development. Improved and unimproved lots can usually be detected from an assessor's tax map. Soil maps and topography maps can help identify fragile areas. The following features should be reserved from development or protected from intensive development:

1. Landslide areas
2. Slopes of greater than 25 percent
3. Floodplains
4. A 100-foot buffer zone next to public water sources
5. Wildlife habitats
6. Wetlands
7. Prime agricultural lands

Next, the planner should conduct site-specific analyses of proposed developments to determine (1) current use of the site and (2) the intensity of the proposed use and hence the environmental impact. The planner should ask these questions: What public services will be paid for by the town and not the developer? Has the developer provided complete basic services, such as roads, a centralized water and sewerage system, or wells and septic tanks? Does the developer intend to sell lots in phases, and, if so, will services be extended? Has the developer used a cluster design, retaining 25 percent or more open space? Does the style of the development fit in with the surroundings?

Planners should avoid allowing congestion and overconstruction; a capital management plan should indicate where development is appropriate. Planners can thus separate conflicting uses and mitigate the conflict between those wishing to develop an area for intensive recreational use (e.g., downhill skiing) and those who prefer to maintain a lighter use (e.g., hiking trails). Planners must be able to delineate the costs and benefits of recreation projects, and must be aware of the experiences of other recreation-oriented towns.

Small towns commonly have only part-time officials who serve with few monetary rewards. The local leadership frequently receives the blame for deteriorating public services. But planning and zoning may be new concepts. The planner will tend to approach a rural recreation community from a regional planning office, so there may be a barrier in being perceived as another outsider. The planner must be sensitive to the needs and values of the local community and must attempt to build a consensus on planning strategies. In this way, the planner can begin to overcome the "crisis mentality" style of planning and look to provide adequate infrastructure and economic diversity far into the future.

Planning in the Single Resource Community: Fisheries

The 1980 census listed just over 83,000 people as employed in forestry and fisheries (U.S. Bureau of the Census, 1980). Less than half of these people were full-time commercial fishermen, comprising under 0.2 percent of the rural labor force. In 1985, U.S. fishermen landed 6 billion pounds of fish worth almost $2.5 billion (U.S. Bureau of the Census, 1986, pp. 662–663). Fishermen as a group contribute very little to the national economy, but may form a significant force in the economic and social livelihood of many coastal towns. In fact, America boasts the world's third largest fishery in terms of pounds of fish caught (U.S. Bureau of the Census, 1986, p. 666). The number of fishermen and the total annual catch of fish today are lower than they were before World War II. Fishermen have been plagued by overfishing, restrictions on fishing seasons, and limited entry into a fishery.

Fish are a unique resource in that the stock of fish is renewable but destructible and constitutes a common property resource, because no one owns the fish (until caught), and limits on access to the fish are difficult to enforce. In an attempt to define property rights, Congress in 1976 passed the Fishery Conservation and Management Act, known as the "200-Mile Bill" (Pub. L. No. 94-265). The act protects American fishermen within 200 miles of U.S. shores from encroachment by foreign fishermen (especially Japanese, Canadian, and Russian boats). The act sought to promote employment and economic vitality in the U.S. fishing industry by extending the property claim from the old three-mile limit to 200 miles.

In addition, the act was aimed at conserving the fishing stocks accessible to U.S. fishermen. The act created regional fishery management councils that have the power to regulate fishing within the 200-mile limit (Yetley, 1982). Each council drafts a management plan for almost all species with either commercial or sport fishing value. To implement the management plan, a regional council may use any or all of the following restrictions: (1) limits on access, with controls on fishing seasons and specific locations; (2) regulations on fishing technology; and (3) limits on the number of boats. In the northwest Atlantic fisheries, for example, conservation measures feature a mix of closed areas, closed seasons, restrictions on technology, and a limit on the number of fishing boats. In the Pacific Coast fishery, the seasons for most fish species were reduced from nine months to three in the 1970s; some species, salmon in particular, had seasons reduced to less than three months in the 1980s. Even with the 200-mile limit, half of all fish and shellfish landed in the United States have been caught within three miles of shore (U.S. Bureau of the Census, 1986, p. 667).

Fishermen have traditionally been staunchly independent. Considerable friction has resulted from the restrictions on commercial fishing, with fishermen arguing that it has become increasingly difficult to earn a living. On top of the regulations, fishermen have faced fluctuating but generally declining fish yields, whereas the work of wresting fish from the sea is as dangerous as ever.

The major obstacles to planning in a fishing community are a lack of data on projected fish stocks and the uncertainty of future government regulations on fishing. Biologists have been unable to provide detailed information about the magnitude, locations, and mortality rates (due to disease and predators) of fish stocks. Data collected on fishermen by federal agencies are confidential. The National Marine Fisheries Service will release only highly aggregated data that are of limited value for planning in a single community. With changing fishing regulations, a planner must be able to forecast who will be affected, the size of the economic impact, what alternative employment opportunities exist, and the likelihood of declines in the local fishing fleet. This forecast enables a planner to estimate future public service needs, tax revenues, and economic base.

An obvious alternative to a community built around commercial fishing is one that relies on sport fishing and tourist facilities. Like recreation areas, fishing communities have peaks and troughs in economic activity during the course of a year, and a fisherman's yields will vary according to the weather. A key aspect of the local fishing economy is whether the fish are sold fresh locally or processed for sale in distant areas. A local fishing industry often can support local restaurants featuring fresh seafood. Processing plants, however, tend to be owned and operated by companies headquartered in far-off urban centers. This creates a situation of direct dependency, in which the local fishermen have little if any control over the price the processing plant is willing to pay for seafood or over the operating season of the plant. This situation should provide an impetus for economic diversity.

Planners working in fishery communities should be in contact with the regional fishery agency, the U.S. Fish and Wildlife Service (Department of the Interior), the National Marine Fisheries Service (National Oceanic and Atmospheric Administration), the state department of fish and wildlife, and the state office responsible for administering the coastal zone management program. Planners must be able to gauge the impacts of proposed developments on the fishing community. Economic diversification should be welcomed, but not at the price of driving the fishing industry into extinction.

Planning in the Single-Resource Community: Mining

There are an estimated 170,000 active coal, metal, and nonmetal mines in the United States (Epstein, Brown, & Pope, 1982, p. 306). Still, mining accounts for only 2.3 percent of America's rural labor force. But because of the highly localized nature of mining, several communities are heavily dependent on the extraction of metal[1] for their livelihood. In general, communities do not rely significantly upon nonmetal mining resources (see Table 8.1). By contrast, a number of towns in the western United States are built around the extraction of gold, silver, lead, mercury, molybdenum, uranium, copper, and iron.

Communities dependent on the extraction of minerals (see Figure 8.1) show similarities both to rural recreation areas and to fishing communities. Like the local fish stock in a fishing community, the size, location, and grade of natural mineral deposits affect the economic growth potential of a town. Like rural recreation areas, mining towns experience an initial boom, a secondary expansion, and then a somewhat uncertain future. Unlike fish, minerals do not regenerate at a pace that is economically meaningful and are considered nonrenewable resources. Unlike rural recreation areas, mining deposits provide a limited livelihood; the western United States is dotted with abandoned mining towns.

Several factors determine the economic lifespan of a mineral deposit. First, a mining company must be satisfied that the size, location, and concentration of the deposit are adequate to justify exploitation. The extent of known reserves will vary as new discoveries are made and current reserves are extracted. The economic supply of a mineral, which

[1]Coal mining is discussed in the following section on the energy community.

TABLE 8.1. Percentage Distribution by Weight and Value of the U.S. Use of Nonfossil Nonrenewable Resources, 1975

Resources	% total weight	% total value
Sand, gravel, rock, clay, etc.	90	12
Iron, aluminum, and magnesium	4	52
Other nonmetals	5	13
All other	1	23

Note. From "The Age of Substitutability: A Scientific Appraisal of Natural Resource Adequacy" by H. E. Goeller, 1979, in V. K. Smith (Ed.), *Scarcity and Growth Reconsidered* (p. 148). Baltimore: Johns Hopkins University Press. Copyright 1979 by Johns Hopkins University Press. Reprinted by permission.

is the portion of the physical supply that can be profitably extracted, changes over time because of (1) new mining technologies; (2) changes in the market price of minerals; and (3) possibilities of substituting other materials. New mining technologies have increased the accuracy of exploration to identify new reserves and have improved the efficiency of extraction methods. This has allowed the use of lower grades of ore and has increased the economic supply. The market price of minerals has a major effect on their economic supply. When the price of gold rose to $800 an ounce in the late 1970s, for instance, it became profitable for companies to open up old mines and extract the more-difficult-to-retrieve gold ore. On the other hand, companies could afford to reduce the extraction of easy-to-mine ore as a form of insurance should the price of gold fall. By the early 1980s, gold prices fell to below $400 an ounce, causing companies to close expensive mines and concentrate on more cheaply mined ore.

The market price of minerals is established in an international market. For example, the more copper Chile exports to the United States, the lower the price of American copper. Thus, the amount of minerals supplied by other countries has a strong influence on price, as does worldwide demand. When mineral prices rise, three responses are likely to occur. First, users will try to conserve on the amount of minerals they use, and greater efforts will be made to recycle mineral products. Second, mining companies have a greater incentive to seek new deposits. And, third, users will seek cheaper substitute materials. A striking example of substitution, noted earlier, is the use of glass fiber optics in place of copper wire in communications. Fiber optics are both less expensive to produce and far lighter and easier to install. As a result, the demand for

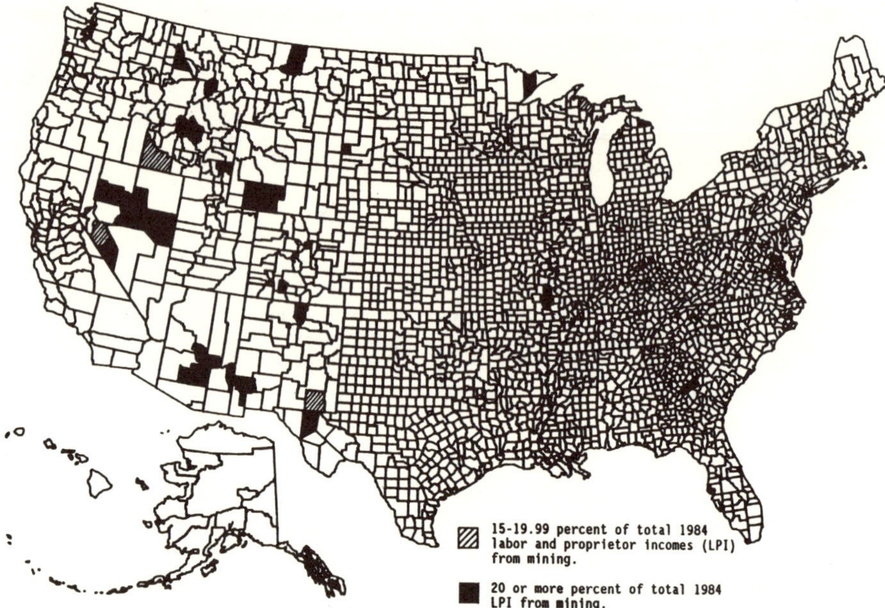

FIGURE 8.1. Mining counties, 1984. From *Rural Economic Development in the 1980s,* by D. L. Brown and K. L. Deavers, 1987, Washington, DC: USDA, Economic Research Service.

copper has fallen in recent years, placing downward pressure on copper prices; consequently, several copper mines and smelters in the western United States have been idled.

The rate of exploitation of a particular mineral deposit has important consequences for a local community (see Figure 8.2). The higher the rate of extraction, the shorter the lifetime of existing mineral stocks; conversely, the slower the rate of extraction the longer the yield of the deposit. The rate of extraction will fluctuate according to the market price of minerals and the operating decisions of mining companies. Mining towns typically exhibit three forms of dependency: direct, trade, and technical. Direct dependency, as noted above, occurs when a distantly based company controls key sectors of a local economy. Mining towns are often remotely situated, and the local mine is likely to be the largest employer within a wide radius. A mining town is often a classic "company town" in which the mining company controls the economic and political life of the town. To develop a mine in the first place, technical equipment, trained personnel, and capital must be imported. If the mine closes, either for a long time or only briefly, the effects are deeply felt in the local community. Local residents have little if any

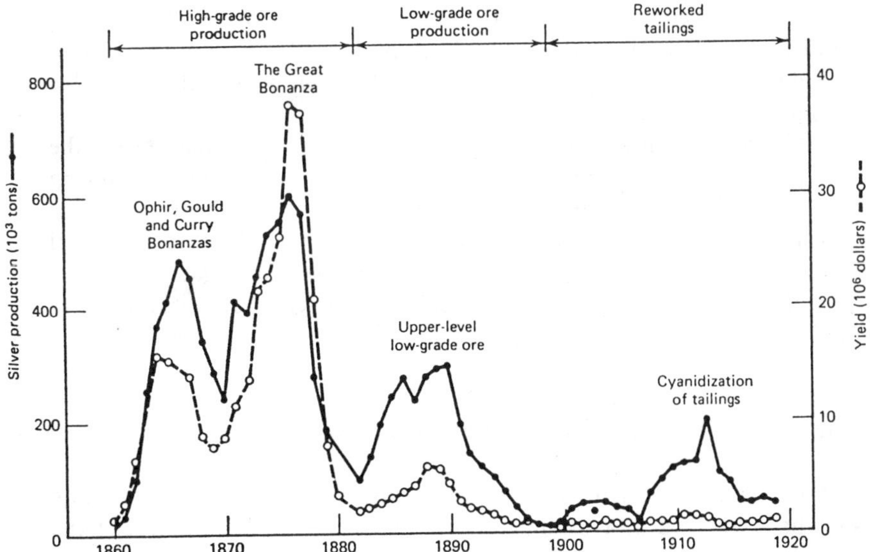

FIGURE 8.2. Production history of the Comstock Lode in Nevada. From "Limits to Exploitation of Nonrenewable Resources" by E. Cook, in *Science* 191 (4228), February 20, 1976, p. 679. Copyright 1976 by the American Association for the Advancement of Science.

control over the price of minerals or the operating decisions of the company. Mining towns must import many goods and services, which are paid for from the export of locally mined ore. When the price of local minerals falls, it becomes more expensive to import goods and services.

The central vulnerability of a mining town is that it may not be self-sustaining, or at least may not be so for very long. Even a secondary boom may not bring enough permanent residents to yield a meaningful diversity to the local economic base. As a boomtown, a mining community will often experience economic, social, and environmental dislocations. Greater economic activity stimulates an increase in the prices of goods and services, especially land and housing. An influx of newcomers causes friction and often brings increased crime, alcoholism, and divorce. The environmental effects of mining often include surface water and groundwater pollution, fish kills, air pollution from smelting, soil erosion, and surface subsidence. U.S. mines generate over 2 billion tons of solid and liquid wastes each year. Some wastes, such as the tailings from uranium mines, are highly toxic. In fact, mining was the first significant source of hazardous wastes. Toxic chemicals in mining wastes include arsenic, cadmium, chromium, copper, cyanide, lead, and mercury. In many mines, the volume of overburden extracted exceeds the volume of ore, and the weight of metals may be a small fraction of the weight of the ore. At some copper mines, one and a half tons of overburden are removed for each ton of ore, and only 16 pounds of copper are extracted. The copper mines become huge open pits, and waste piles resemble human-made mountains.

The planner in a mining town must be aware of the market for the metals extracted in the community. The planner must develop a working relationship with the local mining company; in addition, he or she must seek to promote economic diversity, given the fact that local mineral resources will eventually be depleted. To minimize the negative environmental effects, local development should be located a safe distance from mining activities.

The Energy Community: Special Problems

Not all towns with major energy-processing, -generating, or -excavating sites have evolved in a similar fashion. The energy communities of the Appalachian coal region have been in existence for decades, as have petroleum exploration and processing towns in Texas, Louisiana, and Oklahoma. This section focuses on the new energy communities that sprang up in the West in the 1970s; however, several aspects of these energy boomtowns are relevant to the experiences of other energy communities. (Figure 8.3 shows the U.S. counties deriving a substantial percentage of income from energy.)

The Arab oil embargo in the winter of 1973–1974 marked the beginning of America's concern over the tenuous link between the supply of energy resources and the nation's economic well-being. Since the late 1940s, America had increasingly turned from coal to oil and natural gas as its major sources of fuel. Energy companies generally found it cheaper to produce and import foreign oil than to use domestic reserves. At the time of the Arab oil embargo, foreign oil comprised 35 percent of America's energy supply. In response to the embargo, President Nixon instituted Project Independence, designed to encourage the development of American energy resources and to lessen the nation's dependence on foreign oil. Project Independence set a goal of complete U.S. energy independence by 1985. The development of alternative energy sources, such as solar and nuclear power, and the use of coal and oil shale reserves in the western United States became national priorities. Skyrocketing energy prices (from $2 a barrel of oil in 1972 to $34 a barrel in 1979) sent energy companies into a frenzy of exploration and development. But the collapse of oil prices to less than $10 a barrel in 1985 resulted in the abandonment of many energy projects, the capping of hundreds of low-yielding stripper wells, and sharp cutbacks in new exploration. By 1987,

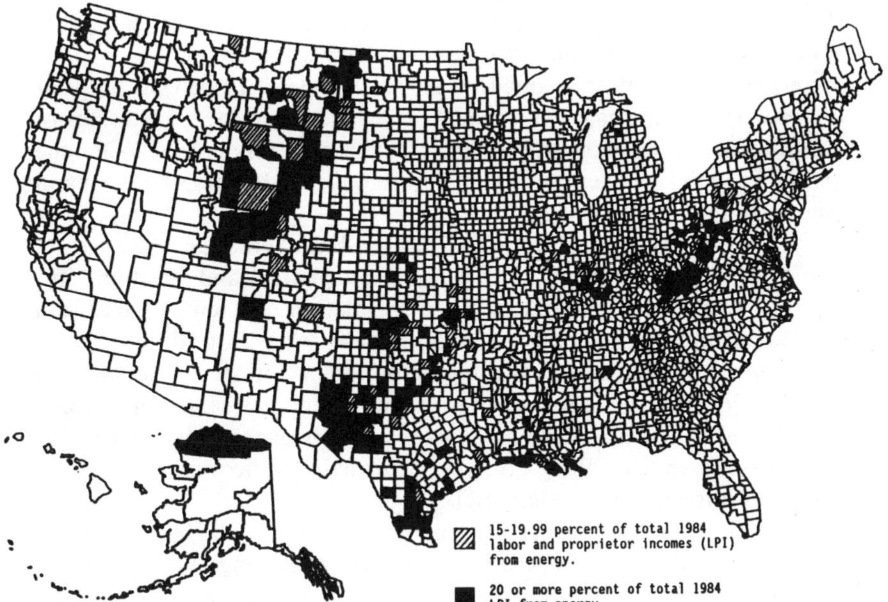

FIGURE 8.3. Energy counties, 1984. From *Rural Economic Development in the 1980s*, by D. L. Brown and K. L. Deavers, 1987, Washington, DC: USDA, Economic Research Service.

oil prices recovered to about $20 a barrel. But the overall effects of lower oil prices on the major oil-producing states, such as Texas, Louisiana, Oklahoma, and Alaska, have been unemployment, bank failures, and stagnant economic growth, particularly in rural areas. Meanwhile, as of 1987, the United States was importing 40 percent of its oil needs ("Data Bank," 1987); the country produced just over 8 million barrels a day and imported about 7 million barrels a day. Domestic oil and natural gas production was worth over $100 billion in 1987.

Roughly half of known American coal deposits are located in the West, along with 90 percent of potential oil shale resources, 95 percent of uranium reserves, and significant oil reserves. Many energy communities sprang up literally overnight in the intermountain West. These communities showed many similarities to mining boomtowns.

The Nature of Boom and Bust

The term "boomtown" applies to small, isolated communities that experience a very rapid rate of growth in a relatively short time span (Weber &

Howell, 1982). An annual growth rate of 5 percent is generally the maximum a small community can absorb without disruptions to its social, political, and economic life. A growth rate of 10 percent per year typically strains the capacity of local government and local businesses to meet the increasing demand for public services and retail trade. When the annual growth rate exceeds 15 to 20 percent, breakdowns in housing and labor markets often occur.

Five phases form the cycle of boom and bust: "preboom," "construction boom," "secondary boom," "operation phase," and "decline." Initially, a small rural community has a population of a few hundred to a few thousand inhabitants; the economic base relies on traditional rural industries of agriculture, ranching, mining, or forestry. The town is typically located 100 miles or more from a regional center of 15,000–20,000 people and up to 200 miles from a major metropolitan area. The local lifestyle and social structure are informal, and because of a limited tax base, infrastructure is minimal.

The decision by an energy company to construct a large new project or to expand an existing operation marks the beginning of the construction boom. However, there may be numerous delays from the time the decision to develop is made until construction actually commences. Considerable uncertainty often exists over the scale and timing of the project, because the energy company must monitor and respond to fluctuations in the national and international economy. The climate of uncertainty also affects nearby towns. The local business community often adopts a "wait-and-see" attitude toward the proposed project. Local governments typically do not begin the expansion of infrastructure to meet increased population demands until construction of the energy project is actually in progress. A significant reason for confusion at the local level is that the major decisions over the future of the community are being made in distant corporate headquarters. Generally there is little advance consultation with local communities. In addition, federal policies and regulations formulated in Washington, D.C., may have a major impact on corporate decisions. For instance, national policies encouraging the development of domestic energy resources through tax incentives and reduced environmental quality standards were important in the corporate decisions to expand coal-mining operations in the Rocky Mountain states.

A local construction boom may last for several years, depending upon the project. Coal-fired electrical generation plants require an average of four to eight years to build, and construction of an oil refinery will take two to three years (see Table 8.2). The construction phase requires the importation of skilled labor and large amounts of capital. About halfway into the construction boom, a secondary boom in retail and service outlets begins. Before the construction boom occurs, the local

TABLE 8.2. Sample Energy Projects

Project	Size	Construction time	Peak construction work force	Operating work force
Electricity-generating plant (including coal mine)	700 MW 2,250 MW	4–6 years 6–8 years	750–950 2,000–3,000	75–100 350–400
Nuclear power plant	1,000 MW	5–9 years	2,500	150
Liquid natural gas conversion plant	1,000 mcf/day	2–3 years	300–400	50–100
Substitute gasification plant (includes coal mine)	250 mcf/day	2½–3 years	3,000–3,500	1,050–1,240
Oil-shale-processing facility (includes mining)	50,000 bbl/day	3–4 years	2,400	1,050–1,450
Coal export mine	9 metric tons/years	2–3 years	175–200	325–425
Oil refinery	250,000 bbl/day	2½–3 years	3,500–4,500	450–900
Offshore oil and gas support	per rig	3–4 years	175	90
Platform fabrication facility	2 platforms per year	5 years	400	1,000–1,500
Deepwater port	2 mooring spaces	3–4 years	1,250	75–90

Note. From *Rapid Growth from Energy Projects: Ideas for State and Local Action. A Program Guide* (p. 3) by the U.S. Department of Housing and Urban Development, 1976, Washington, DC: U.S. Government Printing Office. MW, megawatts; mcf, million cubic feet; bbl, barrels.

business community typically offers only the goods and services that are essential to the everyday activities of the townspeople. Gas stations, a restaurant, a grocery store, a hardware store, a bar, a bank, a general store, and a movie theatre are found in most small Western towns. Local residents commute to urban areas for major expenditures in services and durable goods. But once the secondary boom begins, new types of businesses locate within the community. Additional people settle in the town to operate these new and expanded businesses. The length and intensity of the secondary boom depend upon the diversity of the local economy. In general, the more diverse the economy, the longer the secondary boom will last.

About halfway into the secondary boom, the construction boom ends, signaling a slowdown in local economic activity. As an energy facility enters the operating phase, the majority of construction workers leave the area. The ratio of workers required in the operating phase to the construction phase may be as little as 1/5th to 1/16th of the construction boom labor force. The degree of reduced economic activity will depend upon the strength of the secondary boom. If the secondary boom is weak, then a community will be severely affected by the loss of construction workers. The operating phase may extend for 20–30 years before the natural resource base is exhausted or an energy facility outlives its usefulness. The bust phase occurs when the energy company pulls out, taking much of the local labor force as well. Retail and service outlets are hard pressed to earn a profit, and many businesses are likely to leave. An attitude of failure and decay pervades the town as the exodus continues. Whether a community degenerates into a ghost town will depend upon the degree of diversity of the local economy in retaining population and jobs.

Some booms may be very short-lived, either because initial discoveries prove uneconomical to exploit in depth or because of a decline in the price of a specific resource. In these cases, there is little if any secondary boom, no operation phase, and a severe bust. The boom–bust experience may not be a new phenomenon to communities in the western United States; as a result, townspeople may be reluctant to take steps toward advanced preparation for an anticipated boom. A "wait-and-see" attitude is common. But once a boom reaches the construction phase, planning and the provision of public services occur on a piecemeal basis, attempting to deal with crises in sewer and water facilities, overcrowded schools, and demands for improved roads.

Coping with Rapid Growth and Turnover

The western United States is distinct from other regions of the country in the amount of land that is federally owned. In the Rocky Mountain states, from one-third to four-fifths of the total land area is owned by the federal government. This high degree of federal ownership diminishes the amount of local control over the region. The federal government, not the state or local governments, makes the decisions to allow energy exploration on federal lands. However, most federal land is not available for residential, commercial, or other development that may be necessitated by an incoming energy company. Any development that crosses onto federal property, such as roads or power lines, may require the preparation of an environmental impact statement—a potentially costly and time-consuming process. This situation heightens the pressure to develop privately held land in the vicinity, including ranchlands and timberlands.

Increased speculation in land is likely, since the supply of private land available for development may be limited. Because of a scarcity of housing, particularly during the construction boom, housing costs tend to escalate rapidly. It is not uncommon for the price of residential lots to double and for rents to triple in the course of the construction boom. Overcrowding and unsightly new housing developments (often in the form of mobile home parks) may be disconcerting to long-time residents.

A large number of relative strangers moving into a community tends to break down informal ties of mutual assistance among the townspeople. In a small town where almost everyone knows everyone else, people are more likely to help out their neighbors when faced with unemployment, illness, or family disputes. This may not be the case in a more urbanized environment where people do not know their neighbors, but find more public assistance agencies, health care facilities, and counseling services available. As one might expect, the transition from one social system to another is especially difficult.

Some old-time residents may feel that their traditions or cultural values are threatened. Many individuals experience a change in their roles within the community. Local political power is redistributed as new interest groups emerge; the balance of power may shift from the old-time landowners to the new industrialists as the community becomes more dependent upon energy development for its livelihood. Social pathologies such as crime, alcoholism, juvenile delinquency, and divorce may tend to increase along with the influx of outsiders (for a different opinion, see Krannich, Greider, & Little, 1985). Many of the newcomers will be single male construction workers who have few if any ties to the local community. Married workers often do not bring their families to live with them while they are building the project.

Many areas of the intermountain West have an arid or semiarid environment. Water is a scarce resource that places limits on development and population growth. The pressures of booming growth rates may cause local water demand to exceed local supplies. Competition among ranchers, the energy industry, and local residents can become fierce. And because the town is located far from larger cities, there is no way the town can depend on those communities to supplement sewer and water systems. The deserts and high plains of the West are also environmentally sensitive areas. Once these ecosystems are disrupted, they may require years to return to a balanced state. In 1977, Congress passed the Surface Mining Control and Reclamation Act, and created the Office of Surface Mining to levy a fee on strip mine operators for each ton of ore mined (30 U.S.C. §§ 1201–1328). The act was intended as an incentive for deep mining, which is more costly but less environmentally damaging. At the same time, strip mine operators were to be compelled to replace topsoil and fill in excavations. To date, the reclamation act has been poorly enforced, and an estimated $150 million in fines have not been collected. Some strip-mined areas have been restored to provide grazing lands and forestlands, as well as recreation lakes. But unrestored areas (especially in mountainous terrain) have experienced severe erosion, resulting in increased stream siltation, acidity, declines in fish populations, and poisoned wells. Wastewater discharge from coal mines is estimated at several billion gallons per year (Epstein et al., 1982, p. 307). Acid drainage occurs in both surface and underground mines when strata contain high concentrations of iron sulfides. Minerals in the soil break down into water-soluble products and form acidic solutions that can enter both surface water and groundwater. The pollution of groundwater is a particular cause for concern in rural areas, because a large majority of rural inhabitants obtain their drinking water from wells.

The "Megaproject" versus Planning Problems: Case Studies from the West

A review of case studies of energy boomtowns in the American West provides insight into the boom–bust cycle of development and the magnitude of problems caused by rapid growth. In the 1970s, over 300 towns were targeted for new or expanded energy projects. Some of these projects have yet to be built. Follow-up studies are sketchy as to what happened to boomtowns after the decline in energy prices in the early 1980s. But it is likely that employment and economic activity in many boomtowns have decreased, along with the slowdown in energy exploration and the construction of power plants (see Table 8.3).

Rock Springs and Green River, Wyoming

Rock Springs and Green River are the two principal communities in Sweetwater County, located in the arid southwestern corner of Wyoming. Both communities experienced rapid growth during the 1970s. In 1971, two power companies, Pacific Power and Light and Idaho Power, began construction of the 1,550-megawatt Jim Bridger coal-fired electrical generating plant. The coming of this enormous plant stimulated a surge in local coal mining, followed by the development of oil and natural gas resources. Rock Springs grew from a population of 11,657 in 1970 to 26,000 by 1974. Green River, 12 miles to the west, had just under 4,200 residents in 1970 and tripled to a population of 12,807 in 1980. The provision of public services was a major problem in both communities. In 1974, during the Sweetwater County boom, municipal bond limits were a maximum of 4 percent of the towns' assessed value, plus an additional 4 percent for sewerage systems—all contingent upon a favorable vote of

TABLE 8.3. Population Change in Selected Western Boomtowns

Place	1970 population	Mid-decade population	1980 population	Net percent change, 1970–1980
Alaska	302,583	N/A	400,481	32.4
Fairbanks	14,771	36,975 (1976)	22,645	53.3
Valdez	1,005	3,840 (1975)	3,079	206.4
Arizona	1,775,399	N/A	2,719,225	53.2
Page	506	9,000 (1975)	4,880	864.4
Colorado	2,209,596	N/A	2,888,834	30.7
Craig	4,205	6,657 (1977)	8,133	93.4
Paonia	1,161	1,276 (1977)	1,425	22.7
Montana	694,409	N/A	786,690	13.3
Colstrip	350	3,000 (1975)	N/A	N/A
Wyoming	332,416	N/A	470,816	41.6
Gillette	7,194	7,270 (1973)	12,134	68.7
Green River	4,196	5,246 (1973)	12,807	205.2
Rock Springs	11,657	26,000 (1974)	19,458	66.9

Note. N/A, not available. From *A Descriptive Model of the Socio-Economic Impacts of Energy Boomtowns in the American West* (unpublished masters thesis) by M. E. Holley, 1982. Eugene, OR: University of Oregon.

the people. Housing was hard to find. Almost two-thirds of the newcomers lived in mobile homes, though more than half of these residents preferred permanent structures. Health care was scarce even before the boom, when only ten doctors served the county's 18,000 people. After the construction boom faded in the late 1970's, the population of Rock Springs declined by almost one-third. A contributing factor was the reduced quality of life in the town (Holley, 1982).

Craig, Colorado

Craig lies in Moffat County in the northwest corner of Colorado. The town is situated just north of the Yampa coal field, which contains an estimated 23 billion tons of low-sulfur coal. Traditionally a farming and ranching community, Craig has witnessed limited coal and oil development since the 1920s. The most recent boom began in the mid-1970s, when two large coal strip mines were opened in the hills south of town, and construction began on a 760-megawatt coal-fired electrical generating plant. In 1970, Craig had a population of 4,205 residents. Construction of the Yampa Project by the Colorado Ute Electric Association began in September 1974, and by 1977, there were 6,657 people in the Craig metropolitan area. Some 1,900 construction workers moved to Craig to work on the Yampa Project. Upon completion, the electrical plant was expected to employ only 270 workers. During the secondary phase, one new job in the service and retail sector was created for every two or three construction jobs. But the local business structure and practices changed dramatically. Craig merchants found it necessary to change their open-credit policies (some even ceased giving credit altogether) because of delays in receiving payments and the increase in nonpayments. Some store owners did not like the new style of business and sold out. Large chain stores from Denver established branches in Craig to compete with local merchants; often there were multiple business turnovers as speculators paid too much for a business, failed to turn a profit, and sold out.

In preboom Craig, there was no police force. Between 1973 and 1976, the incidence of crime skyrocketed. While the local population grew by 100 percent, crimes against property jumped 222 percent, and crimes against people soared 900 percent! The number of child abuse, juvenile delinquency, and family strife cases increased substantially as well. In 1977, the high school dropout rate in Craig was three times the Colorado state average (Freudenberg, 1979, p. 143). A survey of old-time residents found that three-quarters felt that change was occurring too fast; two-thirds of the newcomers surveyed agreed (Freudenberg, 1979, p. 152).

Page, Arizona

Page lies in the canyon lands of the Colorado River along the northern border of Arizona. It was established in 1957, during construction of the Glen Canyon Dam and Bridge on the Navajo Indian Reservation, as a federally owned community sponsored by the Bureau of Reclamation. Its peak population of 6,000 fell to less than 1,000 when the dam was completed. Then, in 1970, construction of the Navajo Generating Station, a 2,250-megawatt coal-fired electricity generating plant built by the Salt River Project, brought a new influx of population to Page. By 1975, as construction of the plant neared completion, the town population had soared to 9,000, and 70 percent of the townspeople lived in mobile homes.

A third energy project was proposed for the area but was canceled. In 1964, plans were first made for four Kaiparowits coal mines and a 5,000-megawatt electricity-generating plant about 45 miles north of Page in Kane County, Utah. When projected increases in energy demand did not materialize, and construction costs inflated by more than fivefold, the project was canceled by the energy consortium in the mid-1970s. The town's population fell to less than 5,000 in 1980.

Maintaining Continuity and Mitigating Negative Impacts: The Planner's Mandate

Rapid, uncontrolled growth and the lack of advanced planning have been major causes of the poor quality of life found in most energy boomtowns. Advance knowledge of and planning for the local impacts of large-scale energy development can help minimize the negative effects of growth and can enhance the positive aspects. The stories of towns that had no advance planning prior to rapid development litter the pages of boomtown case studies. Towns that are poorly prepared for rapid growth typically experience shortages of goods and services; serious disruptions of the community social structure; haphazard, sprawling development; fiscal crises; and a general degradation of the quality of life. Although planning cannot provide a panacea for all the problems of boomtowns, it can help to accommodate rapid growth.

Advance planning is critical to successful growth management in a prospective boomtown. The lead time required to expand physical infrastructure (roads, sewer and water facilities) may be as much as two years or more. In order for a community to expand its facilities to match the increase in demand, there must be open communication and coordinated action among local residents, the local government, and the energy company, and up-front financing available to the town. Often there is a lack of technical assistance and funding from the state or federal government. Towns may not even have a planner or adequate expertise to request assistance. A town may be faced with hasty reactions to meet sudden critical needs. But state limitations on municipal bonding capacity prohibit some communities from expanding services in advance of the boom. Increased tax revenues do not arrive until the project is well underway. Towns typically are caught in a vicious circle: Advance preparation is essential for the provision of adequate uninterrupted public services, but advance financing is generally not available to the small, preboom community. Incremental planning decisions are then required to provide for emergency housing, double school shifts, amendments to zoning ordinances, and the lowering of environmental quality standards.

The planning profession brings a comprehensive and interdisciplinary perspective to boomtowns. Public sector planners are needed to help state, local, and federal officials predict the impacts of energy development, formulate goals for guiding growth, and design specific programs to deal with the rapidly occurring changes. Specific functions of the local and regional planning staff might include drafting grant applications, monitoring demographic changes, preparing and administering comprehensive land use plans, implementing zoning ordinances, developing social service programs, and devising a capital improvements program. Coordination among the participants in the boomtown (including state, local, and federal officials; industry; commercial interests; and the general public) is essential. An important part of the planner's role is to facilitate this coordination by maintaining open lines of communication and by working with all parties to speed the transfer of up-to-date information. The creation of an information-sharing and planning group has proven useful in many boomtowns. The Sweetwater County (Wyoming) Priorities Board was established in 1974 on the authorization of the county and the towns of Rock Springs and Green River. The functions of the board are to provide open communication, to identify problems and recommend solutions, to develop proposals for industry assistance in addressing boom problems, and to work toward long-term economic self-sufficiency in the local towns.

The Mercer County (North Dakota) Economic Development Board was formed in 1977 by the county government, its six towns, and its four school districts. The major goals of the board are to plan for and manage energy related growth and to ensure that energy conservation considerations are incorporated into the local planning and development process. The board is authorized to develop a general plan; to negotiate financing; and to monitor the construction of public facilities, housing, and commercial enterprises. The board is staffed with an executive director, three planners, and an administrative assistant. It receives technical assistance from state, regional, and federal agencies as well as from private consultants.

The goals of growth management will vary with each community. These goals, however, should generally include (1) determining how much growth the community or area can absorb and at what rates; (2) deciding where to locate new development; (3) determining how to pay for the public service costs associated with growth; and (4) deciding how to bring together the interested parties in order to manage growth. It is important that local officials take the lead in making decisions and in ensuring that they are carried out. Once the general goals of growth management have been established, specific tasks can be undertaken. These include (1) balancing capital investment by industry with investment in the service sectors; (2) managing the use and conservation of land, air, and water; and (3) accommodating and retaining population by providing a good quality of life.

The need for balancing capital investment becomes evident during the construction boom phase, when the energy industry is pumping millions of dollars into building facilities, while the local goods and services sector has only limited access to new investment capital. The local services sector needs to invest about 5–20 percent of the amount invested by the energy company in order to achieve a balance in which the local goods and services are adequate for the construction worker boom (Gilmore & Duff, 1975, p. 538). One way to attain this balance is to limit the size of individual energy projects or restrict the grouping of facilities, so that the local services can reach a sufficient level of expansion. Alternatively, local investment in services can be stimulated by new taxes, increased bonding authority, state and federal impact grants, and grants or low-interest loans from the energy industry. Most of the taxes from the energy facility will usually not be available until after the project is completed and may go to jurisdictions other than the local boomtown. Some communities have received prepayments of taxes from energy companies, along with corporate or state debt guarantees. Special districts may be created to collect and distribute taxes so that the boomtown, its suburbs, and the energy development are all within a single political boundary.

Control over the use and conservation of local resources can be achieved through land use plans, zoning, and other legislative actions. A land use plan can determine where different types of development should be located. Through zoning ordinances, a land use plan can protect residential, commercial, agricultural, industrial, and environmental sensitive areas from encroachment by other noncompatible land uses and can ensure that adequate land will be available for each use. Land use plans can speed up the development process by determining in advance whether a certain use is appropriate for a particular site. Other planning tools, such as capital improvements plans, growth boundaries, subdivision regulations, mobile home and planned unit development ordinances, and special service districts for sewer and water facilities, are also useful in regulating the location and quality of development. Local governments should take care to afford sufficient flexibility for innovative and temporary facilities, such as mobile home parks, cooperative housing arrangements, modular school classrooms, and package sewage treatment plants.

Although economic development may be a necessary condition for social well-being, it is not a sufficient condition. Assuring a good quality of life requires specific strategies and programs to provide adequate housing, medical facilities, water supplies, sewage treatment, streets, shopping facilities, schools, and recreation. For example, the National Health Service Corps, an agency of the U.S. Department of Health and Human Services, has been helpful in placing doctors, dentists, and nurses in medically underserved areas. The corps recruits the medical personnel and pays their salaries for two years. Several energy companies have recognized the need to keep their workers contented and have provided housing, recreation, and medical services.

Boomtowns present the toughest challenge to a rural planner. The demand for services explodes in a sudden rush, and local funding for services is limited. Local political expertise is often lacking, and an antiplanning attitude is often prevalent. The more advanced planning that a community or region can do, the better the area will be able to cope with rapid growth. Open communications and coordination with an incoming energy developer can help manage the growth process and lead to sources for the funding of health care, housing, and public resources. The social impact of a large number of newcomers is difficult to mitigate or avoid. State human services agencies should be alerted, and staff members should be allocated to the community at the outset of the boom. The planner must strive to be forward-looking, anticipating changes in the boomtown economy and population that will require new and different planning strategies.

CHAPTER NINE

Natural Areas in the Rural Environment

The Importance and Valuation of Natural Areas

Unique Features of Natural Areas

Many rural communities have unique natural features, such as a floodplain, wetlands, important wildlife habitats (for flora or fauna), or geological formations (see Table 9.1). Often the public health, safety, and welfare values of these resources have been taken for granted or severely undervalued. The importance of renewable environmental resources, and the educational and amenity values of natural areas, must be more widely recognized. Wetlands, for instance, provide a buffer against flood damage, serve as groundwater recharge areas and water filtration systems, and are wildlife breeding grounds. Yet over half of America's coastal wetlands have been dredged and filled in the last 200 years ("Protecting Our Wetlands," 1986, p. 7).

Safety from natural hazards, protection of water quality, and public well-being all mandate development patterns that are sensitive to the carrying capacity of a site. "Carrying capacity" is the ability of a site to support development without environmental degradation or the risk of damage to the development. This is not to suggest that all natural areas must be preserved at all costs; rather, the increasing competition over rural lands for a variety of uses must be recognized. It is therefore crucial to identify the most valuable natural areas, utilize them sensibly, and develop techniques to protect them.

The identification of natural areas and determination of the capacity of each area to support development constitute a first step in organizing a community's natural areas. The identification of natural areas should

TABLE 9.1. Types of Unique Natural Features

I. Geological land forms
  A. Mountain peaks, notches, saddles, and ridges
  B. Gorges, ravines, and crevasses
  C. Deltas
  D. Peninsulas
  E. Islands
  F. Cliffs, palisades, bluffs, and rims
  G. Glacial features (moraines, kames, eskers, drumlins, and cirques)
  H. Natural sand beaches and sand dunes
  I. Fossil evidence
  J. Caves
  K. Unusual rock formations

II. Hydrological
  A. Significant and unusual water–land interfaces (scenic stretches of shore, rivers, or streams)
  B. Natural springs
  C. Marshes, bogs, swamps, and wetlands
  D. Aquifer recharge areas
  E. Water areas supporting unusual or significant aquatic life
  F. Lakes or ponds of unusually low productivity (oligotrophic)
  G. Lakes or ponds of unusually high productivity (eutrophic)
  H. Waterfalls and cascades

III. Biological—flora
  A. Rare or unique species of plants
  B. Plant communities unusual to a geographic area
  C. Plant communities of unusual diversity or productivity
  D. Plant communities representative of standard forest plant associations identified by the Society of American Foresters and American Geographical Society

IV. Biological—fauna
  A. Habitat area of rare, endangered, and unique species
  B. Habitat area of unusual significance (feeding, breeding, wintering, and resting)
  C. Fauna communities unusual to a geographic area
  D. Habitat areas supporting communities of unusual diversity or productivity

Note. Adapted from Data Form Guideline, New England Natural Areas Project. Reprinted from Rural Environmental Planning (p. 68) by F. O. Sargent (1976) Burlington, VT: Author. Reprinted by permission.

begin with an inventory of sites, including landownership patterns and information about soil and water quality. Landownership patterns reveal who owns the natural areas, the number and size of parcels, and the proximity of these areas to intensive development. Who owns the natural areas may suggest possibilities for negotiations, conservation restrictions, and general education about the management and protection of natural areas. The number and size of individual parcels will often give an indication of the ease or difficulty that may be involved in protecting sensitive lands. If, for example, there are many landowners, each holding a small parcel, then protection is likely to be less successful and more expensive than if there are only a few landowners holding large parcels. The proximity of a natural area to intensive development signals the likelihood of human encroachment. Areas close to intensive development may require immediate protection to prevent serious damage or destruction. In sum, ownership information is critical to assessing both goals and options for the use of natural areas.

Information about soils indicates a parcel's ability to support buildings, to absorb water, and to produce vegetation (see Table 9.2). Although Soil Conservation Service (SCS) soil surveys and maps primarily reflect agricultural productivity, they contain detailed information that can be used to detect potential limitations to development. Soil features such as steep slope, poor drainage, shallow depth to bedrock, and wetness can be significant barriers to development. The construction of sturdy buildings may be difficult. On-site sewage disposal by septic tanks and leach fields may result in water pollution, constitute a public health hazard, and cause the loss of wildlife in nearby natural areas.

Water resources data (see Table 9.3) should show the location and quality of public water supplies, groundwater, aquifers, and recharge areas in relation to the location and density of development. Intensive development on poor soils near watershed areas indicates a likelihood of soil erosion and runoff from sewage systems, with a resulting decline in

TABLE 9.2. Soil Capability Ratings of the U.S. Soil Conservation Service (SCS)

Class I soils have few limitations that restrict their use.

Class II soils have moderate limitations that reduce the choice of plants or that require moderate conservation practices.

Class III soils have severe limitations that reduce the choice of plants, require special conservation practices, or both.

Class IV soils have very severe limitations that reduce the choice of plants, require very careful management, or both.

Class V soils are not likely to erode but have other limitations, impractical to remove, that limit their use largely to pasture, range, woodland, or wildlife.

Class VI soils have severe limitations that make them generally unsuited to cultivation and limit their use largely to pasture or range, woodland, or wildlife.

Class VII soils have very severe limitations that make them unsuited to cultivation and that restrict their use largely to pasture or range, woodland, or wildlife.

Class VIII soils and landforms have limitations that preclude their use for commercial plants and restrict their use to recreation, wildlife, or water supply, or to aesthetic purposes.

Note. The general SCS ratings are primarily based on agricultural performance. However, more detailed subclass information is available that indicates a soil's erosion, wetness, depth, or stoniness, and thus is useful in determining carrying capacity.

TABLE 9.3. Water Quality Ratings

1. *Class A* waters are waters of a quality that is suitable for public water supply with disinfection when necessary. Character is uniformly excellent.
2. *Class B* waters are waters suitable for bathing and recreation and for irrigation and agricultural uses; good fish habitats, good aesthetic value; acceptable for public water supply with filtration and disinfection.
3. *Class C* waters are waters suitable for recreational boating, irrigation of crops not used for consumption without cooking, habitat for wildlife and for common food and game fishes indigenous to the region, and such industrial uses that are consistent with other Class C uses.
4. *Class D* waters are waters suitable for supporting aerobic aquatic life, and for power, navigation, and certain industrial process needs consistent with other Class D uses for restricted zones of water to assimilate approximately treated wastes.
5. *Class E* waters are waters that carry untreated sewage in such concentrations that they constitute a public nuisance. These waters are unfit for body contact, fishing, or boating.

watershed quality. Similarly, soil erosion and chemical runoff from farming operations can pose a serious threat to water quality. A reduction in public water quality means greater risks to public health and the need for costly water treatment systems. But groundwater, once polluted, is very difficult to purify. As water and sewage runoff from developed areas increases, the natural filtering system of wetlands and aquatic wildlife may be destroyed through sedimentation and pollution. A greater runoff also increases the likelihood of flooding and reduces the capacity of wetlands to absorb floodwaters. On the other hand, local water tables may be disrupted, causing some wells to run dry. Water quality data provide insights into the effect a development may have on water quality. This information can be useful in developing regulations on the siting, density, and design of new developments.

The valuation of natural areas presents a number of complex issues, including methods to determine the most valuable sites, the benefits and costs involved in preserving a site, and the benefits and costs of developing rather than protecting a site. The value of a natural area depends on several factors:

1. Whether the site is irreplaceable or renewable. If irreplaceable, the site is more valuable.

2. The frequency of occurrence. The more common the site, the less valuable it is; the more scarce the site, the more valuable it is.

3. The size of the area. The larger the site, generally the more important it is.

4. Diversity or variety. The greater the diversity of plants, wildlife, scenic views, and other features, the more valuable the site is.

5. Significance. If plant and wildlife species are rare and endangered, or if a site is of national importance, then the site is more valuable.

6. Fragility. This consists of two factors: the quality of the site in its natural state and the threat of human disturbance. For example, the higher the elevation, the thinner and more erosion-prone soils are likely to be. Similarly, the closer a natural area is to human development and the more intense that development is, the greater the likelihood of intrusion and misuse.

These six standards focus on the environmental, educational, and aesthetic performance of natural areas. However, the implementation of these standards for the protection of natural areas requires financial resources and land use controls.

Cost–Benefit Analysis

Only recently have public decision makers begun to assign dollar values to the environmental and educational values offered by natural areas. Although no markets or market prices may exist for the goods and services produced by natural areas, indirect methods such as travel cost studies, surveys, and shadow pricing have been used to derive estimates of how much people are willing to pay to protect a natural area. These estimates have enabled "hard-dollar" comparisons between the costs and benefits of preservation and developments.

Since the late 1950s, cost–benefit analysis has been widely used for improving the allocation of resources and economic efficiency. Cost–benefit analysis has been relied upon for both the evaluation of individual projects and the comparison of competing projects. The general formula for cost–benefit analysis is as follows:

$$NPV = \sum_{t=0}^{n} \frac{(B_{+} - C_{+})}{(1 + r)^{+}}$$

where $NPV =$ net present value; $t =$ the horizon time or "life" of the project up to the nth year; $B_{+} =$ the benefits of the project; $C_{+} =$ the costs of the project; and $r =$ the rate of time preference, also known as the discount rate or rate of interest.

The goal of cost–benefit analysis is to select projects with the greatest net present value, indicating the highest return on an investment. Each year over the life of a project, the annual net benefits (benefits minus costs) are "discounted" by an interest rate that reflects the "time value" of those net benefits. For example, a dollar is worth more today ($1) than a year from now (e.g., $.90). If a person were to agree to accept a dollar a year from now, he or she would want to receive interest—say, 10 percent, or a total of $1.10. Discounting shows the value today of a project's future stream of net benefits, corrected for the value lost in waiting to receive those benefits.

In any comparison of the options to develop or protect a natural area, the better option will have a higher net present value. The benefits of a project are the *social* values in monetary terms of the goods and services created by the project. These benefits include what are known as "secondary benefits," such as income multiplier effects on a community. The costs of a project are the social values in monetary terms of the resources (including environmental services) used in completing the project, and the negative externalities,[1] such as air and water pollution, produced by the development. The benefits of a natural area are the social money values of the goods and services provided by the area. The costs of the natural area are the expenses to society of protecting the area and the lost benefits from not developing a proposed project. These lost benefits are known as the "opportunity costs"; they are *implied* costs rather than expenses involving an outlay of money, but they are very real costs.

If the net discounted social benefits (benefits minus costs) of a project are greater than the net discounted social benefits from preserving the natural area, then cost–benefit analysis recommends the construction of the project. If the net discounted social benefits of preservation are greater, then the natural area should be protected.

But cost–benefit analysis also has certain limitations. First, the true value of the services provided by a natural area is not likely to be accurately reflected in people's willingness to pay. An understanding of the functions of a natural area may not be found among the public at large. Moreover, the services of a natural area appear to be available free of charge. That is, natural areas are public goods, even though they may be privately owned. Everyone benefits from the natural area, but only the particular landowner must actually bear the cost of protection. Thus people are likely to be "free riders" or otherwise to understate the true

[1]An "externality" is a by-product of production that may be positive or negative, but for which there is no market. For example, a bee-keeping operation will have created a positive externality in pollinating a neighbor's apple trees. A steel mill will create negative externalities in the form of air and water pollution in the process of making steel. There is no market for dirty air or water. The challenge to society is devising ways to compel generators of pollution to "internalize" those externalities. In this way, polluters would include the cost of environmental damage in their product prices. One way to internalize externalities is by imposing a tax on polluters. Another way is to enact standards for pollution control technology and acceptable pollution levels.

value of the services provided by a natural area. For instance, the true social value of an acre of wetland has been estimated by one study to be as high as $85,000; yet it is unlikely that people in a rural community would be willing to pay that much (Hufschmidt, James, Meister, Bower, & Dixon, 1983, pp. 210–212).

Second, some of the benefits of protecting natural areas accrue over time in the form of reduced public costs. Limits on developments in floodplains may save the cost of building expensive storm sewers and water purification systems. Yet these future savings are often overlooked in the short run.

Third, the discounting of costs and benefits creates a bias in favor of projects with short-run benefits. Thus, discounting tends to work against natural areas. Moreover, it is likely that with population growth, natural areas will tend to become more valuable over time.

Fourth, in either the preservation or the development option, there will be a certain distribution of costs and benefits both among local residents and between local dwellers and those from far away. Who bears how much of the costs and benefits of preservation or development is often determined by the balance of political power within a community or by a community's lack of political power in opposing state or federal programs. The situation is analogous to that of the creation of wilderness areas in national forests: Local people bear the cost of land removal from the tax rolls, while those visiting from urban areas gain most of the benefits.

Cost–benefit analysis is best seen as a guide to decision making. It is not conclusive evidence. The measuring of environmental benefits and expected costs from development is not as accurate a science as we would like it to be. In addition, there may be irreplaceable features of a natural area and threshold effects to wildlife. If too much development occurs, then pollution tolerance levels may be exceeded; wildlife habitats may be destroyed and the wildlife may disappear. The compatibility of development and natural areas is thus often limited.

Cost–benefit analysis also frequently reflects the biases of the decision makers involved. Conservationists see the landscape as a habitat that must be protected for future generations. Conservationists tend to advocate a controlled-growth ethic, which permits development on sites with suitable carrying capacities, along with certain restrictions on the scale and design of development. In this way, economic growth can be made compatible with environmental protection. Furthermore, the protection of environmental services may actually result in higher land values throughout a community.

On the other hand, developers may see the landscape as their creation and a development as an improvement upon nature. In proposing to build in a natural area, a developer may be unconvinced that the benefits of preserving the area outweigh the costs of changing or abandoning plans. Similarly, local governments may be reluctant to restrict development in natural areas for fear of losing property tax revenues. And an individual landowner may be willing to protect the natural area he or she owns only if public funds compensate him or her for not seeking personal gain in developing the area. Although a consensus may not exist about imposing some kind of order on the landscape, the persistent challenge to any natural area protection program is that it must be both politically acceptable and financially feasible.

Federal, State, Regional, and Local Roles in the Future of Natural Areas

The Federal Role

At the federal level, three agencies—the Army Corps of Engineers, the U.S. Department of Agriculture, and the U.S. Department of the Interior—primarily influence the preservation or development of America's natural areas. Since the 1930s, the Army Corps of Engineers has built over 250 dams, mainly for flood control (Reisner, 1986). In some cases, these dams were erected to protect property built in downstream floodplains. The dams have flooded thousands of acres, including wildlife habitats, wetlands, and geological sites. Under Section 404 of the 1972 amendments to the Clean Water Act, the Corps of Engineers has the authority to grant or deny permits for the dredging and filling of any wetlands for uses other than farming, forestry, and ranching (33 U.S.C. §§ 1251 *et seq.*). The U.S. Congress Office of Technology Assessment estimated that the Section 404 permit process saves about 50,000 acres of wetlands each year ("Protecting Our Wetlands," 1986, p. 3).

Within the Department of Agriculture, the SCS and the U.S. Forest Service are the two agencies that have the most impact on natural areas. The SCS has sponsored stream channelization and ditching programs that have had adverse effects on wildlife habitats along stream banks and on aquatic wildlife. By the mid-1970s, the SCS had planned over 400 projects involving some 12,000 miles of rivers and streams (Natural Resources Defense Council, 1977, p. 33).

The U.S. Forest Service manages the national forest system, which comprises about one-fourth of all public lands, or 191 million acres (U.S. Bureau of the Census, 1986, p. 653). There are 155 national forests in 40 states, although over 160 million acres are concentrated in the Western states and Alaska. Since 1960, the Forest Service has been required to manage the national forests on a "multiple use–sustained yield" basis (see Chapter 7). The Forest Service develops management plans for individual forests for outdoor recreation, livestock grazing, timber harvesting, some mining, watershed protection, and fish and wildlife habitats.

In 1964, the Wilderness Act designated 54 national forest areas containing 9 million acres as the beginning of a National Wilderness Preservation System (16 U.S.C. §§ 1131–1136). As of 1984, there were 327 wilderness areas comprising 80 million acres, of which 56 million were in Alaska (Council on Environmental Quality [CEQ], 1984, p. 256). The Wilderness Act defines wilderness as

> an area where the earth and its community of life are untrammeled by man, where man himself is a visitor who does not remain. And if wilderness is further defined to mean an area of undeveloped federal land returning its primeval character and influence, without permanent improvements or human habitation, which is protected and managed so as to preserve its natural conditions. (16 U.S.C. § 1131(a))

Nonetheless, certain commercial and industrial uses were allowed on wilderness lands between 1964 and 1983, including timber harvesting and the construction of logging roads, livestock grazing, and mineral exploration. Since 1983, only the grazing of livestock has been allowed.

The Department of the Interior and four agencies under its aegis have perhaps the farthest-reaching impact on the regulation and protection of the nation's natural areas. The Secretary of the Interior administers the Wild and Scenic River System, begun in 1968.[2] As of 1984, 65 rivers and adjacent lands, covering more than 7,000 miles of waterways, had been entered into the system (CEQ, 1984, p. 264). These rivers and their banks are protected from water resources projects. In addition, commercial and industrial uses of river banks are prohibited, and other development must meet acreage, frontage, and setback requirements.

To be eligible for inclusion in the system, a river must be a free-flowing, unmodified waterway, and adjacent lands must possess one or more natural values—scenic, geological, fish and wildlife, and/or historic. Each river must be classified as one of the following types, although different segments of the same river may be put into separate categories:

1. A "wild river" is free of dams and is generally inaccessible except by trail. Waters are unpolluted and shorelines primitive, representing America's natural past.
2. A "scenic river" is free of dams but is accessible in places by roads. Shorelines are largely primitive and undeveloped.

[2]The Secretary of Agriculture administers wild and scenic rivers in national parks.

3. A "recreational river" may have undergone some diversion or water impoundment in the past. The river is readily accessible by road, and shorelines may have experienced some development.

A river can become part of the Wild and Scenic River System in two ways: by an act of Congress, or by an act of the legislature of the state or states through which the river flows. This latter method includes state administration without expense to the federal government, but must have the approval of the Secretary of the Interior.

The Fish and Wildlife Service of the Interior Department manages the National Wildlife Refuge System, which consists of over 700 wildlife refuges, game ranges covering some 90 million acres, and about 100 small fish hatcheries (CEQ, 1982, p. 261). Wildlife refuges provide habitats for the protection and conservation of fish and other forms of wildlife that are often threatened with extinction. Game ranges serve as preserves in which herds are managed for present and future public enjoyment.

The National Park Service manages over 23 million acres of national parks and other lands "to conserve the scenery and the natural and historic objects and wildlife therein as will leave them unimpaired for the enjoyment of future generations" (Natural Resources Defense Council, 1977, p. 229). In addition to the national parks, the Park Service also manages national monuments and historic sites, national seashores and lakeshores, national rivers, national recreation areas, and national trails. The Park Service has placed a natural areas designation on all national park lands and national monuments of scientific importance.

In recent years, some of the land and water resources placed under the control of the Park Service have included substantial amounts of privately owned land. The management of these lands has become in effect a partnership among federal, state, and local governments and private landowners. Protection techniques consist of zoning, cooperative agreements, and easements as alternatives to outright federal purchase. Examples of rivers and areas managed by means of this new approach include the Upper Delaware Scenic and Recreational River in Pennsylvania and New York, the New River Gorge National River in West Virginia, and the Santa Monica Mountains National Recreation Area in California (CEQ, 1982, p. 153).

The Bureau of Land Management (BLM) administers roughly 470 million acres, or over 60 percent of all federal lands. About 300 million acres of BLM lands are in Alaska, and nearly all of the rest are located in the 11 Western states. BLM lands are generally unsuited for agriculture, except as rangeland, and have not been included in the national forest or park systems. BLM lands are to be managed on a "multiple use–sustained yield" basis, including outdoor recreation, range, timber, watershed, fish and wildlife, industrial development, mineral production, occupancy, and wilderness preservation. Critics of the BLM argue that the agency has favored ranching interests to the point of overgrazing. The effects on plant ecology, erosion, water quality, and fish and wildlife habitats have been severe. In 1982, over three-quarters of the BLM's grazing lands were in fair to bad condition (CEQ, 1982, p. 254). Original grasses have been largely destroyed, and grazing wildlife has been forced to compete with domestic livestock over dwindling supplies of forage. Also, soil erosion, trampling livestock, and herbicide spraying have damaged wildlife habitats and filled streams with sediment.

The Bureau of Reclamation, like the Army Corps of Engineers, plans and develops water projects that provide flood control, water-related recreation, water for irrigation, municipal and industrial water, and hydroelectricity. These projects may involve the construction of dams, reservoirs, canals, and aqueducts. Unlike the Corps of Engineers, the Bureau of Reclamation operates only in 17 states west of the Mississippi and Hawaii. The Bureau of Reclamation provides irrigation water to about 9 million acres, furnishes municipal and industrial water to some 15 million people, and can generate approximately 8 million kilowatts of electricity (Natural Resources Defense Council, 1977, p. 241). In developing water projects, the creation of reservoirs and the use of heavy equipment along river banks have led to serious disruption and destruction of ecological systems. Areas of scenic, biological, and historical importance have been permanently lost beneath reservoirs behind dams.

In keeping with the National Environmental Policy Act (42 U.S.C. *et seq.*, 1970), all federal agencies proposing projects that involve the development of natural resources must prepare an environmental impact statement citing (1) the adverse environmental effects of proposed projects; (2) alternatives to the proposed action; and (3) the long-term and irreversible use of resources. In addition, the Federal Water Resources Council has issued standards of planning water and related land resources, which apply to federal and federally assisted projects and programs. These standards are intended to supplement the process of preparing an environmental impact statement and call for broad public participation throughout the project's planning stages.

Environmental impact analysis has been used to protect natural areas and rare and endangered species. One of the most famous instances of this occurred in Tennessee, where the proposed Tellico Dam was blocked under the National Environmental Policy Act because of alleged threats to the snail darter, a rare and endangered fish species. In the celebrated *Mineral King* case (*Sierra Club* v. *Morton*, 1970, U.S. Sup. Ct. 405 U.S. 345), Walt Disney, Inc., proposed a large recreation development in Sequoia National Forest, but was denied permission because of potential destruction to the natural environment.

Above all, over the past 15 years, natural areas and wildlife have been granted a status of their own in the American legal system. The importance of natural areas has been recognized at the federal level, and this has served to make the general public more aware of the environmental consequences of development.

State Programs

Every state has a fish and wildlife department and other environmental agencies with some authority over natural areas. More than 20 states require environmental impact statements for state agency projects, and nearly every state has one or more programs aimed at managing specific natural areas (see Table 9.4).

Some states have adopted comprehensive land use planning programs that include authority over the development of natural areas. Since 1961, Hawaii's State Land Use Plan has placed natural areas in special conservation districts (Haw. Rev. Stat. § 205 (1973)). Development within these districts is tightly regulated because of the dangers of flooding, erosion, landslides, and volcanic eruptions. Over the years, the Hawaiian Land Use Commission has added a considerable amount of private land in mountainous terrain and along shorelines to the conservation districts. Since 1970, Vermont's Land Use and Development Law (Act 250) has required permits for all commercial, industrial, or residential development above the elevation of 2,500 feet (VSA 10 § 6001–6039). In addition, proposed developments involving the construction of more than nine units or the creation of more than nine lots must meet environmental quality criteria, involving limits to the burden on water supplies, soil erosion, and impacts on scenic or natural beauty and on rare and irreplaceable natural areas. Maine requires development permits for developments of over 20 acres or the construction of structure or structures in excess of 60,000 square feet. The development must meet certain standards, including no adverse effects on the natural environment.

Florida's Environmental Land and Water Management Act of 1972 concentrates state concern on critical areas and developments of regional impact (Florida Statutes Chapter 380, 1972). Critical areas are defined as "containing environmental, historical, or natural resources of regional or statewide importance." The Division of State Planning serves to identify critical areas subject to the approval of the governor. However, no more than 5 percent of the state's total land area may be designated as critical areas. The Oregon State Land Use Act, passed in 1973, requires the

TABLE 9.4. State Activity Related to Natural Areas

State	Floodplain regulations	Wetlands management	Critical areas	Coastal zone management
Alabama	No	No	No	Yes
Alaska	No	No	No	Yes
Arizona	Yes	No	No	N/A
Arkansas	Yes	No	No	N/A
California	Yes	No	No	Yes
Colorado	Yes	No	Yes	N/A
Connecticut	Yes	Yes	No	Yes
Delaware	No	Yes	No	Yes
Florida	No	No	Yes	Yes
Georgia	No	Yes	No	Yes
Hawaii	Yes	No	No	Yes
Idaho	No	No	No	N/A
Illinois	No	No	No	Yes
Indiana	Yes	No	No	No
Iowa	Yes	No	No	N/A
Kansas	No	No	No	N/A
Kentucky	No	No	No	N/A
Louisiana	No	Yes	No	Yes
Maine	Yes	Yes	Yes	Yes
Maryland	Yes	Yes	Yes	Yes
Massachusetts	No	Yes	No	Yes
Michigan	Yes	No	No	Yes
Minnesota	No	Yes	Yes	Yes
Mississippi	No	No	No	Yes
Missouri	No	Yes	No	N/A
Montana	Yes	No	No	N/A
Nebraska	Yes	No	No	N/A
Nevada	No	No	Yes	N/A
New Hampshire	No	Yes	No	Yes
New Jersey	Yes	Yes	No	Yes
New Mexico	No	No	No	N/A
New York	No	Yes	No	Yes
North Carolina	Yes	Yes	Yes	Yes
North Dakota	No	No	No	N/A
Ohio	No	No	No	Yes
Oklahoma	Yes	No	No	N/A

TABLE 9.4. (*Continued*)

State	Floodplain regulations	Wetlands management	Critical areas	Coastal zone management
Oregon	No	No	Yes	Yes
Pennsylvania	No	No	No	Yes
Rhode Island	No	Yes	No	Yes
South Carolina	No	No	No	Yes
South Dakota	No	No	No	N/A
Tennessee	No	No	No	N/A
Texas	No	Yes	No	Yes
Utah	No	No	Yes	N/A
Vermont	Yes	Yes	No	N/A
Virginia	No	Yes	No	Yes
Washington	Yes	Yes	No	Yes
West Virginia	Yes	No	Yes	N/A
Wisconsin	Yes	No	No	Yes
Wyoming	No	No	No	N/A

Note. N/A, not applicable. Source: U.S. Department of Interior Office of Land Use and Water Planning (1975). Reprinted in *Land Use Controls in the United States* (pp. 254–255) by E. Moss (1977). New York: Dial Press. Reprinted by permission.

adoption and implementation of comprehensive plans by cities and counties (O.R.S. 215.505 *et seq.*). These plans and subsequent local land use decisions must be consistent with 19 statewide planning goals, 12 of which include protection of natural areas. The goals require that inventories be conducted and protective measures be implemented for (1) fish and wildlife areas and habitats; (2) ecologically and scientifically significant natural areas; (3) outstanding scenic views; (4) wetlands and watersheds; (5) wilderness areas; (6) potential and approved federal wild and scenic waterways and state scenic waterways; and (7) areas subject to natural disasters and hazards. Local land use decisions must seek to minimize impacts on natural areas.

Regional Programs

Perhaps the best-known regional program that directly affects natural areas is New York's Adirondack Park Agency. Situated in northern New York State, the Adirondack State Park consists of roughly 5 million acres, with 60 percent in private ownership and 40 percent in state-owned forest preserves. These preserves have been protected since 1895 by an article in the state constitution requiring that they be kept "forever wild." The park contains almost 3,000 lakes and ponds and 46 mountains over 4,000 feet in altitude. It constitutes the major Northeastern airshed and watershed. Its integrity is critical for the well-being of tens of millions of people.

In the late 1960s, an interstate highway joined the Adirondack area to Montreal and New York City. With the greater access came an increased demand for second homes. In 1971, the New York legislature created the Adirondack Park Agency to manage all state lands within the park and to develop land use plans for all private lands. Private lands have been placed into six zones: hamlets, moderate-intensity use, low-intensity use, rural use, resource management, and industrial use. For each of these zones, development densities above a certain number of buildings per square mile are proscribed. Large-scale or potentially disruptive projects proposed in sensitive areas require the approval of the Adirondack Park Agency. Less critical projects are controlled by local government land use programs.

The Adirondack Park Agency has been criticized by local residents and developers for strictly limiting economic development opportunities. For example, almost 2 million acres of private land are zoned for resource management, which allows very few buildings. But in the 1980s, the demand for land inside the park has skyrocketed, especially for vacation home sites. Striking a balance between economic development and conservation will continue to be the major challenge for the agency.

Local Programs

Many local programs involving the protection of natural areas have been established in cooperation with state programs. Typically, a state regulatory agency sets minimum standards for local zoning, subdivision controls, and perhaps even setbacks from water courses and septic tank siting. Similarly, state enabling legislation has transferred some responsibility for the protection of natural areas to local governments. For example, localities have been authorized to develop regulations for critical areas and shoreline protection, subject to review by the state.

The most common means of protecting natural areas is through zoning restrictions, which are made in accordance with a comprehensive plan. The restrictions must relate to the town's health, safety, and general welfare, and may not constitute an "unreasonable" taking of private property without just compensation. Sensitive area zoning restrictions include the following:

1. Restrictions on the types of land uses permitted, such as the banning of commercial uses on floodplains or the prohibitions of filling in wetlands.
2. Restrictions on minimum lot sizes and the location of buildings, including setback requirements.
3. Performance standards that define the maximum acceptable level of disturbance to natural areas, such as alteration of stream flows, floodplains, and vegetation. New land uses require special permits, and the applicant must show that the new use will not exceed permitted disturbance levels.

Variations on traditional zoning have proven successful in minimizing impacts on natural areas. Cluster zoning allows the concentration of development at densities higher than ordinarily permitted so that open space is preserved. The "planned unit development" is a "floating" district in which mixed residential uses and even commercial and industrial uses may be allowed. A floating district is not part of the zoning map and has no definite boundaries until a proposal is made from a developer. However, the floating district must be included in the text of a town or county zoning ordinance. A planned unit development usually features a clustering of development to preserve open space, and minimize environmental impact on a site. Planned unit developments are commonly planned for large parcels of land.

Other local programs that can have significant effects on natural areas include subdivision controls, conservation easements, capital improvement programs, and local conservation commissions. Because the density of development is generally inversely related to the performance of natural areas, large-lot zoning may be an effective means of minimizing intrusion on natural areas. Conservation easements may be purchased from private landowners so that they receive compensation in exchange for the development rights to the natural areas they own.

Capital improvement programs, such as the construction of roads, schools, and water and sewer lines, can have a major influence on the future location and intensity of development. Such programs can be directed away from environmentally sensitive areas. Local conservation commissions have been established in the six New England states, New York, New Jersey, and Pennsylvania to hear citizens' concerns about environmental issues and to pass them on to local government bodies. Conservation commissions serve primarily an advisory function, but exercise regulatory powers over wetlands in Massachusetts and Connecticut. Conservation commissions are charged with preparing inventories of open space and natural resources; advising the local government on resource protections; reviewing development proposals on open space, wetlands, or other critical natural areas; educating the public on environmental issues; and coordinating the efforts of local environmental groups.

An emerging protection technique is the use of private, public, and community land trusts to protect natural areas. The private land trust is a private, nonprofit corporation that accepts gifts of land and donations of money and development rights, with the object of holding land in its natural condition. Two noteworthy models are the Maine Coastal Heritage Trust and the Lake Champlain Islands Trust, which have inventoried the natural significance of many coastal islands and have accepted gifts of land and development rights. Under a public land trust, the state holds public lands in trust for the people of the state. In practice, public trusts have been limited mostly to tidelands and the protection of fishery resources. The community land trust combines the concepts of the public land trust and cluster zoning. Under a community land trust, a landowner is allowed to develop a portion of his or her land at a higher density than normally permitted in return for leaving natural areas unharmed. The landowner deeds over the development rights on the undeveloped land to the town or county. In this way, the landowner receives an adequate economic return on his or her land, and natural areas are protected. Experience with community land trusts, however, is very limited.

National environmental groups involved in the protection of natural areas include the following:

- The Nature Conservancy
- The National Audubon Society
- The National Wildlife Federation
- The Natural Resources Defense Council
- The Izaak Walton League of America
- The Sierra Club
- The Environmental Defense Fund
- The Wilderness Society

Protecting Natural Areas in Rural Communities

Wildlife Habitats

A major responsibility for wildlife protection lies at the local level. The successful protection of wildlife habitats depends on the creation of data bases for day-to-day and long-term planning, the inclusion of identified habitats in a comprehensive plan, and the use of one or more protection techniques. The principal cause of decline in the number of wildlife species is the destruction of their nesting and breeding grounds. Often species are vulnerable because they are highly concentrated in a small area. If a locality can save at least a representative sample of native habitats, most species should be able to survive.

Habitats can be identified by volunteers from local environmental organizations, state universities, and the state department of fish and wildlife. Next, the selected areas can be mapped and added to a local comprehensive plan for protection by zoning and subdivision codes. In the development review process, wildlife can be granted an equal status with other resources, such as air, water, and prime agricultural land. If no comprehensive plan exists, the local elected governing body should at least establish goals for the monitoring and preservation of the most critical sites.

Several tools have been used to protect wildlife habitats. For large projects, cluster zoning can concentrate development on a small portion of a site, thus preserving much of the site for wildlife habitat and open space. Cluster zoning tends to require redesign on the part of the developer, but the economic value of a site is not lost; in fact, it may be increased because of the improved environmental amenities (Hallock, 1984, p. 14). Conservation easements in the form of reduced property taxes may be granted in exchange for development restrictions on private land which support significant habitats. Another option is the outright public purchase of wildlife habitats. The Nature Conservancy, for example, purchases important habitats and in some cases will deed the land to local jurisdictions for limited or appropriate public use. Similarly, a town or county may contain land owned by the U.S. Forest Service or other government agency; cooperative agreements can be arranged for the joint management of wildlife habitats.

Floodplains

An estimated 140–180 million acres, or 6–8 percent of the total land area of the United States, experience occasional flooding (Baker & Platt, 1983, p. 245). Roughly $4 billion in flood damage and the loss of several lives occur each year. Agriculture incurs the majority of flood losses in rural areas, topping $1.6 billion a year. Most other flood damage is inflicted on buildings that have been erected in floodplains.

Three distinct types of flooding exist: shallow-river flooding, flash floods, and coastal flooding. Shallow-river flooding usually occurs in the early spring, with snow melts causing rivers to overflow their banks. Flash floods are produced by heavy downpours from thunderstorms that concentrate large amounts of water into the canyons of the Southwest or the narrow valleys of Appalachia. In fact, about half of the nation's communities that have required federal flood disaster assistance are in Appalachia (Baker & Platt, 1983, p. 248). Coastal floods vary greatly in severity, time of year, and physical impact, depending on the location and type of shoreline. In the late summer and early fall, hurricanes may hit the Gulf states and the Southeast; winter and spring storms cause mudslides on the Pacific coast; and in the Northeast winter storms result in shoreline erosion.

Floodplain management is a pervasive issue in rural America, but rural areas receive less attention than urban areas under federal programs. Campaigns to increase public awareness of flood hazards have not been very effective; this is especially apparent in continued construction in floodplains, even though flood insurance is not included in ordinary homeowner policies. Despite the existence of the National Flood Insurance Program (NFIP), a large amount of property in floodplains is not insured. NFIP provides low-cost flood insurance to over 2 million policy holders owning $100 billion worth of flood-prone property. The average policy costs about $200 a year for $60,000 in coverage. In addition, areas of special flood hazard in 20,000 communities and rural areas have been identified and mapped (Baker & Platt, 1983). If a community does not have a floodplain management program, the NFIP places the community on an "emergency basis." This status carries lower levels of insurance coverage on existing properties and more severe restrictions on floodplain development, including limits on the type, location, height, and design of new construction. If a community adopts a satisfactory floodplain management program, then it is placed on a "regular basis" and property owners are eligible for higher levels of coverage. But if a community fails to adopt a management program, then property owners can no longer obtain flood insurance. So far, about 3,100 communities have achieved a "regular basis" standing. In addition, the Flood Disaster Protection Act of 1973 requires states and communities receiving federal disaster aid to prepare a long-range hazard mitigation plan for the affected area (42 U.S.C. § 4001 *et seq.*).

The federal programs emphasize the need to incorporate floodplain management into a community's general planning process, instead of treating flooding as a separate and infrequent problem. Local jurisdictions can direct public infrastructure away from floodplains and discourage private development in flood-prone areas. Floodplains may be zoned for nonintensive uses, conservation easements may be granted, or floodplains may be purchased for public use. The protection of floodplains from development not only minimizes potential flood damage, but often meshes with community environmental goals of open space, water quality, visual amenities, recreation, and the preservation of natural areas.

Moreover, successful floodplain management can preclude the need for large flood control dams to protect development in floodplains. The reservoirs behind these flood control dams typically cover thousands of acres of valuable land.

Wetlands

Wetlands are areas of transition between dry land and open water and include such formations as marshes, bogs, swamps, and ponds. These areas tend to be low-lying, with poor drainage and standing water. Wetland boundaries may change according to variations in climate and rainfall. Different types of wetlands can be identified by vegetation, fresh or salt water, and average water depth (see Table 9.5).

Wetlands perform three important environmental services:

1. Wetlands act as a water filtration system; they provide a settling basin for silt from runoff, and wetland plants catch inorganic nutrients. As a result, downstream or offshore water resources and wildlife are protected from pollution.

2. Wetlands moderate extremes in water supply. They function as water storage and aquifer recharge areas in dry periods, and act as a buffer against storms and floods. One acre of marsh can absorb 300,000 gallons of water.

3. Wetlands provide wildlife breeding, nesting, and feeding grounds. Plants and animals found in wetlands may include rare and endangered species or commercial species such as fish. The variety of plant and animal wildlife offers recreational and educational opportunities.

Since 1776, America's wetlands have been reduced from 215 million acres to 95 million acres. Wetlands are currently under pressure in virtually every state for conversion to agricultural and urban uses. Although a permit from the Army Corps of Engineers is required before a wetland can be filled, agricultural, forestry, and ranching uses are exempt. The "swampbuster" provision of the 1985 Food Security Act penalizes farmers who drain and fill wetlands to create cropland; a farmer will be ineligible to participate in any federal farm programs (Farm Bill, Pub. L. No. 99-198. 16 U.S.C. § 3801 *et seq.*).

Wetlands are frequently viewed as a source of cheap land (see the discussion of the *Just v. Marinette County* case in Chapter 4). It is often less expensive to buy and fill a wetland than to purchase a nearby piece of dry land. But the private real estate market does not take into account the public benefits of a wetland's services, nor the costs imposed on the public in the form of reduced water quality when a wetland is filled. Moreover, the incremental effects of nearby residential, industrial, and

TABLE 9.5. Wetlands Characteristics

Wetland type	Region of largest acreage	Representative vegetation (north)	Representative vegetation (south)
		Inland fresh	
Seasonally flooded basins or flats	Mississippi north	Varies with flood duration; lowland hardwood trees, smartweed, wild millet, fall panicum, tealgrass, chufa, cypress	Same
Inland fresh meadows (sedge meadow)	Atlantic south, Mississippi north	Carex, rushes, redtop, reedgrasses, mannagrasses, prairie cordgrass, mints	Cordgrasses, paspaluma, breakrushes
Inland shallow fresh marshes	Atlantic south	Reed, whitetop, rice cutgrass, carex, giant burreed, bulrushes, spikebrushes, cattail, arrowheads, pickerelweed	Maidencane, sawgrass, arrowhead, pickerelweed, rushes, cattails
Inland deep fresh marshes	Central north, Atlantic south	Cattails, reeds, bulrushes, wild rice, spikerushes. In open areas: pondweeds, naiads, coontail, water milfoil, duckweed, water lily	Many of the same species, plus water hyacinth and water primrose in some areas
Inland open fresh water	Mississippi north	Pondweeds, naiads, wild celery, coontail, water milfoil, muskgrasses, lilies, spatterdocks	Similar, plus water hyacinth
Shrub swamps	Mississippi north	Alders, willows, buttonbush, dogwood, swamp privet, usually along sluggish streams	Same

TABLE 9.5. *(continued)*

Wetland type	Region of largest acreage	Representative vegetation (north)	Representative vegetation (south)
		Inland fresh	
Wooded swamps	Atlantic south	Tamarack, arborvitae, black spruce, balsam, red maple, black ash. Northwest: Western hemlock, red alder, willows, and thick mosses	Water oak, overup oak, tupelo gum, swamp black gum, cypress
Bogs	Atlantic south, Mississippi north	Leather leaf, Labrador tea, cranberries, carex, cottongrass, sphagnum moss, black spruce, tamarack, insectivorous plants	Cyrilla, persea, gordonia, sweetbay, pondpine, Virginia chainfern, insectivorous plants
		Inland saline	
Inland saline flats	Pacific south	Sparse: seabite, saltgrass, Nevada bulrush, saltbush, burroweed	
Inland saline marshes	Pacific south	Same	Alkali or hardstem bulrushes, wigeongrass, sago pondweed
Inland open saline water	Pacific north, Pacific south	Sago pondweed, wigeongrass, muskgrass	
		Coastal fresh	
Coastal shallow fresh marshes	Mississippi south	Redgrass, big cordgrass, carex, spikerush, sawgrass, cattails, arrowheads, smartweed	Similar, with maidencane
Coastal deep fresh marshes	Mississippi south	Cattails, wild rice, pickerelweed, giant cutgrass, spatterdock, pondweed	Similar, with water hyacinth and water lettuce
Coastal open fresh water	Mississippi south	Similar, excluding water hyacinth	Scarce in turbid waters: pondweeds, naiads, water milfoils, muskgrasses, water hyacinth

Coastal saline			
Coastal salt flats	Central south	Sparse vegetation: glassworts, seablite, saltgrass	Similar, with saltflat and saltwort
Coastal salt meadow	Atlantic north, Atlantic south	Saltmeadow cordgrass, saltgrass, Olney threesquare, saltmarsh fleabane. Pacific coast: carex, hairgrass, jaumea	Same
Irregularly flooded salt marshes	Atlantic south	Dominantly needlebrush	Same
Regularly flooded salt marshes	Atlantic south, Mississippi south	Atlantic and Gulf coasts: salt marsh cordgrass. Pacific coast: alkali bulrush, glasswort, arrowgrass	Same
Sounds and bays	Central south	Eelgrass, wigeongrass, sago pondweed, muskgrass	Southeast: shoalgrass, manateegrass, turtlegrass
Mangrove swamps	Florida only		Much red and some black mangrove

Note. From *Performance Controls for Sensitive Lands* (p. 39) by C. Thurow, W. Toner, and D. Erley, 1975, Chicago: ASPO. Copyright 1975 by American Society of Planning Officials. Reprinted by permission.

agricultural runoff can seriously deplete the quality of a wetland over time.

Wetlands are public goods. There is no market for the various services performed by the wetland, except that the public at large benefits. Regulatory approaches to protecting public goods are justified if the public health, safety, and welfare are being promoted without an unreasonable taking of private property. With wetlands, public health protection involves ensuring water quality; safety relates to reducing the risk of damage from flooding; and public welfare is maintained by keeping a wetland in its natural state for wildlife-related outdoor recreation and educational activities.

As with wildlife habitats and floodplains, special districts exclusively for wetlands or "overlay zones" combining wetlands protection with other uses can be created. Districts and zones should spell out permitted uses, conditional uses, and prohibited acts. Performance standards can be used to evaluate how the functioning of a wetland or other natural area will be affected by a proposed development. These standards should help in designing developments that are compatible with wetlands. Often wetlands and other natural areas can be effectively augmented by the use of adjacent low-density buffer zones. Also, conservation easements and the public purchase of wetlands can be employed to compensate a landowner for development restrictions placed on his or her land.

Coastal Zone Management

The coastal zone consists of coastal waters (estuaries, bays, marshes, and lagoons) and adjacent shorelands that have "a direct and significant impact on coastal waters" (Coastal Zone Management Act, 1972, 16 U.S.C. § 1453(a)). Over half the population of the United States lives in areas adjacent to the coastal zone, and many rural communities earn their livelihoods from commercial fishing and recreation activities along the shoreline (see Table 9.6). Throughout the 1960s, development pressures weighed heavily on the nation's coastal resources. Recreation and tourist facilities, industrial plants, urban development, and mineral exploration proliferated. Massive dredging and landfill operations covered estuarine marshlands, and air and water pollution notably worsened. Entire ecosystems were destroyed or severely damaged, and the environmental value of coastal areas was often overlooked. Coastal areas form the spawning grounds for aquatic organisms that are the basis of the ocean food chain. Salt marshes serve as nurseries for shellfish and game fish, and a single acre of tidal wetlands yields an estimated 535 pounds of

TABLE 9.6. Coastal Zone Resources and Uses

Extractive use/activity	Primary resources
Commercial fishing	Deep ocean fisheries
Aquaculture	Shore zone biota
Surf fishing	Shore zone fauna
Kelp harvesting	Offshore flora
Boat sport fishing	Coastal zone fisheries
Desalination	Ocean chemicals
Mineral production	Marine minerals
Sand/gravel production	Nearshore sediments
Petroleum production	Subsea minerals
Nonextractive use/activity	**Primary resource**
Policing and regulation	Access to coastal areas
Commercial shipping, deep draft	Navigable areas
Commercial shipping, moderate draft	Inshore areas, harbors
Transport right of way	Shoreline
Government reservation	Coastal areas
Underwater parks	Undisturbed sea bed
Housing, real estate development	Shore zone
Swimming, sunbathing, surfing	Unpolluted beaches, water
Waste disposal	Shoreline access to offshore areas
Resorts, parks	Shoreline, beaches
Recreational boating	Protected harbors

Note. From *Coastal Zone Management: Multiple Use with Conservation* (p. 8) by J. F. P. Brahtz (Ed.), 1972, New York: Wiley. Copyright 1972 by John Wiley & Sons. Reprinted by permission.

fish each year! Moreover, some 50 percent of commercial fishing occurs in coastal waters (U.S. Bureau of the Census, 1986, p. 667).

In 1972 Congress passed the Coastal Zone Management Act, the only nationwide land use planning measure ever enacted (16 U.S.C. § 1451 *et seq.*). The act stated:

> The increasing and competing demands upon the lands and waters of our coastal zone occasioned by population growth and economic development . . . have resulted in the loss of living marine resources, wildlife, nutrient-rich areas, permanent and adverse changes to ecological systems, decreasing open space for public use, and shoreline erosion. (16 U.S.C. § 1451(c))

The Coastal Zone Management Act is administered through the National Oceanic and Atmospheric Administration and provides federal funds and technical assistance for the planning and implementation of water and land use management programs in 30 coastal and Great Lakes states. Federal grants cover up to two-thirds of state planning and implementation costs, and 50 percent of the cost (up to $2 million) of purchasing estuary areas for research and educational uses. Federal approval of state plans is required. All subsequent federal actions must be as consistent as possible with state plans.

State plans must include the following:

1. The boundaries of the coastal zone.
2. Permissible land and water uses within the zone.
3. An inventory of sensitive areas of particular concern.
4. An identification of state controls to enforce compliance:
   a. Direct regulation by the state or local regulation with state standards and review.
   b. Acquisition of land, water, and buildings through condemnation, fee simple interest, less than fee simple interest, or other means.
5. The priority of uses in different areas of the zone.
6. Coordination of federal–state–regional and local agencies:
   a. Federal water and air pollution laws.
   b. Permits for filling wetlands.

In addition, states must consider the national interest in siting developments of greater than local impact, such as electricity-generating stations, national seashores, national defense installations, historic sites, harbors and ports, and fisheries. Sensitive areas of special concern facilitate the setting of management priorities. These areas include unique and scarce plant and animal species and their breeding grounds; hazardous areas subject to flooding or landslides; recreation areas; urban development; and commercial and industrial water-related uses.

Summary

Over the past 20 years, substantial federal, state, and local efforts have been made toward the protection of natural areas. The importance of these areas in contributing to society's health, safety, and welfare has been recognized. Experience with natural areas demonstrates that they are complicated resource systems. Placing an economic value on these resources is often difficult, since markets for the products and services of natural areas may not exist. Still, cost–benefit analysis is often used to determine whether a project should be built or a natural area preserved; given the limitations of cost–benefit analysis, however, it is best used as a guide to decision making rather than as conclusive evidence. Several tools exist for the protection of natural areas while allowing for limited but appropriate development. Economic growth and environmental protection need not be mutually exclusive; often careful project design and siting can harmonize human needs with environmental quality standards.

Many rural communities have been lax about the protection of natural areas. Often these areas are privately owned, and communities have been reluctant to restrict the use of private property. Furthermore, because the value of these resources has not been established, people think they are without value. State and federal laws often appear to be aimed at restraining rural communities from spoiling the environment for everyone. The ultimate local goal should be sustainable economic development together with active management of natural resources. Nonetheless, many obstacles exist in rural communities against protecting natural areas and environmental quality. These include the following:

1. Inadequacies in the monitoring and enforcement of existing environmental protection laws and regulations.
2. Extensive poverty, which puts a premium on current income-producing activities, to the detriment of long-term protection of natural systems.
3. Scarcity of financial resources in relation to current needs, which constrains the willingness to protect natural systems.
4. The often perverse distributional effects of environmental quality plans and programs, which may worsen the existing inequitable distribution of income.
5. Difficulty in controlling the environmental quality effects of private sector and public sector development activities; this limits the effectiveness of public programs for environmental quality management.
6. Inadequacies in the technical, economic. and administrative expertise available for the planning and implementation of environmental management programs.
7. Widespread market failures, which require extensive use of shadow prices to replace market prices.
8. Minimal participation in environmental quality planning, either by the general public or by many affected governmental agencies; this reduces the effectiveness of implementation.
9. Inadequacies in environmental, economic, and social data—including difficulties in data collection and processing and lack

of knowledge of past trends and baselines—which limit the quality of analysis.

10. Wide diversity of cultural values, which increases the difficulty of social evaluation of environmental quality effects.

Rural planners must be aware of these limitations and, where possible, work around them to balance growth with a safe, healthy, and diverse environment. Perhaps in no other situation is the "planner as educator" model more appropriate than in the management of natural areas.

CHAPTER TEN

Tools and Skills: Land Markets and Tenure

Planning has long been concerned with the location and density of different land uses. Land use patterns in a town, county, or region have strong influences over the current and future economic base, the cost of providing public services, and the location of future development. Land values—either with or without the existence of land use regulations—to a large extent determine present and future land use patterns.

In drafting comprehensive plans and land use regulations and in reviewing development proposals, the planner must be aware of how the rural land market works and of the local land ownership patterns. Such an awareness will enable the planner to offer more realistic and effective plans, regulations, and technical advice on proposed developments. For example, a comprehensive plan that seeks to maintain a town's compact residential and commercial land use pattern will also tend to discourage those land uses on farmland half a mile beyond the town fringe. The prevention of sprawl will help to keep public service costs down and to maintain nearby farmland in large ownership blocks, and may increase the value of land within the town.

Rural Land Economics

Land has three major features: (1) It is fixed in location, except in cases of natural disaster; (2) it provides renewable flows of such services as housing, crops, and timber; and (3) the quantity of land on the earth's surface is virtually fixed, but the quality of land, in terms of its ability to support development and to produce food and fiber, varies greatly and is subject to change from erosion or human-made improvements. In sum, land is a unique resource: It is essentially nonrenewable, but has some attributes of a renewable resource.[1]

In rural America, the market system has traditionally directed the allocation of land resources among competing uses and users over time. The U.S. rural land market is composed of thousands of smaller, more localized markets; separate land markets also exist for each type of property, which involves different groups of buyers and sellers. The three main types of land that are exchanged in rural areas are farmland/forestland, residential land, and vacation/recreation land.

The rural land market is characterized by (1) infrequent participation in the land market by the average buyer and seller; (2) customary use of credit in purchasing land; (3) common use of a real estate broker's services to execute a transaction; and (4) interaction of supply-and-demand conditions in the determination of land price.

Unlike the stock market, the land market has no centralized exchange; unlike commodity markets, the land market is not an organized futures market. Thus, the value of a parcel of land at any point in time, or for a particular use, is somewhat uncertain. As land economist Raleigh Barlowe (1978) has observed, "There is probably no market in the American economy where as much haggling and 'horsetrading' over price takes place as in the real estate market" (p. 355). Prices are based upon what buyers are willing to pay and what sellers are willing to accept. Still, both buyers and sellers must have some idea as to the value of the land in question.

Land value is determined by a multitude of factors which can be separated into three main groups: use capacity, location, and scarcity. Use capacity encompasses the biological and physical qualities of land, such as soil quality, slope, drainage, and parcel size; these indicate the ability of the land to support different land uses and development densities. Land value per acre typically varies directly with land quality and inversely with parcel size. The smaller the parcel size the more valuable the land and the more likely it is to be developed for intensive uses (residential, commercial) rather than extensive uses (agriculture, forestry).

Location is important because land is fixed, not portable. Locational aspects include accessibility to population centers, markets, public services, and natural amenities. Location affects transportation costs and hence the location of different economic activities and land uses. The more accessible a property is to transportation networks and population centers, generally the more valuable the property is. Also, the higher the population density in an area, the greater the land values will usually be.

[1]For example, under ideal conditions, topsoil is created at a rate of about one inch per acre every 30 years.

Land scarcity measures the supply and demand of land for different uses. Land scarcity incorporates the concepts of land's fixed physical supply, fixed location, and varying land quality. Because the same parcel of land can potentially be put to different uses, the value of land will depend upon the returns to land in each use. Normally, land will be put to the use that commands the highest value as reflected in the market price. Market prices are a convenient way of expressing land scarcity: The higher the price per acre, the more scarce that land is in terms of location and use capacity for that particular use. For example, farmland in rural areas has a relatively low value per acre, compared with that of an improved lot for building a house; farmland is more plentiful than are serviced building lots. The "highest and best" use of land is the use that brings the highest return to the landowner and has the highest value.

Accessibility, location, scarcity, and use capacity also affect the scope of the land market—whether there are a few or many buyers and sellers, and whether these people are local residents or live some distance away. Remote rural areas usually have land markets with limited scope, involving a few local buyers and sellers. Rural resort communities typically have broad land markets with many buyers and sellers who do not reside year-round in the community.

Use capacity, scarcity, and the location of economic activity can be influenced by institutional factors, such as zoning. Zoning ordinances place certain restrictions on the use of land in certain locations and so affect the supply of land for different uses. In an "R-1" zone, for example, only single-family dwellings are permitted. The value of land and the supply of land in this zone are restricted to residential use, even though the land may have a higher value in a commercial use. A landowner proposing a commercial development in an R-1 zone must either seek a zoning change for the property to allow the commercial use, or else look to locate the development in an existing commercial zone. It is important to note that, as this example shows, in rural areas land use planning is not only physical planning, but economic planning as well.

Land Value

The "highest and best" use of land, as noted above, is the use that brings the highest return to the landowner. This use, however, may not be the highest and best use from a community's standpoint. The following formula can be employed to estimate land value for a variety of uses: land value per acre (V) = net annual returns per acre (R)/ an interest rate (i). If the net annual returns (revenues or rents minus operating costs) from an acre of farmland are $50, and the rate of interest is 10 percent, then the land is worth $500 as farmland. But if the same acre could produce $100 a year in net returns from residential use, then the value of the land for housing is $1,000. This method is helpful in describing the changes in land prices, the location of land use, and the competition among uses for a particular parcel.

Those uses producing the highest returns ordinarily have the first claim upon land. For example, land farther away from a city or town will produce lower returns in residential use. This land, by comparison, will produce relatively larger returns in farm use; that is to say, the competition for land between farming and housing in a remote area is likely to be low. If farm commodity prices were to increase, then the value of farmland would rise, and farming uses would become more competitive with housing uses. On the other hand, if a new road were to increase the accessibility of a remote area, housing would become more competitive in relation to farm uses. Changes in technology can also influence land prices by affecting accessibility, the price of goods produced on land, and land quality (use capacity). The point to remember is that land prices and land markets are constantly changing.

Land prices are determined in an imperfect market; they are characterized by a lack of information, time lags in the perception of market conditions by buyers and sellers, and spillovers from one land use that affect other properties. There is often little certainty as to what a certain parcel of land is worth, especially raw land. Normally, there are fewer buyers for raw land than for improved real estate, and the annual rate of turnover of all kinds of real estate is relatively low, except during a land boom. Thus, few comparable data on the land sales are likely to exist. Moreover, the expected future value of raw land is highly dependent on social, political, and environmental trends in the state or community in which the land is located. The prevailing economic climate may also significantly determine land value. For example, during inflationary periods, land may rise rapidly in value and attract investors seeking a hedge against inflation. But as Healy and Short (1981) emphasize, "Borrowed money is the lifeblood of all real estate markets" (p. 96). If interest rates are high, the demand for land will most likely be reduced as the cost of buying land and building structures increases and land values will fall. Federal monetary policy has a direct influence on interest rates and hence on land values. This is another example of the financial dependency of rural areas. Local interest rates and land values depend on what is decided in Washington. An important recent example is the sharp rise in interest rates to historic highs from 1979 to 1982 and the subsequent plunge in farmland values in many rural areas.

Several rural land uses can create "spillovers" (also known as "externalities") that affect the productivity, healthiness, or aesthetic benefits of neighboring properties. The spillover costs generated on a particular piece of land are not included in the price of that property, but may influence the price of affected properties. For example, the noise, dust, and odors that agricultural land use produces do not change the value of that land for farming, but may reduce the value of adjacent properties for residential use. When populations increase in a rural area, spillovers are sure to increase, meaning that the land market will become less accurate in reflecting the true "social" value of land. Another type of spillover occurs when the land market fails to account for the interest of future generations. Land development often creates irreversible effects, either in the destruction of natural areas or in the consumption of land beneath asphalt and concrete. Future generations, of course, have no say in these decisions, and yet they must deal with the consequences.

All of these imperfections in the land market result in a divergence of public and private costs and benefits, and form the justification for public intervention in the land market. Yet there is no guarantee that a public program will improve the fairness or the efficiency of the land market in satisfying people's needs. Public incentives, such as tax policy, may distort the land market by promoting one use over another. For example, the deduction of mortgage interest for income tax purposes has encouraged the proliferation of single-family houses. Zoning regulations also influence land prices by controlling development location and the supply of land for different uses. Farmland protection programs, such as differential assessment and purchase of development rights, seek to maintain the supply of land in farm use (see Chapter 6).

Land Supply

Land is supplied in the market according to economic, social, and institutional factors. Economic factors include natural characteristics (physical location, slope, drainage, soil quality, views, access to water, whether the land is cleared or not, and the relative scarcity of different land uses); technological settings (proximity to highways, recreation areas, population centers, and amenities); the cost of converting land to other uses; and the changing market prices for different land uses. Social factors are comprised of landownership patterns and landowner characteristics. For example, a landowner may sell land because of age, divorce, cash liquidity needs, illness, the need to settle an estate, or a good offer. Institutional factors involve government land use regulations and taxation policies, such as zoning ordinances and property taxes.

Land supply is a function of a seller's "reservation price" (the minimum price at which the seller is willing to sell) and of a seller's "reservation demand" (the enjoyment of the land that the seller is not willing to give up below a certain price). Land supply therefore does not normally refer to the total physical land stock in an area, but rather the "effective supply"—the amount of that stock that landowners are willing to put on the market at a certain point in time. The supplies of most types of land are responsive to price changes. Generally, when prices are high, land supplies move from extensive to intensive uses (e.g., farming to housing; when prices drop, land supplies shift from intensive to extensive uses.

Demand for Land

The concept of demand reflects buyers' needs, desires, and willingness to pay; the demand for land is a derived demand for the services that land provides (e.g., housing, recreation, food production). Changes in what buyers are willing to pay for land and the amount of land demanded have a significant impact on changes in land prices. The changing demand for land involves current and expected future income levels; population and population growth; individual tastes and preferences; the number of buyers; the fertility of land as a factor of production; current and expected future interest rates and land prices; the productivity of land as an investment asset; returns to assets other than land; location; and changes in technology that affect accessibility.

The Rural Land Market and Competition among Land Uses

Rural land is distinguished from urban land primarily by lower settlement density. An incorporated jurisdiction of over 2,500 people is classified as urban, according to the U.S. Bureau of the Census. A broader definition, however, would include a standard metropolitan statistical area (SMSA), which requires a city of at least 50,000 people. According to Healy and Short (1983), "The rural land market is a series of interconnected local markets, segmented geographically by type of land, and joined together by fitful and imperfect flows of information and capital" (p. 115).

The rural land market, as noted earlier, deals in three basic types of land: (1) farmland/forestland; (2) rural residential land; and (3) recreation/vacationland. Although location and accessibility remain important factors, the crucial factor is that land is not a homogeneous good, but, depending on terrain, can be used for either intensive development purposes or for extensive resource-related uses. That is, the market for each of these kinds of land is to some degree interconnected, and each may be a substitute for the other under certain conditions. The sources of this competition are the keys to understanding the ways in which these three land markets are interconnected and the allocation of land resources over time.

The Scope of the Rural Land Market

After World War II, and particularly through the 1960s and 1970s, transportation and communication networks improved in America and helped increase the number of people active in buying and selling rural land. Until the early 1980s the results were (1) rising prices for all types of land; (2) an increased demand for land, along with greater absentee ownership; and (3) a reduction in average parcel size owned—commonly called "parcelation."

As rural lands became more accessible, more actors entered rural land markets, and many of these actors became absentee landowners operating from a regional, national, or international base. These actors typically placed higher values on land than local residents, and the result was that prices for all kinds of rural land rose rapidly. Through the 1970s, population growth rates in rural areas outstripped those in urban areas. Rural residents tended to stay in rural areas, and many people left urban areas to resettle in small towns and in the countryside. Rural economies in many areas continued to shift away from a mainly agricultural base toward an urban type of economy based on manufacturing and services. Although newcomers brought social and economic change to rural areas, they also created new demands for land.

As personal incomes rose through the 1960s, demands for recreational land grew—especially after 1972, when farm commodity prices increased as well. Investment in land also came to be regarded as an attractive hedge against inflation. Inflation caused real interest rates, and hence the true cost of borrowing, to fall; in addition, land values were appreciating faster than the inflation rate, bringing real capital gains. The strengths of these demands no doubt differed in various rural land markets according to proximity to urban areas, population levels, job opportunities, retirement amenities, recreation activities, soil productivity, and farm support services. But in general, the greater scope of rural land markets brought increased competition among land uses over a limited land base. As traditional rural landholders gave way to residential and recreational landowners, land investors, and developers, rural land holdings became fragmented, land uses became more intensive, average parcel size shrank, and price per acre rose.

Between 1970 and 1979, the average price of farmland tripled (see Figure 10.1). Rural land prices for residential, timber, and recreation uses

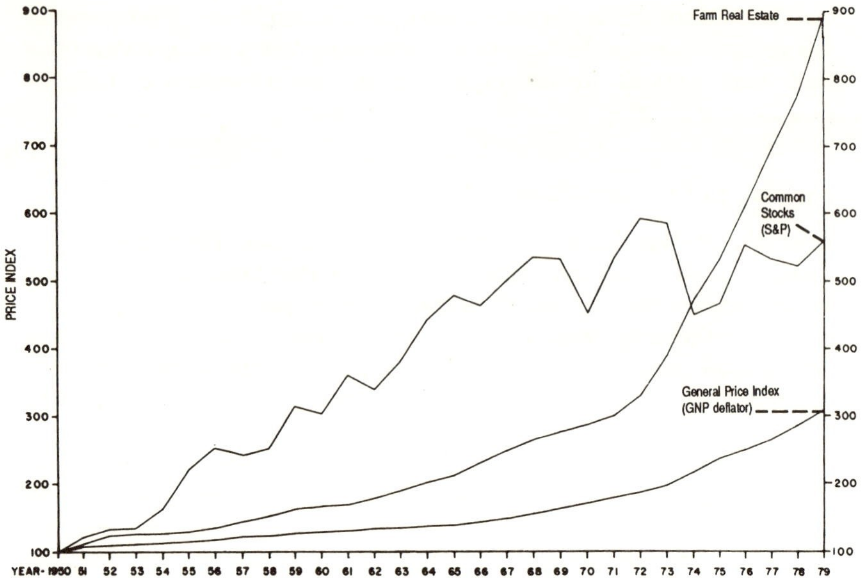

FIGURE 10.1. Comparative record of price change, 1950–1979. (1950 = 100.) From "Changing Markets for Rural Lands: Patterns and Issues" (p. 123) by R. G. Healy and J. L. Short, 1983, in R. H. Platt and G. Macinko (Eds.), *Beyond the Urban Fringe: Land Use Issues of Nonmetropolitan America*. Minneapolis: University of Minnesota Press. Copyright 1983 by Association of American Geographers. Reprinted by permission of the University of Minnesota Press.

rose even faster (see Figure 10.2). Although rural land prices have generally fallen since 1980, especially farmland prices, they remain high compared to just 15 years ago. Land prices per acre tend to increase substantially as parcel size shrinks. Thus, higher land prices, along with smaller parcel sizes and more absentee owners, portend more intensive land uses in those rural places that are accessible to urban areas, as well as a further decline in resource-based industries such as agriculture and forestry.

Land as Investment

The attractions of land as an investment include

> (1) the traditional tendency of people to rate property ownership as desirable, (2) the characteristic durability and long life of land investments, (3) the investor's feeling that he understands land and real estate, (4) the realization that investors can often manage their own investments, and (5) the belief that ownership of real estate and land resources provides an excellent hedge against inflation. (Barlowe, 1978, pp. 343–344)

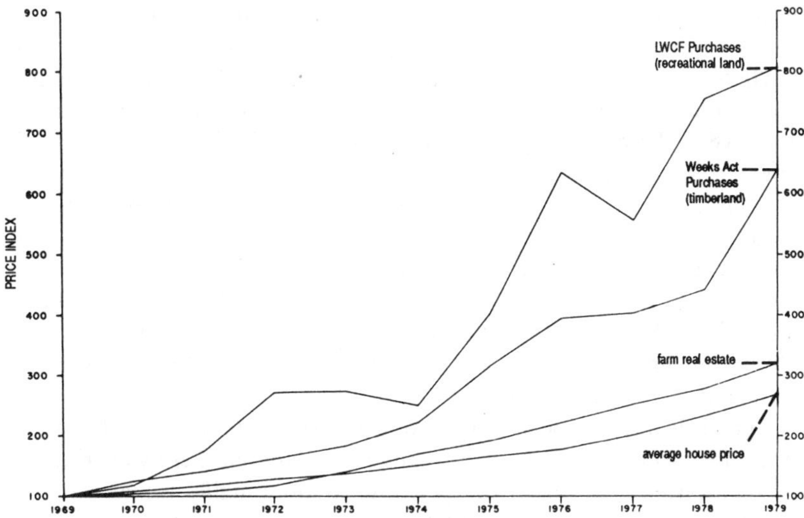

FIGURE 10.2. Prices of farmland, timberland, and recreational land, 1969–1979. (1969 = 100.) From "Changing Markets for Rural Lands: Patterns and Issues" (p. 125) by R. G. Healy and J. L. Short, 1983, in R. H. Platt and G. Macinko (Eds.), *Beyond the Urban Fringe: Land Use Issues of Nonmetropolitan America.* Minneapolis: University of Minnesota Press. Copyright 1983 by Association of American Geographers. Reprinted by permission of the University of Minnesota Press. LWCF, Land and Water Conservation Fund.

Barlowe adds that "real estate–resource investments usually compare favorably with other investment alternatives during periods of stable or rising prices, although they may show up somewhat less advantageously during deflationary periods" (1978, p. 344).

Between 1970 and 1979, inflation averaged an annual rate of 6.6 percent; Standard and Poor's index of 500 stocks rose at an average rate of 4.3 percent a year, meaning that stockholders on average lost money after accounting for inflation. Meanwhile, the average value of farmland increased 16.1 percent each year or 9.5 percent after adjusting for inflation (Healy & Short, 1981, p. 43). The search for a hedge against inflation in the 1970s drew investors toward farmland and real estate in general. As Healy and Short (1981) comment, "Inflation increases the attractiveness of land investment for people whose primary objective is not using the land for productive or personal enjoyment, but as a means of hedging against inflation" (p. 289). Expectations of future price increases caused people to purchase real estate sooner rather than later. In addition, interest rates when adjusted for inflation stayed low (or even negative) for much of the 1970s, thus further increasing the attractiveness of investing with borrowed money. The increased demand along with inflation led to much higher prices, because inflation distorted market signals so that supply became less responsive to changes in demand.

It is important to understand the difference between investment in land and land speculation. An investor tends to hold land for at least five to ten years before resale. Also, an investor will often make improvements in the property—either for personal enjoyment, to earn income from the property, or to enhance the value at the time of resale. A land speculator seeks to earn a quick profit through the purchase and resale of land. Land speculators generally sell land within five years of purchase, and spend little money on improving the property. Speculators hope to gain an "unearned increment" by purchasing land that will increase in value from greater demand, often caused by public improvements in roads and sewer and water lines.

The role of land speculators is subject to considerable debate. On the one hand, speculators provide liquidity in the land market by bearing risk and facilitating the transfer of open land to more highly valued intensive uses. On the other hand, speculators can have a sudden yet long-lived impact on local land markets and land use patterns through the purchase and subdivision of large tracts of land. The parcelation of the land base means that land prices will be higher and that intensive development is likely to follow. This process tends to drive up the demand for and cost of public services. In addition, parcelation can rapidly preclude the use of land for extensive uses, such as farming and forestry. Thus, land speculation, through parcelation of the land base, can bring about major changes in a local economy in a relatively short time. Land speculation has been especially pronounced in the urban–rural fringe land markets and around rural resort areas.

Alternatively, speculators can hold land off the market while they wait for land to "ripen" as public improvements are made nearby and the demand for land in the area increases. This withholding of land can result in awkward development patterns, especially at the urban–rural fringe, where "leapfrog" or "buckshot" development causes a sprawling jumble of open space and development that is expensive to service.

Through the 1970s, farm real estate became a "growth stock," attracting persons of large wealth or high income. These people could easily tolerate a low current return in the first years after purchase and could benefit most in terms of tax advantages from capital gains. Philip Raup (1982) has characterized this change in landowner behavior as a shift from profit seeking through prudent business management to seeking capital gains. Raup argues that the search for capital gains has converted almost all rural landowners into speculators.

Healy and Short (1978) looked at the ratio of cash rent of farmland to farm value as a measure of whether land prices were rising faster than underlying value. They concluded that increased demand for nonfarm uses led to increased land value but did not affect yearly agricultural rent. Moreover, "States where such nonfarm possibilities are greatest, such as Vermont, did tend to show the greatest drop in the rent-to-value ratio" (1978, p. 193). Vermont's ratio fell from 7.7 percent in 1960–1969 to 5.9 percent in 1970–1975. Healy and Short stated, "In the face of very high inflation prevailing in the 1970–1975 period, investors were willing to settle for a low current return (in the form of cash rent) in anticipation that appreciation in land prices would preserve their capital from inflation" (1978, p. 192). This means that agricultural land was seen primarily as an investment asset rather than as a factor of production. Farmland prices in the 1970s were bid up beyond the price that could be paid out of a farm income.

The value of American farmland peaked in 1981. Recall the formula for valuing land: $V = R/i$, where V = land value, R = net annual rent or returns of land, and i = the interest rate (corrected for inflation). In the 1970s, net returns to farmland rose as world grain prices climbed. Interest rates adjusted for inflation were low (and sometimes negative—which renders the land value formula useless). The two factors combined to drive up the value of farmland.

In late 1979, interest rates rose sharply, reaching over 20 percent in 1981. Inflation slowed, leaving adjusted interest rates at historical highs. Meanwhile, the high interest rates began raising the value of the dollar against other currencies. The demand for U.S. farm exports sagged and world grain prices plummeted under the weight of increased world grain output and the rising value of the dollar. Returns to land fell and interest rates rose, and the value of U.S. farmland plunged in the major grain-growing states of the Midwest and Great Plains. Between 1981 and 1986, the value of an acre of farmland in these states fell by about half. As of this writing, farmland values in the Midwest and Great Plains show some promise of increasing, as the drought of 1988 has driven up crop prices.

Farmers who borrowed against the rising value of their land in the late 1970s and early 1980s saw their collateral dwindle and loans become more difficult to repay. For thousands of farmers, the roller-coaster ride of land boom and bust has taken its toll. The recent boom and bust in U.S. farmland show how quickly land value can change and how difficult it can be for land buyers and sellers to anticipate changes in rational monetary policies and global economic influences.

Summary

The competition for rural land intensified between 1960 and 1980. Since the late 1970s, the competition for rural land has abated somewhat, primarily as a result of high interest rates, reduced inflation, and declines in farm commodity prices. Still, many new actors representing rural residential, recreational, and investor demands have entered land markets and added to rising land prices. Because these actors are typically more willing and able to pay, they have often successfully outbid agricultural demands for land. Although farmland prices increased substantially between 1970 and 1979, net farm incomes grew, but not as fast. The farmers' search for capital gains, along with that of other investors during an inflationary period, dovetailed with the greater willingness to pay for rural land on the part of nonfarmers. As a result, acreage was transferred out of agricultural and forestry markets and into rural residential and recreational markets. The plunge in farmland prices in the 1980s indicates a glut of farmland from a national perspective. The shifts in the rural land market have, of course, varied among rural regions. Rural land prices remain high in the Northeast, in rural recreation areas, and within commuting distances of urban centers. In more remote and farming-dependent areas, such as the Midwest and Great Plains states, rural land prices have fallen significantly from the highs of the early 1980s.

The competition for rural land raises questions about the reliability of an imperfect market in allocating land resources over time. The divergence of private and social costs and benefits, and the adequacy of land supplies to meet competing demands, become crucial issues. As populations grow relative to finite land resources, and as incomes and mobility increase, the demand for land uses will increase, raising land prices and reducing the average size of land holdings. The greater density is certain to produce land use conflicts. Legal means have been the primary method used to settle land use disputes. This is not to say that laws employing the police power of a state or locality will ever produce completely desirable or efficient land use patterns and prices. The market retains the appeal of convenience in determining land prices and uses by voluntarily bringing together buyer and seller. But planning and land use regulations can influence the supply, location, and price of different land uses. Planning, we hope, can guide the land market to create the types of development desired by the community in appropriate places.

The Tenure Factor: Ownership and the Definition of Use

For a better understanding of the competition for rural land, it is helpful to examine land ownership as it affects the markets for the three major rural land uses: farming, housing, and recreation. Land tenure is concerned with the rights of landownership, who actually owns or controls land, and how the ownership of or control over land resources is distributed in a particular area. The use of land and its spatial relationship to other resources are crucial elements in rural land management (Lapping & Clemenson, 1983). Yet, perhaps more fundamentally, the characteristics of land tenure and the motives and objectives for holding land determine and define the behavior of rural land markets. According to land economist Gene Wunderlich (1979, p. 319), land tenure is derived from three sources of authority: membership, rank, and use. "Membership" suggests residency in a community. "Rank" refers to kinship or a sense of worth as defined by society and the marketplace. "Use" implies a socially accepted employment of land resources for the benefit of the owner. In America, land tenure is primarily determined by membership and rank.

The key tenure issues in rural areas are the concentration of landownership in the hands of fewer individuals and corporations and the growth in absentee landownership. Perhaps the most prominent issue has been the concern for the future of the family farm as farm consolidation continues. Other issues include the debate over the 160-acre limitation on the irrigation of Western farmland, the Appalachian movement to take lands back from mining companies, and the passage of the Agricultural Foreign Investment Disclosure Act of 1978 to keep track of the foreign ownership of U.S. farmland (7 U.S.C. §§ 3501–3508). In all of these cases, land tenure affects the control and benefits of land as determinants of wealth and power in rural societies.

Those who own and control rural land will continue to shape the rural landscape. Although land use policies can and do have some effect on behavior and tenure systems, the ownership of land will remain the key variable in the use and sustainability of land resources. As Wunderlich (1974) has noted,

> [U]nless there is some major change in our ownership patterns, the future of our rural landscape in many areas will depend on the widely varied motivations of a shrinking number of landowners. And our economic system would still have no effective way of incorporating the values of persons with interest in, but no ownership of, land. (p. 16)

The Market for Farmland

Farmers comprise the largest single group of landowners in the United States. Of the nation's 1.3 billion acres of privately owned rural land, farmers hold title to about 38 percent, or over 500 million acres. An additional 280 million acres of farmland and ranchland are owned by nonfarmers. In the market for farmland as a final use, demanders perceive land as a factor in the production of food and fiber. These demanders seek to combine land and other inputs (e.g., machinery, labor, fertilizer) to achieve maximum output at a minimum cost. Bids are made for farmland according to the land's expected future returns. These returns will be influenced largely by land productivity, technological change, expected crop prices, and the extent to which the land to be purchased will enable a farmer to achieve economies of scale (lower unit costs) in production. Another factor influencing bid price is the farmer's ability to pay or borrow, which in turn is often related to the farm size. Consequently, bid prices for a certain parcel of agricultural land are likely to vary among farmers.

Buyers of farmland fell into the following categories and average percentages over 1972–1980, according to U.S. Department of Agriculture (USDA) data (USDA, Economic Statistics and Cooperative Service, 1981): tenants, 12 percent; owner–operators, 61 percent; nonfarmers, 25 percent; retired farmers, 2 percent. It is important to note that not all demand for farmland comes from farmers or prospective farmers. The final use to which a nonfarmer will put land is not guaranteed to be an agricultural use.

Farmland is usually supplied in the market by long-term landowners who have previously used the land for agricultural purposes. The reasons for offering land in the market will vary among sellers, but for farmland the most common reasons are retirement, a decision to leave farming, or a general need for capital. Sellers of farmland fell into the following categories and average percentages over 1972–1980, according to USDA data (USDA, Economic Statistics and Cooperative Service, 1981): active farmers, 40 percent; retired farmers, 15 percent; estates, 19 percent; and nonfarmers, 26 percent.

Farmland may be sold for other uses (such as residential or recreational) because those uses command a higher price, which nonfarm buyers are willing and able to pay. Also, the demand for farmland in other uses will influence farmland prices: The greater the nonfarm use potential of agricultural land, rather than its potential use in producing food and fiber, the higher and more rapidly its price will rise. This has been the case with farmland in much of the Northeast in the 1980s.

Agricultural land uses typically involve large parcels of several hundred acres; the average size of a farm in the United States was 455 acres in 1984 (U.S. Bureau of the Census, 1986, p. 629). Farmland transactions often carry a large total price, but a relatively low price per acre compared to intensive land uses. Overall, the level of farmland in physical supply is relatively constant. About 900 million acres of varying quality were in farmland use between 1972 and 1980. Of this total, about

25 million acres of farmland were sold each year between 1972 and 1980, meaning that just under 2.8 percent of the total physical supply changed hands in the market each year (USDA, Economic Statistics and Cooperative Service, 1981). Reasons behind this low ratio of economic to physical farmland supply include (1) the lack of farmer mobility (i.e., the difficulty of selling a farm and buying another comparable one); (2) the size of the long-term investment and borrowing associated with a farm operation; (3) the limited number of farmers (2.2 million nationally); and (4) the reluctance of long-term owners of farmland to sell, as the land that supports farming is also seen as supporting a way of life. But when sufficient incentives exist in the form of higher crop prices and greater expected income, additional land can be brought into production, as the experience of the early 1970s has shown.

The characteristics of landowners and their motivations for holding land are thought to provide a key to explaining the level of land in different land markets. Owners of agricultural land tend to be long-term landowners who acquired their land as primarily a factor of production, at a time when expectations of development were remote or nonexistent. Farmland owners typically reside upon the land they own, and are often thought to have a personal commitment to their property. Moreover, the property is frequently the owner's chief financial asset. These "traditional" landowners seldom actively participate in the real estate market, mainly because they are reluctant to place agricultural land on the market.

The Market for Rural Residential Land

In the market for rural residential land uses, demanders view land primarily as providing housing services for personal consumption. Given a budget constraint, the buyer seeks to obtain as much rural land and accompanying amenities as possible to support the housing service; in keeping with this behavior, Healy and Short (1981, p. 19) note that buyers of rural land try to acquire more acreage for their money than they would get in urban areas. Still, the market for rural residential land usually involves smaller parcels and higher prices per acre than the market for agricultural land. For example, a Vermont study found that rural lots under two acres were selling for over 12 times more per acre than parcels greater than 25 acres (Ottauquechee Regional Planning and Development Commission, 1977).

The ownership of rural residential land is dispersed (often for reasons of privacy), but the land must be located within a reasonable distance of places of work and public services. Individual mobility is greater for rural residential landowners than for farmers, because the site of residence is rarely the place of work as well. Suppliers of rural residential land are less likely to be long-term landowners than are suppliers of farmland; speculators, investors, and developers are more likely to supply land in rural residential land markets than in farmland markets.

The supply of rural residential parcels depends on (1) the portion of the total physical land supply that has potential for rural residential use and is offered for sale at prevailing prices at any one point in time; (2) whether the parcel already has a house on it; (3) the cost of improving raw land (clearing, building roads for access) relative to the cost of building a home; and (4) the parcel size. For the purpose of this analysis, the supply of raw land with residential potential is emphasized. Although parcels bearing houses are usually more frequently supplied in rural land markets, the determination of rural residential land value may be difficult to separate from the value of the house it supports.

The market for rural residential land can be further separated into two components: small parcels of less than ten acres, and large parcels of more than ten acres. Small parcel sites are created almost expressly to support housing; as the total price of land in providing housing is likely to be relatively small, the quantity of land demanded is expected to change proportionately less than a change in land price. This, to a large degree, reflects the limited demand that most people have for land. Also, the uses to which a small parcel can be put are more limited than the potential uses of a large parcel.

The demand for large residential parcels will largely depend on prospective buyers' preferences for small versus large parcels and willingness to pay for the large parcels. Generally, the greater the proportion of land price in the total housing package, the lower the demand for land will be. However, one may find that people who purchase large residential parcels have a substantial ability to pay and may desire the additional amenities and recreational opportunities that a large parcel provides.

As noted above in the discussion of the market for farmland, owners of large parcels tend to be long-term, traditional landowners who reside upon their property, which is also their chief financial asset. In addition, these owners often acquired their land at a time when expectations of development were remote or nonexistent. Long-term landowners are likely to be influenced most by family and life cycle factors (e.g., retirement) in offering land for sale, although taxes and good market prices also play a role. Thus, long-term landowners, who also tend to be suppliers of large rural residential parcels, are often reluctant to put their land on the market.

The sale of small rural residential parcels describes a process often associated not only with changes from extensive to intensive land uses, but also with changes in the characteristics of landowners. As Brown, Phillips, and Roberts (1981) point out, "The transfer of land from traditional owners who hold property for its current rural use to investors and developers who value land for its potential urban (or urban-like intensive uses in rural areas) is a cumulative process" (p. 138). These investors and developers are often absentee owners, and the land purchased typically represents only a small portion of their total assets. Brown et al. state that "most landowners in this group intend either to sell their land or develop it in the near future" (1981, p. 138). In rural–urban fringe markets, Brown et al. conclude that "a substantial number (of landowners) would sell immediately if a good offer were forthcoming" (1981, p. 138).

The Market for Recreation/Vacation Land

The overwhelming majority of the literature on rural land market activity has concentrated on the expansion of urban areas into the nearby countryside. But with the advent of interstate highways and increased access to the hinterlands, urban dwellers have sought recreational activities far from home. The impact of these recreational demands on the rural land market has yet to be explored in depth. The market for recreation/vacation land is far more difficult to describe than the other two related markets. Although Healy and Short (1981, p. 106) found that most sales of rural land were comprised of 5- to 40-acre parcel sizes, they also found that sales of vacation land showed the greatest variation in price per acre and parcel size.

In the market for vacation land uses, demanders view the land as providing consumption services, which may include housing as well as recreation and privacy. Demanders are more likely to be temporary visitors than local residents, unlike rural residential and farmland buyers. The recreational activities and amenities that a parcel of vacation land offers are related to parcel size and location. Small parcels are more often associated with an agglomeration of vacationers, as found in resort areas near a body of water or skiing facilities, where there are high prices per acre associated with potential building lots. Large parcels are often more remote, providing such activities as hunting and hiking, and price per acre is usually low.

The demand for large vacation land parcels will largely depend on prospective buyers' preferences for small versus large parcels. As in the case of large residential parcels, the greater the proportion of land price in the total vacation housing or experience package (without a house), the smaller the amount of land that will be demanded. Large vacation parcels may differ from large residential parcels in two important ways. First, vacation parcels need not be near a population center for employment and services; second, large vacation parcels need not support a dwelling (or one that is habitable year-round) to provide a satisfactory vacation experience. It is likely that the value of the land portion of the vacation housing or recreation experience is greater for large vacation parcels than for small vacation parcels or large residential parcels.

The characteristics of suppliers of vacation land will tend to vary according to the parcel size offered for sale. Suppliers of building lots will tend to be short-term landowners, such as speculators, investors, and developers. For the large vacation parcels, suppliers will tend to be long-term landowners. A fairly large proportion of the stock of small vacation land parcels and potential small vacation land parcels is expected to be placed on the market at any given time; also, owners of existing and potential small vacation lots tend to be short-term landowners, indicating a greater frequency of turnover than for long-term landowners. The proportion of the stock of existing or potential large vacation parcels placed on the market at any given time is likely to be small, as is the proportion of the supply of agricultural land. In addition, large vacation parcels are more likely to be held by long-term landowners who are often reluctant to sell and hence cause land supply to be less responsive to price.

How to Determine and Appraise Tenure

There are about ten acres of rural land in America for each man, woman, and child. Of the nation's 2.3 billion acres, about 1.3 billion acres are privately owned. Farmers as a group own the largest amount of private land, roughly 500 million acres. Nonfarmers own another 280 million acres of crop and pasture land. The other half billion acres consist of privately held forestland, urban areas, highways, and wasteland (Wolf, 1981, p. 549). Until recently, few data were available on the private ownership of land in America. The 1978 Landownership Survey of the USDA (Lewis, 1978) provided a data base for an assessment of land tenure characteristics. Although the survey did not detect ownership trends over time, it is useful in pointing out issues of present and future concern. Although at the local level tenure information may be limited to property tax maps, the USDA Landownership Survey suggests several questions that influence the economic, social, and political makeup of rural communities.

From a local perspective, a planner can find out about land tenure and ownership patterns from tax maps in the town or county assessor's office, from land deeds, and from land sales records held by the town clerk or at the county courthouse.

1. *Occupation of landowners* provides information on how many owners use their land as part of their occupation. This is particularly important in determining the stability of land-based indus-

tries. For example, ownership of the land base by family farm units has been considered essential to the long-term survival of the local or regional agriculture industry. If control over farmland is out of the hands of the farm community, then its future stability cannot be guaranteed, and farmers will be less likely to make the investments necessary to maintain soil productivity and to control erosion. Owner occupations include farmers, white-collar workers, blue-collar workers, and retirees. A high percentage of land held by retirees suggests that there will be a considerable turnover in ownership within the next decade, and perhaps major changes in land uses.

2. *Type of ownership* describes how the land is held. The primary concern here is the amount of land held by nonfamily corporations and nonfamily partnerships. These ownership arrangements imply outside control and are thought to be less responsive to local desires and needs. Other forms of landownership include sole proprietorships, husband–wife partnerships, family partnerships and family corporations.

3. *Concentration of ownership* indicates the possibility of exercising political and economic power in a community and the degree of fairness in the ownership of land resources. For example, in the state of Maine, 1 percent of all landowners control 73 percent of the land base. Much of rural Maine is owned by pulp and paper companies whose interests are not necessarily the same as those of the people who live in rural communities. By contrast, the largest 1 percent of all landowners in neighboring Vermont own only 14 percent of the state's land base, which is one-third the size of Maine's.

Land ownership concentration can be conveniently illustrated by a Lorenz curve, which compares the percentages of acres owned to the percentage of owners (see Figure 10.3). Perfect equality of ownership exists when the curve is a 45-degree line. The more bowed the curve becomes, the greater the concentration of ownership; that is, the land ownership pattern depicted by curve b shows a greater concentration than that illustrated by line a.

4. *Age and sex of landowners* may hold important implications for land use and fairness. Most privately held land is owned by individuals over 50 years of age, and a sizeable proportion by people over the age of 65. These facts indicate the likelihood of a large turnover of landownership within the next 10–15 years. Women tend not to own much land, and this may be a problem for those who wish to see a more equitable distribution of wealth.

5. *Land transfer* tells how land is acquired. Most land changes hands through the market place. Other means include inheritance, gifts, and purchase from relatives.

6. *Absentee landownership* brings attention to issues of local control of the land base and changing land uses. Absentee owners tend to be

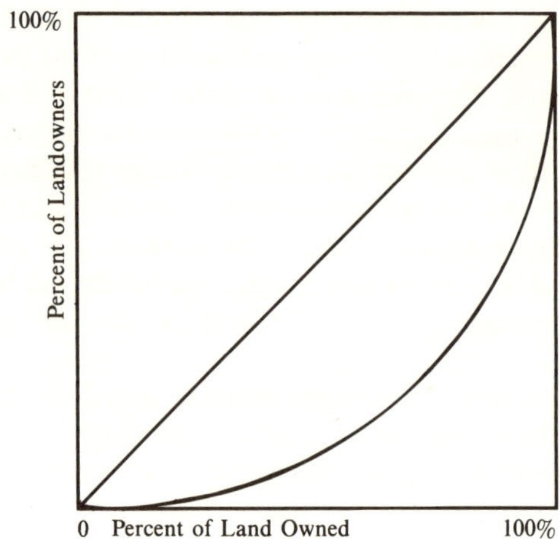

FIGURE 10.3. The Lorenz curve and the concentration of landownership.

less interested in local concerns and to contribute to higher land prices and smaller parcel sizes, which discourage land-based industries. For those occupying absentee-owned farmland, the security of tenure is uncertain, and land management practices may not enhance the long-term productivity of the land. Absentee owners include those who reside within the same county, those who reside within the same state, those who live out of state, and foreign owners.

Tenure data may or may not indicate policy problems, depending on one's concept of a just and equitable society. But tenure data can be used to detect such situations as the loss of ownership of farmland by farmers, the growing concentration of ownership in the hands of a few, the accelerating pace of land transfers, and the presence of both absentee and foreign land ownership. Rural land tenure is intimately related to the character of the social order. When the control of land is no longer widely spread throughout a community, democracy and responsiveness to local needs are not guaranteed. In sum, land tenure is a powerful force in determining land uses, economic growth, the appearance of the rural landscape, and the impact of land use regulations.

Although landowners and land regulators are often at odds with each other, the concentration of landownership can serve to focus political power in rejecting, evading, and otherwise frustrating the intentions of land use controls. Land use planner Frank Popper (1979) warns,

> A planner in such a community who tries to devise or implement a zoning ordinance without taking into account its ownership patterns will inevitably accomplish little. If we conceive of land as a social weapon, it follows that land's power whether exerted offensively or

> defensively, will always advantage—often unintentionally—those who own land at the expense of those who do not. It will generally favor those with more (or more valuable) land over those with less. (p. 131)

Yet little information has been available on who owns the land and why. Ownership data are often available from town or county plat maps or assessors' property tax maps. But gathering this data can be time-consuming. This is especially apparent in the lack of community-level studies on the social and economic effects of concentrated landownership. Nonetheless, if plat maps and assessors' tax maps are not examined for tenure data, various land use regulations may not be designed to achieve desired results. Still, rural land use regulations, like their urban forerunners, will probably continue to emphasize the protection of existing property values and maintaining the environment at the expense of the poor and landless. This need not be the case; however, special funding from the state may be necessary to help provide housing opportunities for the poor. Such a program was initiated in Vermont in 1988 (Vt. Stat. Ann. tit. 10, § 155, 1988). Oregon's State Land Use Program requires communities to provide housing opportunities for all income groups (Oregon Land Conservation and Development Commission, 1980).

Planning Tools for Land Market Analysis and Change

The analysis of rural land markets depends upon a good understanding of how these markets work and of the land use and landownership issues that determine economic and social well-being. The identification of problems will be influenced by what the planner feels are the needs of society or where he or she perceives a gap between the performance of the land market and implied or stated social goals. As a first step in analyzing land markets, land prices offer a key source of information as to what buyers and sellers think land is worth. Land prices indicate not only competition among different uses for a particular parcel, but also the direction of change in land uses. As Healy and Short (1981) have observed:

> The expectations and institutional arrangements affecting land are frequently revealed in the land market long before any change occurs on the land itself. For example, the expectation that agricultural land will be put to urban or other developed use is usually reflected in the land's price years before a single building is constructed. (pp. 1–2)

Rising land prices for forest, recreational, and residential uses may signal increased activity by absentee buyers, speculators, or newcomers. Higher prices may also signal a reduction in average parcel sizes. The growth in the number of small parcels can be detected from plat maps and increases in the local property tax rolls. Smaller parcel sizes portend more intensive land uses and greater pressure on aesthetics, wildlife habitats, natural areas, and the viability of farm and forest operations. Furthermore, intensive land uses, especially residential development, tend to raise the cost of providing public services.

The rate of land turnover provides an indication of how rapidly development may occur in the near future. If many sales of bare land are registered within a short period of time, then a "boomtown" may be in the offing. Generally, rural land does not change hands often, especially large parcels. Closely tied to the rate of turnover are the buyers' and sellers' places of residence. If local people constitute the large majority of buyers and sellers of land, then turnover is likely to be slow. But if locals comprise a large proportion of sellers and out-of-towners are doing a sizeable amount of buying, then changes in land prices and land uses are likely to be in motion.

Such government actions as the building of roads, sewer and water lines, and schools play a significant role in determining the location of future development. Knowledge of state, local, and federal public service investment plans is important in putting together local land use plans and regulations. Case studies of previous public investment decisions should provide insight into the likely impacts of current and future government actions.

There is a definite and growing need for improved information on nonfarm land ownership in rural areas. Also, much research remains to be done on how the distribution of landownership affects land use and land prices. The monitoring of land prices, parcel sizes, rate of turnover, and residences of buyers and sellers can give a valuable perspective on what is happening to rural regions.

An understanding of the workings of land markets provides planners with a valuable skill in anticipating the location and intensity of development. Land prices, parcel sizes, and ownership patterns to a large degree determine land uses and economic activity. The planner must be able to interpret people's expectations about the future course of land use, and yet to steer privately made land use decisions toward socially desirable goals. This function is fundamental to planning.

A plan is an anticipatory document that attempts to prescribe how a community should grow, based on current land uses and expected economic and population trends. In implementing a plan, the planner helps administer one or more land use controls, such as zoning and subdivision regulations; yet the planner must be aware of the potential impacts of certain controls and proposed developments on the local land market. In this way, the planner as educator will be able to recommend effective controls and realistic advice on proposed developments, in order to preserve the integrity of the local plan.

CHAPTER ELEVEN

Rural Economic Development and Community Development

Rural Economic Development and Diversification[1]

Most people talk about "economic development" without defining what they mean. Economic development is a process of change whose goal is to increase the wealth of a community by raising incomes, increasing access to services, and reducing unemployment. Economic development occurs when private entrepreneurs, nonprofit firms, cooperatives, or government agencies make investments in a town; these investments increase the demand for goods and services, and thus create new jobs and sustain existing ones. Small towns rely upon reinvestment for their survival. When the town's population declines or when residents cease to reinvest their wages and profits, homes and storefronts begin to decay, community spirit falls, and visitors wonder whether the townspeople care about their surroundings.

"Economy" literally means "the management of a household." Successfully managing a town's "household" consists primarily of three elements: good leadership, teamwork, and the wise and effective use of a town's resources (the natural setting, people, and human-made structures). It is the planner's role to help private business people, government, and the public work together for the betterment of the community. But to achieve this role, the planner must first understand how to nurture economic development.

[1]This section draws heavily from *Small Town Economic Development* (Planners Advisory Service Memo), by F. Alguire, T. Daniels, J. Keller, & L. Mineur, 1987, Chicago: American Planning Association. Copyright 1987 by the American Planning Association. Used by permission.

Not every town experiences the same type of economy at the same time; some prosper while others struggle. The economic bases of small towns have historically been extractive industries, such as agriculture, wood products, and mining; moreover, many small towns have a majority of their jobs in one or two industries. Extractive industries, unlike manufacturing or service firms, are affected both by natural and by market constraints. Thus, an extractive company's ability to modify its product is limited; for instance, a copper mine cannot produce nickel when copper prices drop. This means that many small towns must both diversify their economic base and retain existing businesses for continued economic growth.

Rural communities face several obstacles to sustaining economic growth (Tweeten & Brinkman, 1976). Small towns have small populations. Towns of fewer than 2,500 people especially appear to lack local markets for many products. In addition, public services are often limited; the labor force is largely unskilled; and a remote location may mean high transportation costs for importing raw materials and shipping products to urban markets. In recent years, there has been an accelerating trend toward the consolidation of rural markets, and retail and service businesses have been increasingly captured by regional shopping malls and regional centers. As a result, many small towns have been weakened and share common problems, such as young people leaving town, businesses failing, and the need for a greater property tax base.

Before launching an economic development program, the planner must realize that there are no panaceas and a number of risks. Efforts may or may not reap rewards; hard work over a long period of time is essential. Success in economic development is rarely accomplished in a year or two, but instead is achieved over decades. A final warning is that the increased exposure of America to international competition has meant that small towns need to take risks and innovate. The traditional rustic advice of "If it ain't broke, don't try to fix it" makes little sense in a highly competitive global economy.

The first step toward economic development is to understand the private sector. Many planners naively overlook the fact that most of their training has been grounded in the public sector. To be effective, the planner must understand and come to respect the private business person who bears the risk in creating the new jobs on which economic development depends. To understand the private sector, the planner must participate in it. It is important for the planner to become a part of the local business community through organizations such as the chamber of commerce.

The second step is organization. The key ingredients of good organization are leadership, commitment, and patience. Economic development cannot be mandated. Only through local consensus can a town marshal the coordinated effort necessary for economic development. This point cannot be overemphasized. The private sector achieves its success by reducing and managing risk. Investors may perceive rural communities as risky for many reasons. The additional risk created by the lack of consensus on the part of a community often sends investors off to look for another town. The intense competition for entrepreneurs has compelled many rural communities to search for effective ways to attract new employers. For these reasons, a town may form a nonprofit economic development corporation funded by the local government or by a public–private cooperative agreement. Economic development corporations for profit may also be formed if local investors are willing, and if such corporations are in compliance with state and federal securities regulations. Development corporations seek to draw new employers from outside the community, as well as to foster new entrepreneurial activity from within. The corporations typically look to package retail and industrial sites, to obtain loans and grants from government agencies, and to provide advice to businesses. They should be willing to assist a variety of business ownership arrangements: community-owned businesses, companies or cooperatives in which the community holds an equity position, or employee-owned firms, not just privately owned and operated businesses.

Unless influential leaders and the public at large are willing to work for community economic development, not much will happen. Often a town will establish an economic development committee, but the committee loses momentum after a few months. Economic development takes time and patience. Here the planner can play an important leadership role by helping organize community participation and support through public meetings and items in the local news media. Keeping people informed and involved sustains interest and continuity in economic development efforts. An especially useful technique is a needs assessment survey that can be distributed to the townspeople, soliciting their opinions on (1) what kinds of businesses are needed (retail, service, industrial, or other); (2) what kinds of businesses the town can support; (3) what barriers exist to developing new businesses; and (4) what the community can do to help existing businesses grow.

Next, the planner can survey existing businesses, compare the results with the needs assessment survey, and make recommendations about attracting new businesses and retaining existing ones. Here, state agencies involved with economic development and local utility companies can offer valuable advice.

The planner should also collect and evaluate data that businesses and industries seek in making a location decision. This information includes the availability of building sites, buildings, and infrastructure; transportation services; education and job training programs; the local labor market; access to customers and suppliers; capital availability; and local taxes and land use regulations. The planner should also be able to evaluate whether the property tax concession and infrastructure demands of a prospective business are reasonable for the community.

The third step in local economic development is forming an overall strategy and plan of action. The strategy should express in a few pages what the community wants to accomplish, how, and when. The core of an economic development plan consists of (1) a baseline projection; (2) projection of a desired level of economic activity; and (3) a description of ways of closing the gap between the two.

The baseline projection is an assumption of what the community's economy can be expected to do in the future, based on historical trends; it assumes no major deviation from those trends in the future. The baseline projection serves as a "best guess" of how the town's economy will perform without organized economic development efforts. The projection of the desired level of economic activity can be expressed in quantitative terms. For instance, a 6 percent growth in employment over five years equals 120 new jobs.

There are three main economic development strategies that small towns can follow. They can try to attract manufacturing firms, promote the expansion of local businesses, or "hunker down" and hope for better economic times. These strategies are known as "search development," "incremental adjustment," and prayer!

Many towns have made enormous efforts to attract large companies that are not appropriate in terms of available local resources and public services. Much of the traditional economic development literature emphasizes courting manufacturing plants from somewhere else. This process often means spending large amounts of money over a long period of time for local improvements in public services and for advertisements in numerous development-related magazines. About 1,000 industrial firms look to relocate or open new plants each year. Yet, there are 10 to 15 or more cities and towns actively competing for each of these plants. Thus, a town's chances for attracting a new plant are slim.

There is a strong temptation to think that economic development involves only the recruitment of large new firms into a community. Certainly this has been the focus of much activity on the part of local development commissions and chambers of commerce. Recent research, however, indicates that most new jobs are created by small local businesses employing fewer than 20 people. Most of these new jobs come not from manufacturing, but from service industries such as recreation, tourism, wholesale trade, restaurants, finance, and real estate.

Although planners should be prepared for out-of-town inquiries from firms looking to locate manufacturing plants, often the greatest success will be in assisting local businesses with expansions or helping local entrepreneurs who want to start businesses. It is vitally important to put into place programs to keep good, productive businesses in the community and help them realize expansion potential. In fact, about half of all new jobs come from the expansion of small businesses. A strategy that promotes the creation of eight jobs here, six there, and another five elsewhere is often more realistic and appropriate to the size and needs of small towns. Local firms understand the community and local customs, and do not completely change the town's lifestyle as larger, "outside" companies frequently do. The Federal Small Business Administration's data indicate that only one out of two new businesses makes it through the first four years and that only one out of five makes it through the ninth year. Failures are as much a part of the economic landscape as are startups and expansions. Thus, it is not wise to foster just one firm.

There are three important points to consider in putting together a successful business venture. First, the firm must have a marketable product. Second, entrepreneurs must have competence in management, production, and marketing. And third, there must be at least a 25–30 percent up-front equity investment. A deficiency in one or more of these categories needs to be addressed if an enterprise is to have a chance to succeed.

The planner can assist local business development in several ways. The planner can offer advice on how to meet regulatory standards and obtain regulatory permits, or he or she can work to change existing zoning and building regulations that stifle appropriate development. Care must be taken here, because some firms actually seek a community with rather strict land use regulations to protect their development from a conflicting land use next door.

The planner should know the many sources of development financing and should understand the financial needs of existing and prospective businesses. The planner should also be aware of time frame constraints. Local financing sources are likely to respond in one to six months to funding requests. Proposals to state agencies normally take three to six months for consideration, and federal agencies often take six months or more to respond to funding requests.

The role of commercial banks in promoting local economic development has often been overlooked. There are currently over 4,000 rural banks throughout the United States, and most of these have assets of only \$15–\$35 million (Daniels, 1986b). Small rural banks serve as a critical source of financing for local inhabitants by channeling local savings into local businesses and purchasing locally issued municipal bonds. The recent relaxation of federal banking laws to allow interstate banking and remove the ceiling on deposit interest rates is likely to have a profound influence over the future availability of capital in rural America.

At the same time, problems with the traditional rural industries of farming, forestry, and mining have led to a rise in the number and size of troubled loans and bank failures. In 1984, 79 banks failed in the United States; in 1985, the number soared to 120; in 1986, 138 banks failed; and three-quarters of the way through 1987, 141 banks had failed—the highest rate since the Great Depression of the 1930s ("Bank Failures Increase," 1987). Most of the banks that have closed in recent years have been small rural banks. Many local commercial banks have little chance to diversify their loans, either geographically or outside of agriculture and energy, because of existing restrictions on branch banking. For example, in many Midwestern states, a bank can only open branch offices in counties adjacent to the home office. Given the large number of sparsely populated counties, this restriction locks small rural banks into a very limited area (Comptroller of the Currency, 1984).

Although much of rural America continues to experience economic restructuring toward service industries, local efforts to diversify the economic base may be hampered by an ailing banking community and by bankers' cautiousness in making only "high-quality" loans.

Federal subsidies, revenue sharing, loans, and public works projects have been reduced or eliminated under the Reagan administration. State economic development funds are limited as well. The overriding message is that local communities will be forced to become more self-sufficient in generating economic development programs and in finding the capital to finance those programs. To do this, new public–private partnerships will have to be formed. Low-interest loan pools and venture capital will be essential elements for stimulating new and expanding businesses. The involvement of local commercial banks could be crucial, and the degree of involvement will probably depend on the health of the local banks.

Various state agencies accept proposals for grant and loan dollars for local economic development projects (see Figure 11.1). These sources can usually be identified through state departments of commerce or economic development.

Federal dollars for economic development are available from many sources, including the Small Business Administration (with programs to assist towns in forming community development corporations, providing financial advice to small business, and offering small business loans); the Economic Development Administration (providing technical assistance and money for projects primarily in towns that qualify as suffering from extreme economic distress); the U.S. Department of Housing and Urban Development (featuring community development block grants, which are administered by the states); the Farmers Home Administration (offering business and industrial loans and community facility loans, especially for sewer and water projects); the Bureau of Indian Affairs (for Native American tribes); and other agencies and programs too numerous to mention here. Planners should see Chapter 5 ("The Federal Government as a Center of Influence and Power") for a more extensive discussion, as well as for a list of sources publishing information on federal grants.

Public funds can often be used for indirect leveraging of private investment dollars, such as providing needed roads to an industrial park. Public money can also result in direct leveraging by adding to a firm's operating capital, bringing down interest rates, or guaranteeing loans. A number of small towns have even raised private money through fund drives to pay for the relocation of new businesses. A particularly popular program for towns and counties has been the sale of industrial revenue bonds for the purchase of land and the financing of buildings. Companies then lease space from the local government and pay rent, which is used to retire the bonds. As of 1988, industrial revenue bonds are tax-exempt only for manufacturing projects.

The most important part of dealing with the many financing sources is communication. First, the planner should take time to read a book about economic development finance or to attend a course at a local college to learn the jargon of the financial world. He or she should also talk to local bankers about what they look for in development financing. Various workshops—such as those offered by the National Development Council, based in Schiller Park, Illinois—are also available across the country to train individuals in small business finance and deal structuring. Better yet, the planner should arrange to visit a town that has achieved a measure of success in economic development.

In determining the financial feasibility of private projects, planners should also be aware of the public benefits and costs of development. Benefits include payroll and income taxes and property tax revenue; net sales, inventory, and corporate income taxes; and spinoff development. Costs include the direct costs of the incentives offered, such as property tax breaks, the preemption of future opportunities, an increase in public service demands and costs, and environmental impacts.

One of the most overlooked sources of funds in a small town is energy conservation. Most of the money spent on energy leaves the local community. By reducing energy usage, a town can keep more money within the community and make it available for other uses. An active conservation program by the local electric or natural gas utility can have far-reaching effects on the financial health of the community. A conservation strategy requires organization and perseverance, and results may not appear immediately. But the returns can be considerable. For exam-

	I. Agriculture-Related Development	II. Transition Tools	III. Rural Business Assistance	IV. Rural Community Assistance
Alabama	C,F	H,J,K	P,R,T,U,W	Z,CC,EE,GG
Alaska	A,C,D,E,F,G	J	M,N,O,R,S,T,U, W,X	Z,AA,CC,DD, EE,GG
Arizona	A		M,N,O,P,T,U,X	Z,AA,BB,CC, DD,EE,FF,GG, HH
Arkansas	A,B,F		P,T,U,V	CC,EE,GG
California	B,F	H	M,N,T,R,U,V,W	Z,AA,BB,CC, EE,GG
Colorado	D		N,O,P,Q,T,U,V, X	
Connecticut	F			DD,HH
Delaware			M	AA,BB,CC
Florida	A,B,E,F	I,K	M	Z,BB
Georgia	A,B,F,D	K	N,Q,S,T	
Hawaii	A,B,C,D,E,F	H,I,J	P,V	Z,CC,GG,HH
Idaho	A,E,F,G	H,J,K	T,U	AA,CC
Illinois	A,B,C,D,E,F	H,J,K	M,N,O,P,Q,R, S,T,U,V,W,X	Z,AA,CC,DD,EE, FF,HH,GG
Indiana	A,B,C,D,F	J,K	N,P,T,U,W	CC
Iowa	A,B,C,D,E,F,	H,I,J,K	M,N,O,P,Q,R,S,T, U,V	Z,AA,CC,DD,EE, FF,GG
Kansas		H,J,K	M,P,R,T,U,V,W	CC,GG
Kentucky	B,C	J,K	M,N,Q,P,T,U,V,W	Z,BB,CC,EE
Louisiana	A,B,C,D,E,F	H,J	N,P,Q,T,U,V,W	Z,AA,BB,CC,EE, GG
Maine	A,B,C,D,E,F	H,I,J,K		GG,HH
Maryland			T,U,V	Z,CC,EE,GG
Massachusetts			M,P,Q,T,U	BB,CC,EE,HH
Michigan	A,B,C,D,E,F	J,K	M,N,P,R,T,U,V,W	Z,CC,EE,GG,HH
Minnesota	D,E,F	H,J	M,N,O,P,R,T,U, V,W	AA,CC,EE,GG, HH
Mississippi	A,E,F	H	P,Q,T,U,V,X,W	AA,CC,DD,GG, HH
Missouri	A,B,E,F	J,K	V,T,Q	CC,EE

I. Agriculture-Related Development
- A. Agriculture Export Development
- B. Attracting Value-added Business
- C. Beginning Farmer Programs
- D. Biotechnology & Technology Transfer
- E. Crop Diversification
- F. Marketing Ag/Rural Products
- G. Other

II. Transition Tools
- H. Ag & Rural Development Commissions, Agencies, Etc.
- I. Assessing Competitive Advantages
- J. Farmer & Agribusiness Financial Programs
- K. Farmer Retraining and Counseling
- L. Other

III. Rural Business Assistance
- M. Economic Development, Comprehensive
- N. Entrepreneurship, Business Incubators, Etc.
- O. Job Creation/Training
- P. Location of New Business/Industry
- Q. Marketing and Export
- R. Plant/Military Base Closing
- S. Procurement Assistance
- T. Retention and Expansion of Existing Business
- U. Small Business Assistance
- V. Rural Enterprise Zones
- W. Tax Incentives for Private Investments
- X. Technology Transfer
- Y. Other

Montana	B,C,D,E,F,G	H,J,K,L	M,P,T,U,W	AA,BB,CC,DD, GG,HH
Nebraska	A,B,C,D,E	J,K	M,N,O,P,T,U,W,X	GG,HH
Nevada			M,N,O,P,R,T,U,W	Z,BB,CC,DD,EE, GG
New Hampshire			M,U	DD,EE,GG,HH
New Jersey	A,B,C,F	I,J,K	P,Q,T,U,V,W	CC,GG,HH
New Mexico	A,B,D,E,F		M,P,R,T,U,W,X	Z,AA,CC,EE,GG
New York	D,F	H,I,J,K	M,N,O,P,T,U,V, W,X	AA,BB,CC,GG
North Carolina	D	H	M,P,T,U	CC,DD
North Dakota	A,B,C,D,E,F	H,I,J,K	M,N,P,T,U,V,Y	Z,CC,DD,GG,HH, EE
Ohio	C,F	J	P,R,T,W	CC,DD,GG
Oklahoma			U,V,W	EE,GG
Oregon	D,F	H	N,P,R,T,U,V,W	Z,CC,EE,GG
Pennsylvania			M,N,O,P,R,T,U,V, W,X	
Rhode Island	F		R,T,U	CC,DD,GG
South Carolina	C,F	H,J	M,N,P,U	EE,GG
South Dakota	C	H,J,K	O,U,X	
Tennessee	A,B,C,D,E,F	H,I,J,K	N,P,R,T,U,V	Z,CC,DD,EE,GG
Texas		H	M,T,U	AA,CC
Utah	D,E,F	H,K	Q,T,U,X	Z,AA,BB,EE,GG
Vermont	A,D,F	H	M,N,P,Q,T,U,V,X	AA,CC,GG,HH
Virginia	A,B,C,D,E,F	H,J,K	M,N,P,T,U,V,X	AA,DD,EE,GG, HH
Washington	A,B,D,E,F	J	N,O,P,T,U,W	Z,CC,GG
West Virginia			P,T,U,W	CC
Wisconsin	A,B,D,E,F	H,I,J,K	P,O,R,T,U,W	Z,DD,EE,GG
Wyoming	A,B,C,D,E,F	H,J,K	M,N,O,P,Q,T,U	Z,GG

IV. Rural Community Assistance
- Z. Culture and Arts
- AA. Financial
- BB. Housing
- CC. Infrastructure
- DD. Land Use
- EE. Parks and Recreation
- FF. Quality of Life
- GG. Tourism
- HH. Other

This chart is the result of a 50-state survey conducted by the Center for Agriculture and Rural Development in the fall of 1986. For additional information on the survey or the Center's Rural Development Database, contact the Center for Agriculture and Rural Development, The Council of State Government, P.O. 11910, Iron Works Pike, Lexington, KY 40578, (606) 252-2291.

FIGURE 11.1. State rural development programs. Source: *The Book of the States 1988–89* (1988) Vol. 27 (Lexington, Ky.: Council of State Governments, 1989 edition), p. 434. Copyright 1989 by the Council of State Governments. Reprinted by permission.

ple, the town of Osage, Iowa (population 3,800) established an energy conservation program in 1974. Thanks to a community-wide effort, the town reduced its energy demand and is saving several hundred thousand dollars each year. These savings have helped keep jobs in the town, and the low energy rates have attracted industry to Osage (Daniels & Farley, 1986).

In addition, nearly every state has established an energy department in response to the sharp increase in oil prices in the 1970s. These departments offer technical assistance for devising community energy conservation programs and the development of alternative energy sources.

The planner can serve as a valuable source of information in a community and is in a unique position to coordinate economic development activities in a town. The planner should establish good communication with local business leaders, prospective entrepreneurs, citizens at large, and state and federal funding agencies. Communication and continued commitment to economic development are crucial in recruiting new business and industry from outside the community, as well as in stimulating local business activity.

The planner has two main jobs: development promotion and land use regulation. Both tasks need to be balanced to ensure that a community will continue to be a desirable place to live, work, and grow. This juggling act is often the most difficult aspect of the economic development process. Still, most towns would prefer too much development to too little, especially when the search for economic diversification and business retention requires an investment of time and money with little guarantee of success.

Planning as Community Development; Community Development as Planning

"Community development" is a process through which a community attempts to improve the social, economic, and cultural situation (Christenson, 1982, p. 264). Community development encompasses a broad range of issues, such as the quantity and quality of public services, the impact of migration in and out of a community, and the alleviation of poverty. In the past, rural development has all too often been equated with economic development, and economic progress has been narrowly defined as an increase in local income per capita. Yet the benefits of economic growth may not be evenly distributed throughout a community. In fact, many social problems may persist. As rural sociologist Kenneth Wilkinson has noted:

> Economic development without community development can increase the gap between social classes and reduce the expression of natural human tendencies toward interpersonal warmth, cooperation, tolerance, and respect. Community development as a purposeful activity is needed to realize the potential social well-being benefits of economic development. (Quoted in Christenson, 1982, p. 266)

Community development concerns how people bring about change and how those changes affect them. Although publicly directed change should contribute to the community's well-being, changes always generate social and economic costs and benefits. Therefore, it is important to have an idea of both the size and the distribution of the costs and benefits that change may bring.

Community development strategies seek to organize and stimulate local efforts by improving communications among different interest groups and by promoting partnerships between local government and private citizens. Community development strategies fall into three general categories: "self-help," "technical assistance," and "conflict." Each of these strategies can be used independently or together to achieve community betterment. At a time when federal grants to communities are declining rather than increasing, it is crucial that communities foster a greater ability to solve their own problems. For example, the National Main Street Program (discussed in more detail later in this chapter) has helped establish self-help programs to revitalize over 100 small cities and towns in 16 states. The Main Street approach emphasizes organization, promotion, design, economic restructuring, and quality. These five facets have combined to promote both the economic well-being of a community and the community's morale, self-image, and cohesiveness.

Technical assistance involves the use of trained personnel and financial resources to solve a specific problem. A planner provides technical assistance when giving advice on land use regulations or when seeking state or federal grants to help finance a project. Often local officials judge a planner's community development efforts according to the amount of grant money the planner is able to obtain. Technical assistance may also include bringing in consultants from private firms or from state and federal agencies. But the more a community relies on outside money and expertise, the more the sense of community self-help is reduced.

Certainly the most drastic strategy for community development is the engineering of conflict to achieve change. Conflict may be necessary to challenge the existing power structure of the community and to force a transfer of some decision-making authority to previously powerless people. Conflict is most likely to occur in communities where political and economic power are concentrated in a few hands, such as a company town or recreation-based community. The use of conflict must be handled very cautiously, because it is potentially divisive. Given the goal of promoting community morale and cohesiveness, conflict may actually defeat that goal. Conflict is therefore best viewed as a last resort to instigate change when all other efforts have failed. Examples of conflict include workers' strikes; boycotts of certain establishments; and demonstrations and marches against employers, stores, or elected officials. In such cases, the planner must decide whether to take sides or to try to remain neutral. It is an awkward position, because often the planner's employer, the elected governing body, will oppose change through conflict. Yet the planner may feel that siding with the powerless is worth the risk of losing his or her job.

Success in Community Development

The success of a community development program depends on three related factors: leadership, consensus, and planning. Leadership combines a vision of what a community can be with a willingness and ability to organize human and financial resources into active programs. These programs need not bring great sweeping changes; rather, small visible changes, such as a community garden or a farmers' market, can bring people closer together and improve a town's morale. These results serve to convince the community of the planner's worth and help boost the planner's own confidence.

The wider the public support, the more effective community development efforts will tend to be. A leader must be able to involve a diversity of individuals and interest groups and to encourage a variety of opinions. From this interaction, the leader must orchestrate a consensus on what problems the community faces, on community goals and objectives, on what action can best achieve these goals, and on how to implement the action.

Planning serves to help direct change toward public goals, and presents a comprehensive process for community developers to follow:

1. Gathering of information and data
2. Problem identification
3. Analysis of the problem
4. Development of goals and objectives
5. Identification of alternative solutions
6. Selection of a solution
7. Implementation

8. Enforcement
9. Monitoring and feedback
10. Readjustment of the solution

Planning and community development share the common purpose of discovering what changes the public needs and desires, and of determining which changes are feasible, given constraints of time, money, and personnel. Because all feasible changes may not be achievable at once, changes must be prioritized. Then as programs are implemented, they must be monitored and evaluated in the light of public needs and desires, and perhaps adjusted.

Planning and Awareness in Community Development

Information must first be gathered and analyzed to determine what problems exist. Next, community developers, together with input from the public, should set out general goals along with specific objectives. For example, a town may have the goal of preserving its rural character and the objectives of protecting open space and ensuring that new growth harmonizes with the natural setting. The articulation of goals and objectives creates a policy statement that acts as a guideline for future development. In the present example, a proposal for a 100-unit apartment complex would not be in keeping with the goal of preserving the town's rural character; therefore, the proposal should be denied. On the other hand, the clustering of 25 housing units on 5 acres to preserve 20 acres of open space would probably meet the town's open-space objective.

Several possible solutions to a problem may exist. Solutions may involve a simple approach or a more complex response. Ideas on potential solutions should be solicited from a group. Brainstorming sessions bring out a diversity of ideas and involve people in the planning process. The challenge to a planner is to sift through the potential solutions and recommend the best course of action. This does not mean that the planner should search endlessly for the ideal solution; rather, the planner should take into account limitations of time, money, and personnel and select an adequate solution. Above all, the planner must realize that the solution must be agreed upon by the group. Agreement need not be unanimous, but neither should the solution entrench those not in agreement. The planner should be prepared to have his or her own solution face opposition or even rejection, and should avoid allowing feelings of personal pride and emotion to hinder his or her ability to function in the public interest.

Once a solution has been selected, an implementation strategy must be devised. This strategy should answer the following questions: (1) Who is responsible for taking the action? (2) When will the action be taken? (3) Where will the action occur? and (4) What financial needs must be met? At the same time, there should be a means of enforcing the implementation. Enforcement is necessary to "buy time," to give the action a chance to take effect and to consolidate gains from implementation. Enforcement can occur through moral persuasion, peer pressure, legal sanctions, or fines.

Monitoring and feedback offer information about how well a plan of action is performing. The planner should assume the responsibility of information coordinator and should obtain public and private reactions to the outcome of the plan implementation. Finally, the planner should make recommendations to the planning commission, elected officials, and municipal employees about potential changes in goals, objectives, choice of solution, style of implementation, enforcement, or monitoring.

The planning process helps not only to explain what community development is, but also to induce greater community well-being. A major function of planning and community development is to inform and educate the public as to the benefits of controlled change. Planning and community development must address the issues of the distribution of wealth and power, both within a community and between the rural region and urban areas. In this way, the planner can publicize the balancing of local needs with external constraints. Also, it will become clearer which decisions can be made effectively at the local level and which decisions are better made at the regional, state, or federal level. Although most planning and community development actions are geared toward day-to-day problems, changes may be swift and drastic: The local mill may close, or a shopping mall may be built in a neighboring town, drawing away local business. Although a crisis can be a catalyst for comprehensive planning and community development, public reaction may be too late. Planning and community development must strive to be forward-looking yet sensitive to the community's needs and desires.

In practice, community development will mean something different to each rural community. In the rural–urban fringe, community development will tend to emphasize protecting the traditional rural industries of farming, forestry, and mining, as well as providing adequate public services for growing populations. In rural recreation areas, community development will aim to diversify the local economy and increase the number of year-round residents. In growing rural areas, community development efforts will seek to accommodate the social and economic changes brought about by the influx of newcomers. In stagnant rural areas, community development will attempt to stem the outflow of population and diversify the local economic base.

Planners must also be aware of the forces that have tended to reduce the cohesiveness of communities. Improved transportation networks have enabled people to live in dispersed settlement patterns. In fact, most newcomers to rural areas have chosen to settle outside of established population centers. Many rural people commute farther to work than do urban dwellers. Wider access to the media, especially television, has increasingly become a substitute for personal interaction. The popularity of outdoor recreation often results in people leaving their local area for day trips or weekends. All of these forces direct individual energy and attention away from the community. Planners must devise ways to overcome public apathy, elicit public interest, and involve people in community development activities.

Organizing Human Resources: The Planner as Community Developer and Animator

The first step in putting together a community development program is to get to know the people in the community and the local physical and economic features. Conversation and personal observation are two valuable yet inexpensive ways of gathering information. Back issues of the local newspaper are a good source of insight into the local power structure and local concerns. Books on local history and census data can suggest community values, trends, and needs.

A more systematic means of learning about a community is to conduct a needs assessment survey. The survey fulfills three important roles. First, local residents are asked for their ideas, and their involvement in the community development process is encouraged. Second, the survey provides data on what local residents perceive as the major needs in the town. This identification of problems forms the basis of planning and community development action. Third, the survey data indicate issues of consensus and disagreement among local residents. This may help the planner to address problems that will not cause friction within the community and to avoid, at least initially, problems that are divisive. Group action is often more effective than individual action, and it is easier to promote group action on issues of common agreement.

Needs assessment surveys should cover a wide spectrum of topics, such as schools, health services, sanitation, community centers, police and fire protection, new businesses, transportation, aesthetics, and housing. Surveys can be designed and conducted in several different ways. Conducting a survey is an art; if it is improperly done, it can do more damage than good. The planner must use extreme care in designing and structuring survey questions. Questionnaires can be sent to local residents. Telephone interviews can be conducted, or personal interviews can be conducted on the street. A variation of the personal interview is to survey just the community leaders. The main benefit of personal communication is that it establishes a basis of trust; face-to-face interviews tend to make a planner visible, as opposed to the stereotyped "faceless bureaucrat" tied to an office desk.

In putting together a plan of action, the planner should determine which problems can be solved quickly and effectively. This enables a planner to show that he or she can get things done; such a record of achievement will increase public confidence and morale. These problems are likely to be small and simple, but they provide a very important foundation for tackling larger, more complex problems. Furthermore, in the process of solving a simple problem, a planner will often develop contacts for technical and financial assistance at the regional, state, or federal level.

It is likely that a town will not be able to solve all the problems that its citizens identify. Towns typically face major limitations in time, money, and personnel. Thus there is a need to establish priorities among needs. Two additional constraints are often severe, particularly in towns of less than 5,000 inhabitants. First, towns may lack a "critical mass" of people necessary to support new businesses. Small towns tend to provide day-to-day kinds of goods and services, such as groceries and gas; to purchase specialty goods and services, such as washing machines and advanced health care, townspeople must travel to a regional center. This lack of critical mass is often a major barrier to local economic diversification and growth. Second, towns are often at a disadvantage in providing public services because of a lack of economies of scale. In urban areas, where there are more people to absorb the cost of a new sewer or water line, the cost per person tends to be fairly low. In rural areas, there are fewer people; consequently, the cost per person tends to be high. For example, in Des Moines, Iowa, residential water rates are about $7 per 5,000 gallons of water, but in rural Iowa, water rates are generally greater than $20 per 5,000 gallons and range as high as $60.

Realizing these limitations, a planner must be able to discern which community needs can be met with limited resources. One way of determining the feasibility of a possible solution is to conduct a market survey. Through this process, a planner can identify a potential market area and assess the demand for retail, commercial, or housing development. If a project appears feasible, a planner can put together a proposal to present to a developer. In essence, the planner is marketing his or her community to developers and indicating that the community will be a good client and a good partner to work with. In the past, towns have waited for developers to arrive, often without results. But when developers appeared, they would tend to present their project proposals on a take-it-or-leave-it basis. Towns then responded by either accepting a proposal or else turning it down if it did not meet local desires. Still, the style was one of confrontation rather than of negotiation and mutual interest in the community.

Today, the planner must be involved in actively seeking new business and development projects, and must be able to act as a negotiator for the community's interests. The planner must look to create public–private partnerships that both meet the community's social and economic needs and fulfill the desire of developers and entrepreneurs to make profitable investments. Local governments will attract more development if they show a willingness to share in the risks of a project. Municipal governments, together with local lending agencies, can help provide needed financing for a project. Local government may also assist a developer by providing a construction site, changing zoning ordinances, offering a partial or total property tax abatement, or securing the necessary permits. The developer, on the other hand, must be prepared to make compromises on the size of the project, the mix of uses, the location, and project design.

Public–private partnerships can help organize public resources and create a positive attitude in the community. The more active a community becomes, the more it will attract attention as a desirable place to live and work; also, the more economic activity will be generated through the "multiplier effect" as the dollars from new jobs circulate through the community, creating additional employment. The planner can play a vital role in community development efforts and public–private partnerships by serving as a coordinator of information and human resources and as a manager of local interests. At the same time, the planner should work to preserve those features that make a small rural community a desirable place to live. This includes maintaining physical appearances, social interaction, environmental quality, and economic health.

Consensus and Conflict Management

The role of the planner is often seen as one of providing technical advice to decision makers. In community development, the planner assumes a more active position, initiating change rather than merely responding to development proposals or crisis situations. In bringing about planned change, the planner relies upon the approval of elected officials and the local planning commission. In building a consensus of opinion on community development efforts, much depends upon the planner's style and ability in presenting ideas. Many rural communities resist change and are slow to accept new ideas and new programs involving increased local budgets. The building of consensus is a three-step process: information gathering, learning, and decision making. A needs assessment survey, together with other data sources, can provide insight into community well-being and suggest projects with potential community support.

The planner can help to interpret the information and suggest a variety of policies and projects to decision makers. The planner's advice will to some degree be influenced by what he or she believes will promote "the good community." Still, the important task is to present choices and their consequences, rather than to appear as one who has all the solutions. The goal is to form agreement on the community's problems and on potential solutions.

Conflicts may arise for several reasons. First, an issue the planner presents may not be perceived as a need by the decision makers, nor may that issue be adequately supported by available data. For example, a planner may argue for a municipal water system in a town where wells and water quality are satisfactory. Second, an issue may not be perceived as being in the public interest. For example, decision makers may feel that conservation easements are not appropriate for inducing a large and wealthy landowner to refrain from developing his or her land. Third, a planner may advocate change without having an adequate knowledge of local needs and values. In this case, a needs assessment survey would increase a planner's understanding of local wishes. Also, the greater the scope of the change proposed by the planner, the more conflict is likely to arise. Often, large changes both are more difficult to understand and present more risks. Fourth, planners must avoid presenting untested, "ideal" solutions. Fifth, some townspeople may feel threatened by change, especially change involving large projects. Sixth, matters related to economic and community change can be perceived as creating "winners" and "losers" or as upsetting the existing social structure. A planner should be aware of the existing distribution of wealth and power, both within a community and between the community and outside forces (see Chapter 5). Finally, a planner should ascertain the sources and magnitude of opposition to potential changes before presenting definite proposals.

Local attitudes toward economic and social change can define five general types of communities (Blakely, 1983):

1. Entrepreneurial communities
2. Analytical communities
3. Defender communities
4. Destroyer communities
5. Desperate communities

Entrepreneurial communities have strong local public and private leadership; ample financial resources; and an aggressive program of seeking new business, industry, and community development projects. By contrast, an analyzer community is cautious about change and may have only limited resources to promote community development. Often the analyzer community suffers from "paralysis by analysis" as local leaders and officials study development proposals but are reluctant to commit themselves. Such caution may or may not ultimately benefit the community. A defender community features a leadership that is generally opposed to change. Defender communities are often desirable places to live and work and community leaders feel that they can choose what sort of change they want. In a destroyer community, the local leadership actively strives to block economic and social change. A desperate community has recently been the victim of a natural disaster or a severe economic setback, such as the closing of a local factory. Such communities often have few resources to attract new businesses or initiate community development projects. However, the crisis situation provides a forceful push for community organization and activity. Shifts in local attitude and development effort can alter a town's character. For example, a defender community may become a desperate community if a natural or economic disaster strikes.

A planner can learn much about a community from polling local leaders and residents about which of the five attitudes toward change is dominant in their community. Such a poll, coupled with a needs assessment survey, can provide insight into potential planning actions or possible pitfalls to community development efforts.

A planner should follow some general guidelines to avoid or minimize conflict and to promote a cooperative spirit. A good first step is to form a citizens' group that represents a wide diversity of interests and organizations. Next, the results of a needs assessment survey and other information can be discussed and interpreted by the group. Brainstorming should be encouraged, because the more people participate, the more they will feel useful. The planner must be able to direct the brainstorming sessions toward agreement on community problems and potential solutions. Group decision making gives clout to decisions and raises feelings of trust and purpose.

The planner functions as a moderator, an information source, an educator, and a cumulative group memory. In addition, the planner must carefully control the use of outside experts, since community development efforts should emphasize local decisions and self-help. Finally, it is virtually impossible that everyone will agree on a community's problems, prospects, and development programs. As noted earlier, a planner must take care not to entrench those in opposition. Changing personal attitudes, values, and behavior is at best a slow process. The most certain way to convince those in opposition is through successful activities and programs. A planner must be sensitive to the concerns of those not in agreement and should look for ways to accommodate conflicting opinions. Yet the planner should not allow minority opposing views to cancel the consensus of the majority of the community. This may be difficult when a local power elite is opposed to change. Nonetheless, a planner should continually strive to educate people that unless a community attempts to control change, uncontrollable forces will tend to control the town.

The National Main Street Program: A Case Study

Between the end of World War II and the mid-1970s, the downtown areas of many towns and small cities suffered major declines in economic health and population. Improved highway networks and changes in personal preferences caused many people to abandon living in downtown areas, and the construction of suburban and regional shopping centers attracted shoppers away from deteriorating downtown environments. Some towns tried promising remedies, such as pedestrian malls or parking garages, only to see their momentary benefits fade into further decline. Other towns spent thousands of dollars on volumes of planning studies that resulted in no action.

Downtowns have long symbolized the identity of communities. Failures to revitalize downtowns have sapped the energy and spirit of elected officials, business leaders, planners, and town residents. By contrast, a healthy downtown can help to recruit new businesses and jobs and can promote civic pride.

In 1977, the National Main Street Program was founded under the auspices of the National Trust for Historic Preservation to provide technical assistance and training in how to revitalize downtowns in small cities of 5,000 to 50,000 people. The Main Street approach emphasizes creating a positive image of downtown—not just fashioning an attractive physical appearance, but also having townspeople work together to improve and maintain the economic and cultural qualities of downtown. Although money is an important ingredient of successful revitalization, community cooperation and spirit are crucial.

Four elements comprise the total image of downtown: organization, promotion, design, and economic restructuring. A key feature of the Main Street formula is the hiring of a full-time project manager. This person is responsible for organizing the separate groups in the downtown area: merchants, bankers, civic groups, local government officials, the media, and individual citizens. The project manager is an information source and a promoter of events, and serves to provide a more centralized style of management for downtowns. Funds for the project manager's position may be made up of private money, public money, or a combination of the two.

The project manager can initiate and coordinate a variety of promotional activities to market downtown as a place to shop, work, and live. Promotional efforts should be aimed at providing greater visibility and awareness of downtown resources. Activities may vary from distributing shopping bags with a special downtown logo to publishing a directory of downtown businesses or sponsoring special events such as craft fairs, farmers' markets, and sidewalk sales.

High-quality design and visual appearance are essential ingredients in downtown revitalization. Design includes a wide range of physical features from window displays to the size and positioning of signs, building facades, landscaping, and public spaces. Some towns have enacted design standards to remove eyesores and to ensure that new construction will blend with existing buildings. But townspeople should avoid phony themes or gimmicks that give a false impression.

Until the 1970s, old buildings were often considered out of style, and many handsome antique facades were covered with modern materials. (For example, one town in Kansas covered its buildings with shiny aluminum!) Today, generous income tax credits exist to promote the rehabilitation of commercial buildings more than 30 years old, especially for buildings listed in the National Register of Historic Places. Many investors who are seeking the tax shelter afforded by historic buildings can be attracted to Main Street if they are made aware of the opportunities there. Furthermore, a town with a significant collection of historic buildings can be appealing to tourists. Above all, an attractive downtown area reflects and stimulates civic pride and advertises a town as a pleasant place to live and work.

America's economy is moving more toward service industries and away from manufacturing. Towns waiting for factories to arrive may be disappointed. This trend toward deindustrialization is forcing communities to look to other sources of economic growth. Small businesses are now the primary generators of new jobs, not large corporations. New retail stores in rural areas can lend a significant boost to the local economy, but competition is likely to be stiff among communities.

The Main Street approach strives to diversify the downtown economy by recruiting new stores for a balanced retail mix, converting unused space into apartments or offices, and improving the competitiveness of downtown merchants. For example, many older buildings along Main Street have three stories; the first floor is a storefront, but the upper stories are often unused. Developing vacant spaces for housing and offices can both improve the appearance of the downtown and increase the number of people who live, work, and shop there.

Revitalization is an incremental process, with many small improvements occurring over a long period of time. In most towns, major results are visible after about three years. Redevelopment often happens in stages, building by building, block by block. This process not only is less expensive than a large one-shot project, but tends to involve a greater number and diversity of people over a longer period of time. Local leadership can be strengthened, along with a commitment to the downtown as a place to work, live, and shop. A major implementation strategy of the Main Street approach is the creation of public–private partnerships. Planners and project managers need to know how to be good clients for developers and how to present proposals that will appeal to developers. After performing a needs assessment survey, planners must be able to perform marketing studies to determine which of the retail, commercial, and industrial needs can be met by private developers. In attracting developers, planners can identify sources of public funding; they can also involve local lending institutions and perhaps induce the town to offer incentives such as low-interest loans or property tax breaks. The public must be willing to take more risks and to negotiate with developers. Too often in the past, the planner–developer relationship has been based on confrontation, rather than exploration of mutual interests with the aim of reaching acceptable compromises.

The results of the National Main Street Program have been impressive. Currently, more than 200 communities in 16 states are part of the Main Street network. Local rehabilitation and investment have been strong. There have been over 650 facade renovations and over 600 rehabilitation projects in Main Street communities. Over 1,000 new businesses have been started, compared with only 500 business failures. Between 1981 and 1984, some $130 million were invested in just 30 small towns, and the National Main Street Program estimates that each public dollar spent in a rejuvenation program has resulted in $11 of private investment. For example, in McKinney, Texas (population 16,000), there have been 26 renovation and rehabilitation projects totaling $1.4 million, 34 business openings, and just 13 closings or relocations, resulting in a net gain of 33 jobs. But, perhaps more important, there has been significant improvement in the cohesiveness and cooperative spirit of communities. These towns have become better places to live. Some states, most notably Oregon, have established statewide nonprofit organizations to provide technical and financial assistance to local downtowns.

The Main Street approach has not been successful in every town. The program has worked well in only about three-quarters of the communities, and failures may be attributed to several factors. When a local Main Street program is initiated, about 20 percent of the townspeople are in favor of it, 20 percent are opposed to it, and 60 percent are undecided. Local leadership must be strong and committed to implementing the four main elements of the program: organization, promotion, design, and economic restructuring. The hiring of a full-time project manager is crucial for the management of the downtown area. Some communities have been reluctant to hire a project manager, and some merchants and elected officials have resisted giving the project manager much authority. Without a full-time project manager, few people are likely to become involved in redevelopment efforts, and promotional efforts are likely to be haphazard. Some communities have neglected to invest in improving their physical appearance, or projects have been developed without adequate attention to quality, opportunities to attract tourism, and the need to create a pleasant new atmosphere. There is a tendency among rural communities to emphasize attracting new business and achieving economic diversification. But unless the other three elements of the program are enacted, economic diversification will be very much a hit-or-miss occurrence.

Small-Town Triage and the Regional Planner[2]

In devising economic development and community development programs, the planner must be realistic about what small towns can hope to do. Local resources in terms of capital, labor skills, consumer dollars, and infrastructure may be severely limited—especially in towns of fewer than 2,500 inhabitants. At a time when federal expenditures for local economic development are declining, many small towns will be hard pressed to avert economic and social stagnation. Competition among hundreds of towns for relatively few new firms is likely to be an expensive and, indeed, a wasteful process. Even so, larger towns are more likely to win out in the long run.

Most planners will work with small towns from a county or regional planning office. In nonmetropolitan areas, planners may work with four different kinds of towns, defined according to the degree of economic reliance and trends in the local population and economy (see Table 11.1). Economically independent towns have more or less self-sustaining economies, as in the case of growing towns, or lack exposure to economic

[2]This section draws heavily from "Small Town Triage: A Rural Settlement Policy for the American Midwest" by T. L. Daniels and M. Lapping, 1987, *Journal of Rural Studies*, *3*, 273–280. Copyright 1987 by Pergamon Journals Ltd. Used by permission.

TABLE 11.1. Small-Town Typology

	Population and economic trends	
Economic reliance	Growing	Stagnant
Independent	Economically diversified communities	Remote, declining communities
Dependent	Commuter communities	Natural-resource-based communities

Note. From "Small Town Triage: A Rural Settlement Policy for the American Midwest" by T. L. Daniels and M. B. Lapping, 1987, *Journal of Rural Studies, 3,* 273–280. Copyright 1987 by Pergamon Journals Ltd. Used by permission.

activity, as in the case of remote, stagnant towns. Economically dependent towns either rely on regional centers for employment and services, or depend on distant markets to sell their raw natural resources.

Growing towns typically feature a diversified economic base and a population of more than 2,500 people; such a town is often the county seat, which is the traditional political, legal, and retail center of the county, or a commuting town within reach of a regional center. Stagnant towns have usually experienced a loss of population over the past five to ten years, along with negative or negligible economic growth. These towns commonly have fewer than 2,500 inhabitants, are remotely situated, and are heavily dependent on a single land-based industry, such as agriculture or energy extraction. These towns rely especially on the importation of goods and services, and because the price of locally produced commodities has fallen in recent years, the purchasing power of these townspeople has also declined.

Towns of fewer than 2,500 people generally are able to support only businesses offering day-to-day goods and services. Stagnant towns tend to show a large leakage of retail trade dollars to regional centers, where consumers can find a greater variety of goods and services, often at lower prices. In addition, both the quantity and quality of public services in stagnant towns are frequently low; local residents are either reluctant or financially unable to raise property taxes to defray the costs of maintaining or upgrading services necessary to attract new businesses and industry. Thus, stagnant towns seeking to revitalize themselves must struggle against a vicious circle: They do not have enough people to attract new retail and service establishments, nor do they have the financial capacity to construct the infrastructure thought necessary to attract manufacturing firms.

The challenge to regional planners in nonmetropolitan areas is how to influence the allocation of public funds and resources to create a sustainable pattern of rural settlement. Much of the rural United States was first settled in the horse-powered era of the 18th and 19th centuries. Small towns have long shown a remarkable resiliency, but ghost towns are not unfamiliar in the American landscape.

Under a policy of small-town triage, planners would seek to provide public funds to the growing towns of 2,500 to 5,000 (see Table 11.1), which could achieve sustained economic growth with some financial assistance. Second priority would be given to larger growing towns of 5,000 to 15,000, which have the ability to generate new economic activity, but may require additional state and federal funds to help build new and expanded infrastructure. Stagnant towns without much prospect for rejuvenation receive the lowest priority. In essence, the triage strategy seeks to promote growth centers, which can provide appropriate services at a reasonable cost and employment opportunities to the surrounding hinterlands. A secondary goal is to slow down regional out-migration.

State highway, education, and economic development programs have a major influence on small towns. Roads are the lifeblood of rural areas, and many small towns have survived because of road systems that make them accessible. But the maintenance of the interstate highway system has largely fallen on the state and has a high priority; county local roads and bridges are receiving a low priority. The state-mandated consolidation of school systems has tended to favor larger towns at the expense of smaller towns, but the result is often a lower cost per pupil. The leading state economic development program is the allocation of community development block grant monies. These funds are federal grants, which each state then distributes to communities on a competitive basis. But few small towns, especially stagnant towns, have the expertise or organization to put together competitive grant applications.

A common theme in state, regional, and county planning agencies seems to be that each town has an equal right to compete for growth and development grants. This democratic principle works fairly well in robust economic times, but is likely to result in a thin and ineffective dispersal of public resources in a depressed economy. The triage strategy aims to make public spending achieve economies of scale in the provision of public services and build up population centers that can be self-sustaining in the long run.

Economic Development: A Summary

Where people live and work influences the production of goods and services, the cost of providing public services, and the ability of a community or region to sustain or increase its population. People live where they can earn a living, or at least can survive. People are usually attracted to a place by other factors besides jobs, but the ability to earn a living is key.

The nurturing of economic development—increasing local incomes, employment, and access to services—is an essential task of the planner in more remote rural areas. The challenge to these mainly natural resource-based communities is how to diversify the local economy to provide adequate employment opportunities and reverse out-migration, especially of young people. At the same time, the planner must try to help existing businesses expand or at least remain in the community.

Community organization and promotion are two important elements in economic development that must be maintained over the long term. Economic development is a long-term process, not a quick fix. America's rural areas and small towns have become part of a global economic system. Competition is often very stiff, and a community's ability and willingness to innovate may decide its growth or decline.

Planners should strive to achieve public–private partnerships to foster economic development. Although much of the planner's work has traditionally involved land use regulation, such an emphasis seems of secondary importance to towns and regions that are struggling to survive. Applying for a variety of infrastructure and economic development grants and helping to organize joint public–private development committees and loan funds are activities that will make a planner useful and needed.

In the rural–urban fringe, ability to earn a living is less of a problem. Most inhabitants have the choice of working in the suburbs or in the inner city at good-paying jobs. Here, the challenge to the planner is how to maintain the environmental quality of an area and the natural-resource-based industries. The directing of growth away from commercial farming, forestry, and mining areas is an important activity.

Planners should recognize that some rural areas and small towns have very limited potential for economic growth. Thes areas are usually sparsely settled and have rather few natural assets.

Economic development is not a panacea and in many cases may not be possible. It is a major part of a community's planning efforts, but a community must not emphasize economic development to the detriment of the community's appearance, health, and safety, or of community facilities (schools, parks, streets, etc.) that have much to do with the local quality of life. The planner should always strive to be comprehensive, explaining the costs and benefits of economic development proposals, so that decision makers can make informed choices about the future of their community.

CHAPTER TWELVE

Special Populations: The Poor, the Elderly, and Native People

The Poor

In America, rural poverty has two faces: isolation and underdevelopment. A number of U.S. regions remain uncompromisingly rural and isolated from the mainstream of American life. Examples of such regions are parts of the intermountain West, Appalachia, the interior of Alaska, the Mississippi delta, and the Southwest–Mexican borderlands. The other face of rural poverty is underdevelopment. The greater the lack of regional development, the poorer the area is likely to be. Underdevelopment is not the result of any one deficiency, or even the lack of people. Though the list is somewhat arbitrary, the interrelated causes of underdevelopment include, among other factors, high proportions of subsistence-level jobs, a lack of infrastructure, aging industries, heavy outside control of banking and industry, poorly developed social services, and a historic poverty of the spirit.

All rural planners must deal with the ubiquitous nature of rural poverty. Though perhaps less visible than urban poverty, the problems of the rural poor are pervasive, vicious, and of long duration. A 1975 presidential commission noted in its report that "rural poverty is so widespread, and so acute, as to be a national disgrace" (President's National Advisory Commission on Rural Poverty, 1978). Little has changed in this regard. Sometimes the problems of the rural poor have been complicated by the nation's history of race and ethnic relations; sometimes they reflect specific issues of age and gender (the "feminization" of American poverty). In still other cases, a long-standing absence of economic opportunity is at the root of rural poverty.

According to U.S. Bureau of the Census data, over 18 percent of all rural Americans live at or below the official poverty level (O'Hare, 1988). This rate is 50 percent greater than in urban areas. The gap between urban and rural incomes, one important indicator of relative wealth and poverty, continues to grow. Between 1979 and 1986, real incomes for rural Americans actually declined by 10 percent, and the difference between urban and rural incomes was over 25 percent. In 1979, there were 242 "persistent poverty counties," which had per capita incomes in the lowest 20 percent of all U.S. counties from 1950 to 1979 (Bender et al., 1985) (see Figure 12.1). Chronic rural poverty areas tend to show a sparse settlement pattern, long-standing low income, and a disproportionate level of disabilities that affect employment participation.

Although rural poverty can be found in all regions and states, and among all age and ethnic groups, some places and some people seem particularly hard hit. Persistent poverty counties are mainly found in the South, where a far greater proportion of the rural population lives in chronic poverty than in any other region of the country. Many of the nation's poorest states are also located in this region, and poor states find it increasingly difficult to aid poor rural areas. Greater federal involvement may prove necessary, but recent cuts in federal programs are not encouraging. Likewise, rural blacks, Hispanics, and Native Americans are more likely to be among the poor than are rural whites, though whites comprise fully 75 percent of those living in rural poverty. Also, young families, households headed by single parents (especially women), the elderly, and children are disproportionately represented among the rural American poor. For example, one recent study noted that "as of 1986, a quarter of all children in rural America were living in poverty" (O'Hare, 1988, p. 2).[1] The picture of the rural poor is a disquieting one that rural planners cannot ignore.

Planning strategies to alleviate poverty must be specific and especially sensitive, though, in reality, a local planner's effectiveness is often limited. First, many of the industries upon which rural America has historically depended are in stagnation or decline. Extractive industries, such as mining, timber harvesting, and energy production, are being buffeted by international shifts in supply and demand. Because so many rural areas are dependent upon one industry or one firm, the narrow economic base of such towns often leads to significant levels of unemployment during the "bust" periods in the boom–bust cycles all too familiar to rural people. Even in those counties where agriculture and agribusiness are dominant activities, growth in the number of jobs has

[1]Much of the statistical material in this chapter is derived from O'Hare (1988) and from Rodgers and Weiher (1986).

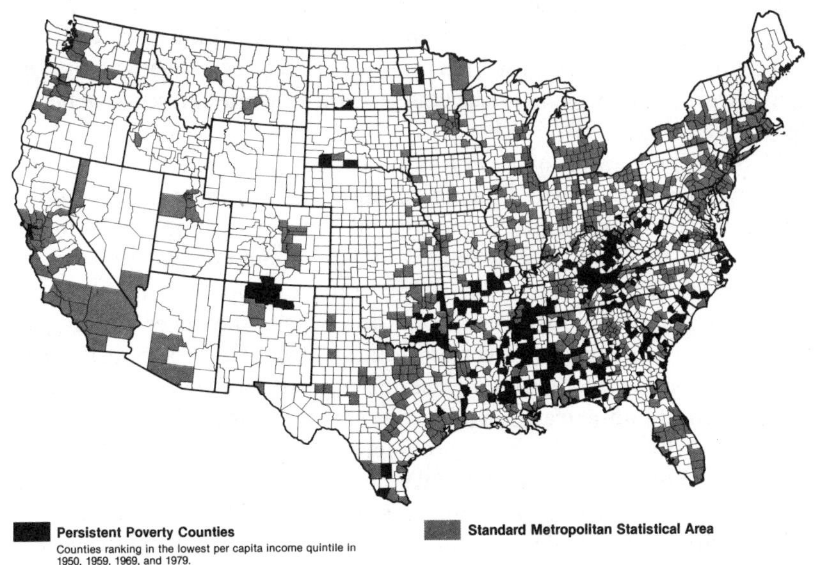

FIGURE 12.1. Persistent poverty counties. From Bender, L. D., et al., 1985, *The diverse social and economic structure of nonmetropolitan America* (Rural Development Research Report No. 49) (p. 14). Washington, DC: U.S. Department of Agriculture, Economic Research Service.

lagged considerably behind that of urban areas. Indeed, according to the U.S. Department of Agriculture, agricultural employment actually dropped by 200,000 jobs in the period 1983–1986. In those rural counties where manufacturing was the major economic activity, total employment also fell in the same period (Beale, 1988). This has further fueled migration from rural areas to cities, and often people with better educations and skills are among the first to leave for metropolitan-based opportunities. Those remaining in the rural areas tend to have minimal job skills and lower levels of educational attainment.

Second, many of the rural poor are among the ever-expanding number of Americans who comprise the "working poor" (Levitan & Shapiro, 1987). They may have employment, usually in the service sector, but their incomes are extremely modest and their package of fringe benefits (e.g., health insurance) is substantially thinner than for urban workers. The lack of generally available employment opportunities is exacerbated by lower wage levels for those who do find jobs. Furthermore, many industries attracted into rural areas are precisely those that find lower wage scales and smaller fringe benefit packages genuine incentives. The lack of unionization in many rural-based industries tends to further diminish pressure for higher wages and better fringe benefits. As many rural Americans can attest, it is possible to work and still live below the poverty level.

Third, the rural poor have limited access to services and programs that have been developed to assist them. Small, isolated rural populations often lack sufficient "economies of scale" to support a well-balanced social service support system for the unemployed and working poor. Unlike urban areas, where public transportation has increased the accessibility of the poor to services, and where the sheer number of people helps to guarantee a certain minimum degree of support, many rural areas still suffer from a lack of accessibility and a paucity of services. Rural public transportation, in particular, is inadequate; several bus routes, for example, have been terminated as part of the federal deregulation of private bus companies. As a result, existing services must be carefully coordinated—perhaps on a regional level—to avoid duplication, maximize efficiencies, and fill what voids exist in service provision.

But access is only part of the problem. Indeed, the "safety net" to support the rural poor has also deteriorated. Food stamp programs, Medicaid, Aid to Families with Dependent Children, nutrition programs, and health care delivery programs have been reduced or eliminated altogether. The result has been a significant worsening in the physical health and well-being of the rural poor. As one study has noted, the rural poor are "more likely to suffer from hunger, malnutrition, health problems and learning difficulties" (Rodgers & Weiher, 1986, p. 285).

Janet Fitchen (1981), who has conducted extensive studies of rural poverty and marginality in New York State and elsewhere, has argued that a particular set of problems explains the persistence of rural poverty over the generations. These include the following:

- The impact of history and how it shapes the future.
- Inadequacies of the social structure to permit movement across class lines and upward mobility.
- The existence of "corrosive stereotypes" that further erode self-esteem.
- The cumulative nature of being poor ("too many problems at the same time").
- An individual and collective syndrome of failure, which reinforces psychosocial deficits originating in the earliest experiences of childhood.
- The "closing-in of horizons": "[A]s an individual experiences years of failure, frustration, struggle, and disappointment that are inher-

ent in poverty and marginality, his [or her] horizons close in and he [or she] becomes locked into a world of limited hopes and bounded environments" (Fitchen, 1981, p. 199).

Clearly, the rural planner cannot attack all of these issues. Moreover, some solutions to these problems do not lie at the level or scale at which the typical rural planner operates. But the rural planner must understand these matters and should attempt, wherever possible, to seek the amelioration of these realities. How the persistence of poverty affects two special populations in rural America will be the focus of the remainder of this chapter.

The Elderly

An ever-increasing number of those living in rural areas are elderly. In many nonmetropolitan counties, 15–20 percent of the population is over 65 years of age, compared to the national average for all counties of 12 percent (see Figure 12.2). Though the rural elderly are not a homogeneous group, certain common characteristics distinguish them. Generally speaking, older rural Americans tend to have lower incomes, to have fewer and less accessible services available, to live in less adequate housing, and to suffer poor health more frequently than their urban counterparts.

It would appear that few issues are as important to the rural elderly as the maintenance of an independent lifestyle. Living arrangements for the rural elderly may vary along a continuum from total independence (living in one's own home and being entirely mobile) to total care (living in a nursing facility or with one's family, having little or no personal mobility, and being totally dependent). The latter condition, often occasioned by deteriorating health and/or falling economic status, can be the cause of great anxiety, since it clashes with the time-honored rural value ideal of rugged individualism and independence.

The vast majority of the rural elderly own their own homes, the predominant type being single-family. A large percentage of these housing units contain one or more serious deficiencies, according to the U.S. Department of Housing and Urban Development (1980). Indeed, such deficiencies can be found among the rural elderly at a rate twice that for the elderly in urban settings. There is often a reluctance to attend to such deficiencies because of low incomes or the expectation that one may not live very much longer. Planners can enhance the quality of the home environment by initiating and managing programs to address two areas of deficiency of greatest concern to the rural elderly: space conditioning

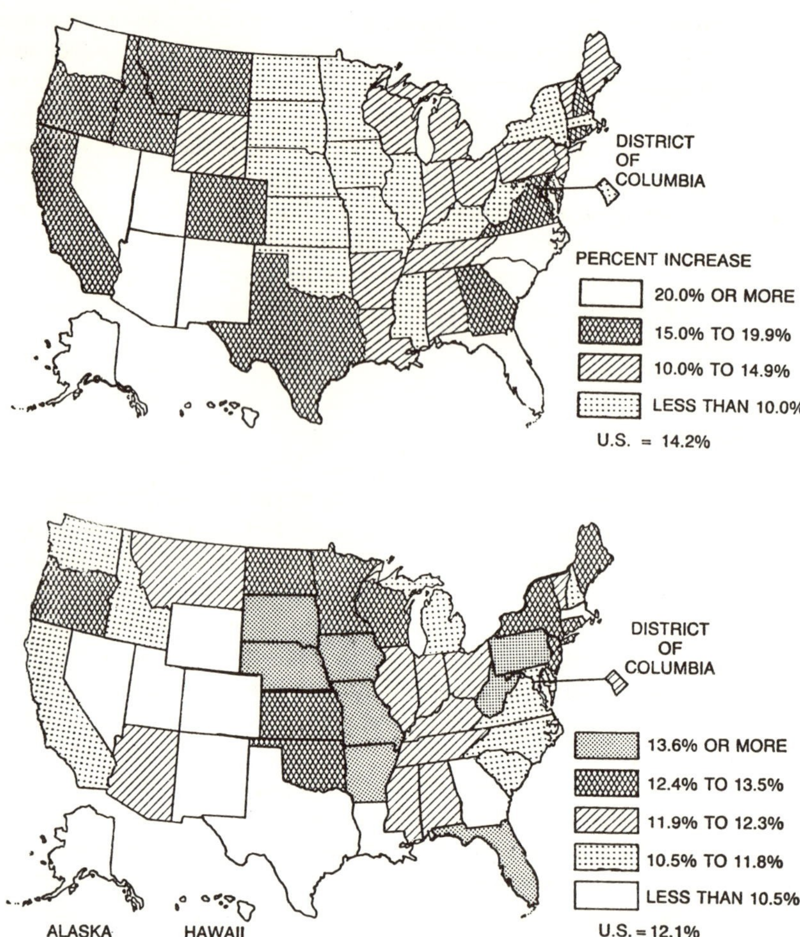

FIGURE 12.2. Percent increase in population of person 65 years of age and older, 1980–1986 (top), and persons 65 and older as percentage of total population, 1986 (bottom). From "State Population and Household Estimates with Age, Sex, and Components of Change: 1981 to 1986" by U.S. Bureau of the Census, September 1987, *Current Population Reports*, Series 1-25, No. 1010.

(i.e., heating/cooling) and lighting. In his work on this problem, Lawton (1980) has noted that the presence of central heating is among the most important variables in housing satisfaction. Windley and Scheidt (1983) have extended this to include the provision of adequate heating in winter, adequate cooling in the summer, and proper natural and artificial lighting that meets the changing needs of the elderly. Windley and Scheidt's work further suggests that satisfaction with one's dwelling environment is one of the most critical elements in the maintenance of the mental health of the rural elderly. This appears to be a pervasive theme throughout rural America and may help to explain why the rural elderly tend to be long-term residents of their homes. In a social context that prizes the maintenance of roots within the community, this is surely a major asset. Increased satisfaction can, then, result from some timely and modest interventions in the areas of home weatherization, energy efficiency, and lighting.

Although the rural elderly spend a great deal of their lives in and around their home communities, they will make occasional trips to larger towns to secure services not available locally. Among the most important will be visits to medical facilities and allied services (e.g., pharmacies). At the very time that the elderly are living longer, the availability of medical support services to maintain and enhance the quality of life has drastically decreased in much of rural America. This is not the result of a shortage of physicians, for example. Rather, it is a function of the unequal spatial distribution of practitioners (Lucas & Himelfarb, 1977). Health personnel in rural areas tend to be "older, work longer hours, earn less money, and are more likely to be [in] solo practices than their urban counterparts"; the urban practitioner, on the other hand, "earns more and works less" and "has increased access to professional peer relationships, coverage arrangements, and continued medical education" (Rosenblatt & Moscovice, 1978, p. 755). These are all vulnerabilities within the rural health care system that planning strategies must seek to address.

If the plight of rural health practitioners is a difficult one, then too the existence of rural and small-town hospitals is in jeopardy. Increased regulation has created a complex and costly environment in which such hospitals must exist. Rural hospitals receive lower reimbursement under Medicare and Medicaid than urban hospitals for similar services performed, for example. Many hospitals are undercapitalized, and this has made it very difficult to support modern facilities and equipment. This leads to a situation in which the hospitals cannot provide the range of sophisticated services that patients require; this, in turn, forces patients away from local institutions to ones that can meet their needs. The underutilization of bed space creates a drain on a hospital's financial health. In many cases, rural hospitals receive subsidies from the counties they serve, but these grants are becoming increasingly inadequate. Practitioners without ready access to hospitals are among the very first to leave rural areas. The lack of a hospital creates a genuine economic impediment to a small-town practice and further inhibits the type of health care most physicians would like to make available. The absence of appropriate and consistent health care services works particular hardships on the rural elderly, who are forced to range ever farther from their home communities for the services and health specialties they require. Moreover, their situation may be made all the more difficult by relatively low or fixed personal incomes, poor physical mobility, and a fear that confinement in a medical facility may not be a temporary situation.

It has been suggested that any planning strategy for rural health care ought to aim at the creation of group practices that include not only traditional health care providers, but also paraprofessionals and skilled ambulance teams capable of providing critical emergency aid. Furthermore, these services should be made available through intercommunity cooperation. Such a program would avoid duplication and maximize efficiencies, and would speak to practitioners' legitimate needs for peer support and mutual learning. In terms of the specific needs of the rural elderly, visiting nurse programs, congregate meal and day care centers, temporary housekeeping and homemaker services, and transportation to and from health facilities can be woven into the fabric of a well-planned regionalized rural health care system. Such approaches must stress nutritional, social, and mental health objectives as surely as they must meet health care needs if they are to address the true needs and concerns of the rural elderly.

The planner should also view the elderly as a potential source of economic and community development. The elderly bring important transfer payments (Social Security and pensions) that help support the local economy. Some rural communities have prospered as places of retirement. In fact, there were some 515 nonmetropolitan counties in 1980 with significant retirement populations (Bender et al., 1985) (see Figure 12.3). Retirement counties had a net in-migration of people aged 60 or over from 1970 to 1980, equal to 15 percent or more of that age group in the country. Retirement counties are found in all areas of the nation, but are especially concentrated in the Southwest, the Ozark Mountain region, Florida, the upper Midwest, and the Pacific Coast. Retirement counties feature high per capita transfer payments, and destination retirement complexes have brought some economic development to these counties. And many communities have benefited through the participation of the elderly in government and community organizations. Rural New England, for example, has seen a significant increase in the expertise base available to local communities as growing numbers of the elderly have retired to the area.

Retirement communities have attracted an increasing number of the urban elderly into rural areas. Some communities are literally "new towns," nearly self-reliant and built upon the premise that total support for the elderly can be provided throughout the lifespan. Emphasis within

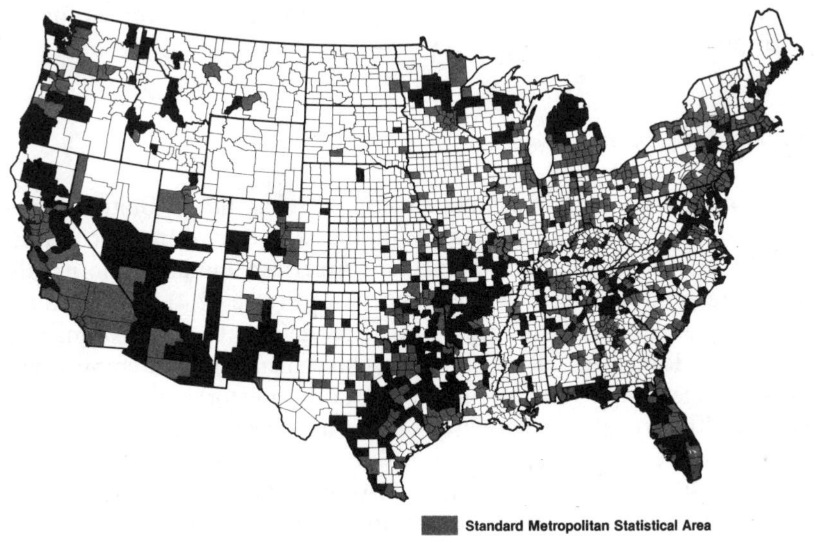

FIGURE 12.3. Retirement counties. From Bender, L. D., et al., 1985, *The diverse social and economic structure of nonmetropolitan America* (Rural Development Research Report No. 49) (p. 19). Washington, DC: U.S. Department of Agriculture, Economic Research Service.

these communities is upon meeting the initial needs of the elderly, as well as those during the last periods of life. Others are only subdivisions in existing communities, and it is assumed that necessary services exist or will exist in the host communities (Marans, Hunt, & Vakalo, 1984). Although both types of communities can be substantial generators of tax revenue, retirement centers or subdivisions are commonly built by non-profit entities (e.g., churches) and may be exempt from certain taxes. Furthermore, both types of settlements will generate a large number of consumer purchases, provide banks with deposits, and create new local labor market demands, especially in the service sector. Some areas, such as the Ozarks, have been transformed by elderly retirement communities and developments.

The rural elderly, then, represent a very special population group within the countryside. Planning strategies must reflect the differences among the elderly, with a special emphasis upon those with the fewest resources and the least access to support services. By the same token, the rural elderly constitute one of the greatest pools of expertise and "wealth" available to rural America, and nurturing the involvement of the elderly in community affairs represents one of the great opportunities for the rural planner.

Native Americans

There are over 1 million Native Americans living in the United States. The majority reside in rural areas. With the exception of Inuit or Eskimo people in Alaska, the vast majority are American Indians, who usually live on reservations in the Western states (see Figure 12.4). Rural Indians have long experienced high rates of poverty defined by a lack of marketable skills; low levels of educational achievement; limited employment opportunities; high rates of infant mortality and other indications of poor health care; and inadequate housing. The American Indians are the poorest of the nation's poor.

The Bureau of Indian Affairs (BIA), created in 1849, is the federal agency charged with the responsibility for protecting the interests of Native Americans. The BIA recognizes over 250 tribal units and "holds in trust" over 50 million acres of tribal lands. The largest reservation, belonging to the Navajo Nation in Arizona, is approximately the size of the state of West Virginia. Though some reservations are rich in energy and mineral resources, few have been developed to date. The smallest reservations are little larger than parks and contain little or no natural wealth, save for the land itself. The vast majority of Indian lands in the United States are sparsely populated, inaccessible, and bereft of resource attributes. Reservations often lack the most basic elements of infrastructure and support services. Employment opportunities are limited, and the majority of jobs that do exist are tied, directly or indirectly, to the federal government (through such agencies as the BIA, the Indian Health Service, etc.). Local tax bases are meager, and the private sector is very small. Unemployment is rampant; on many reservations, the majority of the population exists on one form of income transfer or another.

On those reservations where resource wealth has been established, no clear consensus among Indians exists as to whether or not to exploit this wealth. Some "traditionalists" have long stressed the need to conserve the land, and with it, those spiritual values that are central to the Indians' way of life. This viewpoint has been well expressed in a Canadian government document:

> For Indian people, land is much more than a source of profit and wealth creation or a place in which to reside temporarily. Their attachment to the land is part of a spiritual relationship with the universe, its

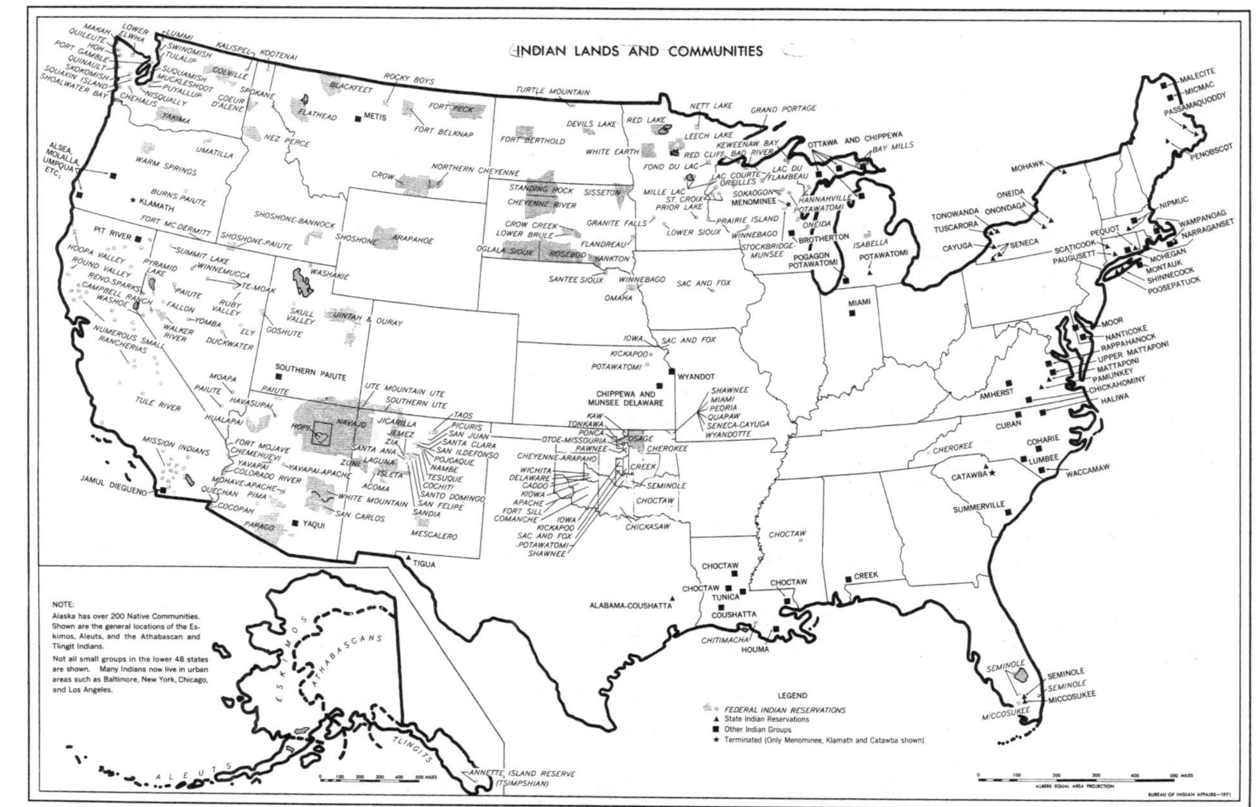

FIGURE 12.4. Indian lands and communities. Source: Tyler, S. L. (1973) *A History of Indian Policy*. Washington, D.C. Bureau of Indian Affairs, U.S. Department of the Interior, U.S. Government Printing Office, p. 280.

> elements and its creatures. Indian people see themselves as caretakers of the land and its resources. Land is thus a prerequisite for and vital to self-government. Indian people assert that their rights flow from their relationship with the land. (*Indian Self-Government in Canada: Report of the Special Committee*, 1983, p. 105)

Others within the community have stressed the need for development as the mechanism to raise the standard of living among Native Americans. Still others have argued for a middle course, which would include modest development within an overall context of resource conservation through wise planning and management.

Because land is the most abundant resource that Indians control, decisions related to economic and community development often involve decisions about the future of the land. A great deal of the Indian land resource is arid; without massive irrigation investments, little agricultural output can be realized. Over a third of the land is currently leased to non-Indians for farming, ranching, timber production, and mining activities. Certain rents accrue to the tribes through these leases, and some jobs are created for Indians from these activities. Also, royalties are often paid to the tribes on the basis of the productivity of mining activities, for example. Leasing land to non-Indians is often the result of the lack of internal tribal capital investment pools. Therefore, tribes must turn to the non-Indian community for the necessary capital to develop resources. The results of such arrangements have not always been beneficial. As anthropologist Nancy Owens (1979) has written:

> In both reservation–bordertown areas whites had gained control of Indian resources, and Indians had shifted away from reliance on their own sources to reliance on wage labor. As wage laborers they were primarily dependent on white employers for their livelihood. The principal implication of these findings is that white control and development of reservation resources will result in the economic betterment of whites, not Indians. Development packages which hold out the promise of increased job opportunities for Indians are likely to result in low-paying or temporary jobs for Indians and higher-paying jobs for a new community of whites whose presence accentuates Indian poverty. (p. 326)

Recently a number of tribes formed the Council of Energy Resource Tribes to combat this situation by bargaining for higher rents, larger royalties, preferential treatment for Indians in hiring, job training opportunities, and the creation of a pool of financial resources to stimulate internal tribal development activities.

Over the generations, a dependency relationship developed between Native Americans and the larger society, represented by the federal government. This created a certain level of tension and often frustrated attempts and movements toward greater Indian self-reliance. Both communities came to understand just how destructive this dependency had become. Many in Congress argued that this historical relationship simply had to be dissolved. Several tribes in Oregon, Wisconsin, and Utah were "terminated," though it soon became apparent that they lacked the resources and background to compete successfully in the marketplace. The result was another reordering of policy: "Termination" was replaced by the goal of "self-determination" for Indian tribes.

With the passage of the Self-Determination Act of 1975 (P.L. 93-638, 88 Stat. 2203, Title 1), the accent was on support of those efforts that would engender local economic development on reservations by Indian-owned and -controlled corporations and entities. Like other Americans, Indians entered the grants economy. And, again, as in much of America, little of this took place within the context of well-planned, comprehensive, well-financed programs. Indeed, much of the BIA funding earmarked for tribal and reservation economic development programs ended up supporting other efforts. For example, the attempt of the Economic Development Administration (EDA) to establish industrial parks on reservations was very largely a failure. According to a General Accounting Office (GAO) audit, this was the result of a cutback in BIA staff whose job it was to attract firms into such developments. Of the 45 industrial parks created by the EDA, 38 had two or fewer firms in operation (U.S. Congress, GAO, 1978). The lack of comprehensive programs and planning led to waste, duplication, and the creation of tribal bureaucracies devoted almost entirely to grantsmanship. Very little in the way of industry or internal economic development resulted. The Reagan years exacerbated problems through indiscriminate budget cuts, which greatly reduced or eliminated programs that had a genuinely favorable impact upon reservations (Hertzberg, 1982). One observer has summed up recent experience thus: "The change from dependency on the federal government to become completely independent was a major transition for the tribes. Federal funding reductions and a lack of a national Indian policy had made this transition difficult for many tribes" (Kirst, 1987, p. 108).[2]

Rural planners working with tribal organizations soon realize that although overall BIA policy affects all tribes, the need exists to treat each tribe and each reservation as separate planning opportunities. No two reservations share the same resource endowments or obstacles, and no two tribes have similar organizational structures. Though planners will have many tasks, few are as important as the provision of expertise, information, and management knowledge, so that tribal organizations

[2]For a more extensive account of this situation, see Levitan and Johnston (1975).

can bargain on a more equal footing with the sophisticated planning and management apparatus of corporations and governments. Unless tribes are effective in defining the terms of development, codes, and performance standards, a policy void will be created. Where this has been the case, the interests of tribal members have become secondary to those of the larger, more dominant community (Goldberg, 1976). It is this uneven nature of power and its distribution that has come to characterize the relationship beween Native American and white society. Planners must seek to redress this imbalance, difficult as this task may be. Planners working for tribal organizations, as well as those employed by counties and regions where reservations are located, must be committed to the enforcement of nondiscriminatory hiring practices. Likewise, enhancing the flow of information concerning opportunities is also important to the job generation process, for Indians and non-Indians alike (Snipp, 1980).

Summary

As rural areas become increasingly "older," it is critical that planners become more sensitive to the changing needs of people as they mature through the stages and cycles of life. Older Americans do not represent problems as much as they constitute opportunities. And though the United States is still far from the elaboration of a humane and credible policy for older Americans, local planners must resolve to understand the needs of the elderly and to meet the challenges of growing old in small towns and rural areas. Sensitivity is also of the greatest importance when working with other special populations, such as Native Americans. As a "people-centered" profession, planning must seek to address the needs of the total community *and* its constituent parts. Planning theory has historically concerned itself with issues of equity, justice, and fairness. These values are timeless ones for the planner. Although rural planners must seek to address the core issues that exist between urban power and rural needs, they cannot ignore those cleavages and problems that exist and persist within the rural community itself.

Local planners may feel powerless to turn back the rising tide of rural poverty. While programs to ameliorate rural deprivation are consistent with the ways in which planners have traditionally functioned, this is not enough. Through the creation of coalitions among groups and classes in the community and region, the provision of information and knowledge to resist inappropriate plans and policy agendas, and consistent clarification of the true nature of problems, planners can help create a society that is at once critical, progressive, and oriented toward change. Few things can be more important.

CHAPTER THIRTEEN

Conclusion: The Future of Rural America and the Role of Planning

Changes and Challenges in the Coming Years

As rural America approaches the 21st century, many challenges exist for planners, policy makers, and rural dwellers. The planner faces a diversity of competing demands for rural lands, a variety of communities, and regions with different needs. At the same time, a decentralizing federal bureaucracy is forcing more planning, public services, and economic development efforts down to the state and local levels.

The planner must understand that the competition for rural land has long-term implications for the local and regional economic base, social structure, and environment. Above all is the issue of who owns and controls the land base. The planner must educate policy makers about this situation and help them to make land use decisions that will result in a diversified economic base, a pluralistic social order, and a healthy environment. The choices involved, however, are likely to become more difficult. Should a community with high unemployment allow or prohibit the construction of a low-level nuclear waste dump, which would provide dozens of jobs and broaden the tax base? Should a county be rigid about enforcing large-lot agricultural zoning when many of the farmers in the county are going out of business and the farmers need to sell land for "top dollar" in order to pay off debts? Should a community resist the development of a vacation resort, which could broaden the tax base but drive up the cost of real estate beyond the means of local residents? In sum, the pros and cons of various development and conservation proposals are likely to be difficult to measure or predict in both the short and the long term. Nonetheless, the planner must attempt to provide decision makers with the best possible information and advice so that thoughtful decisions will be made.

How a planner functions often depends on the setting. Generally, there are two rural Americas: declining places and growing places. This is not to say that all declining places will never recover or that all growing places will never stop expanding. One of the curious features of economic life is that prosperity often moves in cycles. For example, rural New England was a region of struggling hard-scrabble farms from the late 1860s to the 1960s. In 1988, New England had the highest per capita income of any region in the nation, in part because of the development of new recreation, manufacturing, and service industries in the hinterlands. Whereas rural New England was formerly desperate to maintain and attract economic activity (and some pockets of underdevelopment remain), today and in the future the emphasis is on preserving farmland and open space, protecting groundwater supplies, and ensuring high-quality development.[1]

The pressure for growth is expected to continue on both the East and West Coasts, and the competition for rural land in these areas will increase as well. Thus, growth management and environmental protection will be major planning tasks. Planners will be called upon to draft comprehensive plans; to write ordinances; and to advise in directing, denying, or modifying developments. Planners should find an increase in public funds available—for the purchase of land or development rights, for example. Such funds have been authorized in half a dozen coastal states since 1985. In sum, on the East and West Coasts, rural planning and planners should enjoy greater importance in the coming years. Quite simply, as more people congregate on finite land, the need for planning the location, type, scale, and timing of development also increases. Greater government intervention in the land market is likely to be necessary in order to protect the public health, safety, and welfare in such areas as water supplies, solid and toxic waste disposal, aesthetics, and recreation. Although there is sure to be some resistance to planning, given the potential fortunes in real estate development, local planning will increasingly occur within a framework of state goals and regulations.

In the Midwest, the Great Plains states, the intermountain West, and parts of the South, many rural areas have experienced net population out-migration in the 1980s. Farmland values have plunged, and mining,

[1]Vermont's Land Use and Development Law (Act 250), passed in 1970, regulates large developments to ensure that they meet ten environmental criteria (Vt. Stat. Ann. tit. 10 § 6001–6039). In 1988, Vermont enacted Act 200, which mandates regional plans and calls for town plans that are in accord with the regional plans (Vt. Stat. Ann. tit. 10, § 155). Maine in 1988 passed law L.D. 2317, requiring local comprehensive plans throughout the state by 1996 (Maine, L.D. 2317; 1988). Both Vermont and Maine have significant land-purchasing programs as well.

timber, and energy activities have declined sharply. In short, the energy boomtowns and soaring farmland prices of the 1970s are things of the past. These regions are badly in need of economic diversification and growth. The role for planning in this situation is quite different from the role of planning in growth management. Yet planners must be wary of the temptation to sacrifice environmental quality for the short-term "quick-fix" of economic gain. It is important for people to be able to earn a decent living in rural areas, but these places must also be safe for human habitation. Furthermore, the projected drop in federal grants will mean a greater competition among rural communities for remaining grant monies. There will also be increased competition over attracting and promoting new businesses to bolster the local economy and tax base. Not all communities will survive, especially the smaller and more remote towns. Planners at the town, county, or regional level will play an important role in applying for grants, assisting in economic development programs, and negotiating with prospective new firms.

A major question for those areas curently in decline concerns the future of federal farm subsidies. Half of the farm population lives in the Midwest, and one-quarter resides in the South. The Reagan administration proposed the elimination of all federal farm subsidies by the year 2000. Such a drop in federal support could, in turn, cut net farm income, especially in the farm-dependent counties. Even a reduction in federal farm support would be felt by many farmers and neighboring communities. But, given the need to reduce the federal budget deficit, lower farm subsidies are likely. This may involve less support for all farmers or a targeting of farmers based on size. For example, under a decoupling program, those with medium-sized farms would qualify for federal deficiency payments, but those with large farms would not. On the other hand, federal rural policies could be shifted to address needs in transportation, housing, and job training for the large majority of rural inhabitants who are not employed in farming.

The price and availability of energy supplies could have a significant impact on the future location of economic activity. Cheap, plentiful energy has enabled Americans to live quite some distance from work and shopping. Higher energy prices could make a dispersed settlement pattern more costly to maintain and could make access to recreation and tourist sites more expensive. On the other hand, higher prices for oil, natural gas, and coal could help revive those rural areas dependent on energy development, especially in the Great Plains states and the intermountain West.

America's dependence on foreign oil stood at over 40 percent of total national consumption in 1988. Proven domestic supplies of oil can satisfy current demands for only eight years. Thus, the United States is very vulnerable to a disruption in oil supplies from overseas. Unlike the situation in the 1970s, however, the United States imports relatively little oil from the Middle East, relying instead on the United Kingdom, Norway, Mexico, and Venezuela.

Rural social problems of poverty, lack of education, substandard housing, and an aging population are likely to continue well into the next century. Rural America has proportionately more people living below the poverty line and in substandard housing than does urban America. Moreover, poverty nationwide has been rising since 1978, especially in rural areas. Education provides a way for people to acquire the skills and knowledge to increase productivity and gain better-paying jobs. But all too often, trained personnel and college graduates leave rural areas for better-paying jobs in the cities. The out-migration of the young contributes to the aging of many communities. Most senior citizens live on fixed incomes, are not entrepreneurs, and are not actively employed. They do bring in transfer income, but they also require special services, particularly in health care.

If current demographic trends continue, urban areas will gain political power, and rural areas will continue to lose power at the state and federal levels. Yet, if America's already crowded metropolitan regions begin to experience drastic reductions in air and water quality, then perhaps migration to rural areas will result. Arguably, one of the most important services that rural planners can perform is to help make rural areas livable, so that out-migration to crowded cities does not occur and that rural areas maintain the potential to absorb urban populations if city life becomes unbearable for one reason or another. During the Great Depression, for example, America registered its greatest number of farms ever, 6.5 million. People could not make a living in the cities, so many turned to the country for survival. And during the 1970s, despite the rise in energy prices, purchases of rural land by urban dwellers remained strong; the countryside represented not just a hedge against inflation or a vacation home, but also a place of refuge. This idea of the hinterlands as a "safety valve" for urban populations has existed since the early 19th century and persists today (see Smith, 1950).

It is not possible to predict how the major factors discussed above will interact in the future. Still, planners must be aware of the importance and the direction of these factors, and the ways in which they will affect individual planning areas. One of the purposes of planning is to anticipate change, and planners should develop contingency plans for a variety of potential circumstances. The worst situation occurs when a community or region is caught unprepared for change. Then local control is lost, and residents are vulnerable to changes beyond their control, and these changes can have long-term impacts on the local economy and environment. Although communities may not be able to foresee some changes, such as a shift in interest rates by the Federal Reserve Board, many changes are the cumulative result of several small alterations. For example, land use changes from open space and agricultural uses to residential and commercial uses do not happen all at once, but evolve over a period of several years. When land use changes are pending or begin, that is precisely the time to take action to influence the location and mix of land uses in the community.

Above all, the planner's role is that of a public servant. Learning public needs and wants, advising public officials, and reckoning what is best for the community or region are all tall orders. Within the framework of the law and political and economic systems, the planner should strive to discover what is possible and what is not. That is, the planner should not be afraid of innovation in economic development or regulation. At the same time, the planner should never forget that being useful to the community or the region in shaping and improving the future is the ultimate goal.

Rural Planning as a Field of Study

Rural planning and development strategies are gradually coming to constitute a distinct field of study. Rural planning requires an interdisciplinary approach to problem solving, combining elements of agricultural economics, law, rural sociology, planning, community development, historic preservation, ecology, and political science/public administration. The purpose of this field of study is to help explain how rural areas and small towns function and to predict what will happen under different circumstances. The study of rural planning also presents the planner with the tools and skills of several disciplines. Community development and historic preservation merge to form a Main Street revitalization effort. Planning and agricultural economics join in issues of farmland preservation. Ecology and planning interact in the protection of natural areas. Rural sociology and political science combine to suggest ways of alleviating poverty through government action.

In this book, we have identified patterns of change, development, growth, and decline in rural areas, as well as patterns of action that the planner can take. The description of the different types of rural counties in Chapter 1 summarizes the economic and population changes observed since the mid-1960s. The theories of dependency discussed in Chapter 5 provide an explanation of why rural communities may lack control over their present and future economic situation. Dependency theory lays the foundation for the major goal of rural planning and development: the creation of more self-reliant communities with more power to determine their own destinies. This is a difficult challenge in an era of global economies. Nonetheless, all rural planning and development strategies can be evaluated according to whether or not they promote (1) greater local self-reliance, (2) more options for rural people, and (3) a healthier environment.

For example, in the case of agriculture and mining, the theory of "comparative advantage" advises a rural region to specialize in commodities it can produce more cheaply than other regions and to rely on trade to improve the local economy. Dependency theory suggests that pursuing comparative advantage makes a rural region more vulnerable by relying on distant markets and fluctuating terms of trade over which the rural area has no control. Although the comparative-advantage strategy may be successful in the short run, a rural region is vulnerable in the long run to (1) increased supplies from other regions, which cause lower prices and greater competition for markets; (2) the creation of substitutes (e.g., polyester for cotton); and (3) changes in political and economic arrangements, such as the loss of a major U.S. wheat market after Iran overthrew the Shah, or the 40 percent increase in the value of the dollar on world exchanges between 1981 and 1985, which made U.S. farm exports less competitive.

The ultimate goal of the field of rural planning and development is to aid the planner in deciding upon courses of action. A planner first needs to gain knowledge and understanding of the rural jurisdiction. This involves not only identifying land use patterns and economic activities, but also discerning how wealth and power are distributed *within* the planning area and *between* the jurisdiction and urban centers. This kind of analysis will enable the planner to gauge the extent of the region's self-reliance, which in turn will largely define the limitations and opportunities for economic development and other programs to improve the local quality of life.

A planner faces the challenge of establishing and implementing priorities that either harmonize with outside forces or else mitigate their impact. In effecting change, the planner must rely on negotiating and organizing to bring together diverse interest groups, in order to orchestrate a consensus for dealing with the problems and potential of rural and small town environments. This consensus provides the planner with the necessary public backing and direction in working toward solutions.

The planner's dilemma, simply stated, is how to turn planning activities into beneficial action. Unfortunately, planning theory has been preoccupied with the creation of plans to the detriment of plan implementation. But, in reality, the planner does not operate from a position of strength; his or her professional recommendations are subject to review by planning commissions and elected officials. Moreover, a planner's attempts to change the distribution of wealth and power within a community, or between a community and outside forces, are likely to be controversial and to meet resistance. As a result, rural planners have often adopted a fragmented and incremental style. This planning approach may be adequate to handle day-to-day land use matters, but it is not sufficiently forward-looking to anticipate and accommodate broad economic and social changes. A planner must strive to maintain a comprehensive view of a jurisdiction, with an awareness of how the different elements of the community (land use, economic base, housing, transportation, etc.) are interrelated and affected by change in any one element.

In sum, planning offers a way of systematically identifying problems, establishing goals, organizing resources, and applying solutions. A comprehensive perspective enables a planner and the community to anticipate and shape the cumulative impact of individual planning decisions. This is particularly important when the community is attempting to achieve greater autonomy and to reduce situations of dependency. The more self-determining a community is, the more purposeful and effective its planning can become.

References

Ada v. Henry, 688 P.2d 994 (1983).
Agricultural Adjustment Act of 1933, 7 U.S.C. §§ 601–624 (1976 & Supp. III 1979).
Agricultural Adjustment Act of 1938, 7 U.S.C. §§ 1281–1393 (1976 & Supp. III 1979).
Agricultural Foreign Investment Disclosure Act of 1978, 7 U.S.C. §§ 3501–3508 (Supp. III 1979).
Alguire, F., Daniels, T., Keller, J., & Mineur, L. (1987). *Small town economic development* (Planners Advisory Service Memo). Chicago: American Planning Association.
Allen-Deane Corp. v. Township of Bedminster, No. L-28061-71 P.W., N.J. Super. Ct., Somerset County (1979).
American Land Resource Association. (1987, September–October). *Land report*. Washington, DC: Author.
Anthan, G. (1988). A grass-roots movement to save farmland. *Governing*, *1*, 52–54.
Appalachian Regional Act, Pub. L. No. 89-4, 5 Stat. (1965).
Ausness, R. (1983). Water rights legislation in the East: A program for reform. *William and Mary Law Review*, *24*, 547–590.
Bailey, L. H. (1911). *The Country-Life movement in the United States*. New York: Macmillan.
Bailey, L. H. (1915). *The holy earth*. New York: Scribners.
Baker, E. J., & Platt, R. H. (1983). The management of floodplains in nonmetropolitan areas. In R. H. Platt & G. Macinko (Eds.), *Beyond the urban fringe: Land use issues of nonmetropolitan America*. Minneapolis: University of Minnesota Press.
Baker, L. (1975). Controlling land users and prices by using special gains taxation in the land market: The Vermont experience. *Environmental Affairs*, *4*, 427–480.
Baldwin, S. (1968). *Poverty and politics: The rise and decline of the Farm Security Administration*. Chapel Hill: University of North Carolina Press.
Bank failures increase. (1987, October 2). *Kansas City* [MO] *Times*, section B, p. 2.
Bankhead Act of 1934, 7 U.S.C. §§ 1010–1012 (1976 & Supp. III 1979).
Barlowe, R. (1978). *Land resource economics* (3rd ed.). Englewood Cliffs, NJ: Prentice-Hall.
Batie, S. S., & Healy, R. G. (Eds.). (1980). *The future of American agriculture as a strategic resource*. Washington, DC: The Conservation Foundation.
Beale, C. L. (1988). Americans heading for the cities, once again. *Rural Development Perspectives*, *4*(3), 2–6.
Belle Terre v. Borass, 416 U.S. 1 (1974).
Bender, L. D., Green, B. L., Hady, T. F., Kuehn, J. A., Nelson, M. K., Perkinson, L. B., & Ross, P. J. (1985). *The diverse social and economic structure of nonmetropolitan America* (Rural Development Research Report No. 49). Washington, DC: U.S. Department of Agriculture, Economic Research Service.
Beuscher, J. (1960–1961). Appropriation water law elements in riparian doctrine states. *University of Buffalo Law Review*, *10*, 448–458.

Blakely, E. (1983, October). *Community attitudes toward change.* Paper presented at the meeting of the American Collegiate Schools of Planning, San Francisco.

Blauvelt v. County Commissioners, 227 Kan. 110 (1980).

Bowers, W. L. (1974). *The Country-Life movement in America: 1900–1930.* Port Washington, NY: Kennikat Press.

Bradbury, D. A. (1983). Agriculture law: Suburban sprawl and the right to farm. *Washburn Law Review, 22*(2), 448–468.

Bradshaw, T. K., & Blakely, E. J. (1987). Unanticipated consequences of government programs on rural economic development. In D. L. Brown & K. L. Deavers (Eds.), *Rural economic development in the 1980's: Preparing for the future.* Washington, DC: U.S. Department of Agriculture, Economic Research Service.

Brahtz, J. F. P. (Ed.). (1972). *Coastal zone management: Multiple use with conservation.* New York: Wiley.

Brooks, P. R. (Ed.). (1980). *Proceedings of the New York State Forest Resources Planning Symposium.* Albany: New York State Department of Environmental Conservation.

Brooks, P. R. (Ed.). (1981a). *The forest resources of New York: A summary assessment.* Albany: New York State Department of Environmental Conservation.

Brooks, P. R. (Ed.). (1981b). *Issues report on the forest resources of New York.* Albany: New York State Department of Environmental Conservation.

Brooks, P. R., Hendren, K., & McCann, B. (1982). Grand experiment: State forest resource planning. *Journal of Forestry, 80*, 585–587.

Brown, D. L., & Deavers, K. L. (Eds.). (1987). *Rural economic development in the 1980s: Preparing for the future.* Washington, DC: U.S. Department of Agriculture, Economic Research Service.

Brown, H. J., Phillips, R. S., & Roberts, N. A. (1981). Land markets at the urban fringe. *Journal of the American Planning Association, 47*, 131–145.

Burdge, R. J. (1982). Needs assessment surveys for decision makers. In D. A. Dillman & D. J. Hobbs (Eds.), *Rural society in the U.S.: Issues for the 1980's.* Boulder, CO: Westview Press.

Carey, J. C. (1977). *Kansas State University: The quest for identity.* Lawrence: Regents Press of Kansas.

Carney, M. (1912). *Country life and country schools.* Chicago: Row, Peterson.

Carter Administration (1980). *Small community and rural development policy.* Washington, D.C.: U.S. Government Printing Office.

Chokecherry Hill Estates v. Devel County, 294 N.W.2d 654 (1980).

Christenson, J. A. (1982). Community development. In D. A. Dillman & D. J. Hobbs (Eds.), *Rural society in the U.S.: Issues for the 1980's.* Boulder, CO: Westview Press.

Clavel, P. (1983). *Opposition planning in Wales and Appalachia.* Philadelphia: Temple University Press.

Clawson, M. (1972). *America's land and its uses.* Baltimore: Johns Hopkins University Press.

Clawson, M. (1975). *Forests for whom and for what?* Baltimore: Johns Hopkins University Press.

Clawson, M. (1979). *The economics of U.S. nonindustrial private forests.* Washington, DC: Resources for the Future.

Clawson, M. (1981). *New Deal planning: The Natural Resources Planning Board.* Baltimore: Johns Hopkins University Press for Resources for the Future.

Clean Air Act Amendment, 42 U.S.C. § 1857 *et seq.* (1970).

Clean Water Act, Pub. L. No. 95-576, 95 Stat. 2467–2469, 33 U.S.C. § 92 (1977).

Clean Water Act Amendments, Pub. L. No. 95-217, 95 Stat. 1566 (1977).

Coastal Zone Management Act, 16 U.S.C. § 1451 *et seq.* (1972).

Cole, L. A. (1982). *Forest resource management: Meeting the challenge in the states.* Lexington, KY: Council of State Government.

Collins, R. A. (1982). Federal tax laws and soil and water conservation. *Journal of Soil and Water Conservation, 37,* 319–322.

Comprehensive Environmental Response, Compensation and Liability Act ("Superfund"), Pub. L. No. 96-510, 94 Stat. 2767 (1980) (codified as 42. U.S.C.A. §§ 9601–9675, West 1987).

Comptroller of the Currency. (1984). *Community banks and rural markets: Case studies in economic development.* Washington, DC: U.S. Department of the Treasury.

Cook, E. (1976). Limits to exploitation of nonrenewable resources. *Science, 191,* 677–682.

Council on Environmental Quality (CEQ). (1982). *1982 annual report.* Washington, DC: U.S. Government Printing Office.

Council on Environmental Quality (CEQ). (1984). *1984 annual report.* Washington, DC: U.S. Government Printing Office.

Council of State Governments. (1989). *The book of the states (1988–1989).* Lexington, KY: Author.

County of Lake v. Cushman, 40 Ill. App. 3d 1045 (1976).

Cramer, G. L., & Jensen, C. W. (1982). *Agricultural economics and agribusiness.* (2nd ed.). New York: Wiley.

Cromwell, D. A. (1984). Strategies for dealing with the urban/forest interface: The recent California experience. In G. A. Bradley (Ed.), *Land use and forest resources in a changing environment.* Seattle: University of Washington Press.

Crop estimates. (1987, September 27). *Manhattan* [KS] *Mercury,* p. 3.

Crosson, P. R. (Ed.). (1982). *The cropland crisis: Myth or reality?* Baltimore: Johns Hopkins University Press for Resources for the Future.

Crosson, P. R., & Brubaker, S. (1982). *Resource and environmental effects of U.S. agriculture.* Washington, DC: Resources for the Future.

Danbom, D. (1979). *The resisted revolution: Urban America at the industrialization of agriculture, 1900–1930.* Ames: Iowa State University Press.

Daniels, T. L. (1985, September). *Trends in U.S. agriculture with an emphasis on the Midwest: Implications for planning.* Paper presented at the American Planning Association Gateway Regional Conference, St. Louis.

Daniels, T. L. (1986a). Hobby farming in America: Rural development or threat to commercial agriculture? *Journal of Rural Studies, 2,* 31–40.

Daniels, T. L. (1986b, March). *Rural banking trends: Implications for economic development in nonmetro areas.* Paper presented at the Conference on the Small City and Regional Community, Stevens Point, WI.

Daniels, T. L. (1988). America's Conservation Reserve Program: Rural planning or just another subsidy? *Journal of Rural Studies, 4,* 405–411.

Daniels, T. L., Daniels, R., & Lapping, M. B. (1986). The Vermont land gains tax: A variation of Henry George's "single" tax. *American Journal of Economics and Sociology, 45,* 441–455.

Daniels, T. L., & Farley, M. W. (1986). Energy conservation means big savings in Osage, Iowa. *Small Town, 17*(1), 29–30.

Daniels, T. L., & Keller, J. W., with Lapping, M. B. (1988). *The small town planning handbook.* Chicago: American Planning Association Planners Press.

Daniels, T. L., & Lapping, M. B. (1984). Has Vermont's land use control program failed? *Journal of the American Planning Association, 50,* 502–507.

Daniels, T. L., & Lapping, M. B. (1987). Small town triage: A rural settlement policy for the American Midwest. *Journal of Rural Studies, 3,* 273–280.

Daniels, T. L., & Nelson, A. C. (1986). Is Oregon's farmland preservation program working? *Journal of the American Planning Association*, *52*, 22–32.

Data bank. (1987, October 4). *New York Times*, section 3, p. 18.

Davis, C. (1978). *Water and water rights* (Vol. 7, Supplement). Indianapolis: Allen Smith.

Davis, C., Coblentz, H., & Titelbaum, O. (1976). *Water and water rights* (Vol. 7). Indianapolis: Allen Smith.

Deaton, B. J., & Weber, B. A. (1985, August). *Economics of rural areas*. Paper presented at the 75th annual meeting of the American Agricultural Economics Association, Ames, IA.

DeCapu v. Cella, No. 19-85-59, slip op. (New Haven Super. Ct. 1982).

Depository Institutions Deregulation and Monetary Control Act, Pub. L. No. 96-221 (1980).

Dillman, D. A., & Hobbs, D. J. (Eds.). (1982). *Rural society in the U.S.: Issues for the 1980's*. Boulder, CO: Westview Press.

Doherty, J. (1984). *Growth management in countrified cities*. Alexandria, VA: Vert Milon Press.

Domestic Allotment and Soil Conservation Act, 16 U.S.C. § 590 *et seq.* (1935).

Drabenstott, M., & Barkema, A. (1987). U.S. agriculture on the mend. *Economic Review* (Federal Reserve Bank of Kansas City), *72*(10), 28–41.

Duncombe, M. (1977). *Modern county government*. Washington, DC: National Association of Counties.

Ebeling, W. (1979). *The fruited plain: The story of American agriculture*. Berkeley: University of California Press.

Economic Opportunity Act, 7 U.S.C. § 2701 *et seq.* (1964).

Eichbaum, (1984). The Chesapeake Bay. *Environmental Law Reporter*, *14*, 10237.

Ellefson, P., & Cubbage, F. (1980a). State forest practice laws. *Environmental Policy and Law*, *6*, 133.

Ellefson, P., & Cubbage, F. (1980b). *State forest practice laws* (Bulletin No. 536-1980). St. Paul: University of Minnesota, Agricultural Experiment Station.

Endangered Species Act, 16 U.S.C. § 1531 *et seq.* (1973).

Environmental Protection Agency (EPA). (1984, August). *Groundwater protection strategy*. Washington, DC: U.S. Government Printing Office.

Epstein, S. S., Brown, L., & Pope, C. (1982). *Hazardous waste in America*. San Francisco: Sierra Club Books.

Farm Credit Act, 48 U.S. Statutes at Large 31 (1933).

Farmland Protection Policy Act, Pub. L. No. 97-98 (1981).

Federal Insecticide, Fungicide and Rodenticide Act, 7 U.S.C. § 136 *et seq.* (1971).

Federal Water Pollution Control Act Amendments, 33 U.S.C. § 1251 *et seq.* (1972).

First English Evangelical Lutheran Church of Glendale v. County of Los Angeles, 96 L.Ed.2d 250 (1987).

Fischer v. Bedminster, 93 A.2d 378 (1952).

Fishery Conservation and Management Act, Pub. L. No. 94-265 (1976).

Fitchen, J. (1981). *Poverty in rural America: A case study*. Boulder CO: Westview Press.

Flood Disaster Protection Act, 42 U.S.C. § 4001 *et seq.* (1973).

Florida Environmental Land and Water Management Act, Fla. Stat. Ch. 380 (1972).

Food for Peace Program of 1954, Pub. L. No. 83-480, 7 U.S.C. §§ 1691–1736(e) (1976).

Food Security Act (Farm Bill), Pub. L. No. 99-198, 16 U.S.C. § 3801 *et seq.* (1985).

Forest and Rangeland Renewable Resource Act of 1974, Pub. L. No. 93-378, 16 U.S.C. §§ 1601–1610 (1976).

Forstall, R. L., & Engels, R. A. (1984). *Growth in nonmetropolitan areas slows*. Washington, DC: U.S. Bureau of the Census.

Franklin, B. A. (1985, August 11). Despite 20 years of federal aid, poverty still reigns in Appalachia. *New York Times*, pp. 1, 19.

Freudenberg, W. R. (1979). *People in the impact zone: The human and social consequences of energy boomtown growth in four Colorado communities.* Unpublished doctoral dissertation, Yale University.

Friedmann, J., & Weaver, C. (1979). *Territory and function: The evolution of regional planning.* Berkeley: University of California Press.

Garcia, P. J. (1987, August 30). Farm credit system struggles to reorganize. *Washington Post*, p. H3.

Gates, F. T. (1912). The country school of tomorrow. *World's Work*, *24*(4), 464.

Gilmore, J. S., & Duff, M. K. (1975). *Boom town growth management: A case study of Rock Springs–Green River, Wyoming.* Boulder, CO: Westview Press.

Goeller, H. E. (1979). The age of substitutability: A scientific appraisal of natural resource adequacy. In V. K. Smith (Ed.), *Scarcity and growth reconsidered.* Baltimore: Johns Hopkins University Press.

Goldberg, C. E. (1976). *The prospects for Navajo taxation of non-Indians* (Bulletin No. 19). Los Angeles: University of California at Los Angeles, Institute of Geophysics.

Goldschmidt, W. (1975). A tale of two towns. In P. Barnes (Ed.), *The people's land.* Emmaus, PA: Rodale Press.

Gregory, G. R. (1972). *Forest resource economics.* New York: Ronald Press.

Hagenstein, P. (1984). Economics and land allocation at the urban/forest interface. In G. A. Bradley (Ed.), *Land use and forest resources in a changing environment.* Seattle: University of Washington Press.

Hagman, D. G. (1980). *Public planning and the control of urban and land development: Cases and materials* (2nd ed.). St. Paul, Minnesota: West Publishing.

Hallman, H. W. (1977). *Small and large together: Governing the metropolis.* Beverly Hills, Calif: Sage Publications.

Hallock, D. (1984). Boulder County makes room for wildlife. *Planning*, *50*(9), 12–14.

Harrington, D. H. (1987). Agricultural programs: Their contribution to rural development and economic well-being. In D. L. Brown & K. L. Deavers (Eds.), *Rural economic development in the 1980's: Preparing for the future.* Washington, DC: U.S. Department of Agriculture, Economic Research Service.

Harrington, M. (1962). *The other America: Poverty in the United States.* Baltimore: Penguin Books.

Hatch Act, 24 Stat. 440 (1887).

Hauser, P. M. (1981). The census of 1980. *Scientific American*, *245*(5), 53–61.

Haw. Rev. Stat. tit. 11, § 3165.1–3165.4 (Supp. 1982).

Haw. Rev. Stat. § 205 (1973, 1983).

Hazardous and Solid Waste Amendments, Pub. L. No. 98-616, 98 Stat. 3221 (1984).

Healy, R. G. (1976). *Land use and the states.* Baltimore: Johns Hopkins University Press.

Healy, R. G., & Short, J. L. (1978). New forces in the market for rural land. *The Appraisal Journal*, *4*, 185–197.

Healy, R. G., & Short, J. L. (1981). *The market for rural land: Trends, issues, policies.* Washington, DC: The Conservation Foundation.

Healy, R. G., & Short, J. L. (1983). Changing markets for rural lands: Patterns and issues. In R. H. Platt & G. Macinko (Eds.), *Beyond the urban fringe: Land use issues of nonmetropolitan America.* Minneapolis: University of Minnesota Press.

Henry, M., Drabenstott, M., & Gibson, L. (1986). A changing rural America. *Economic Review* (Federal Reserve Bank of Kansas City), *71*(7), 23–41.

Hertzberg, H. W. (1982). Reaganomics on the reservation. *The New Republic*, *187*(20), 15–18.

Hexem, R. (1980). *Agriculture districts and land use: A pilot study.* (A. E. Report 80-19, Department of Agricultural Economics). Ithaca, NY: Cornell University.

Holley, M. E. (1982). *A descriptive model of the socio-economic impacts of energy boomtowns in the American West.* Unpublished master's thesis, University of Oregon Graduate School of Planning.

Holmes, R. H. (1912). The passing of the farmer. *Atlantic Monthly, 110*(4), 217–223.

Homestead Act, ch. 75, 12 Stat. 392 (1862).

Housing Act of 1937, 42 U.S.C. § 1401 (1958).

Housing Act of 1949, 42 U.S.C. §§ 1471–1490 (1976 & Supp. III 1979).

Housing and Community Development Act, Pub. L. No. 93-383, 42 U.S.C. § 5304 *et seq.* (1974).

Hufschmidt, M. M., James, D. E., Meister, A. D., Bower, B. T., & Dixon, J. A. (1983). *Environment, natural systems, and development: An economic valuation guide.* Baltimore: Johns Hopkins University Press.

Ill. Rev. Stat. ch. 5-1101 (1981).

Indian Self-Government in Canada: Report of the Special Committee. (1983). Ottawa: Queen's Printer of Canada.

Ind. Code Ann. § 34-1-52.4 (Burns, 1982).

Iowa Code Ann. §§ 1729.1–1729.4 (West Supp. 1982–1983).

Jackson, K. T. (Ed.). *Atlas of American history.* New York: Scribner.

Johnson, J. A. (1984). Wisconsin's farmland preservation program. In F. R. Steiner & J. Theilacker (Eds.), *Protecting farmlands.* Westport, CT: AVI.

Just v. Marinette County, 210 N.W.2d 761 (1972).

Kan. Stat. Ann. § 19-2921 (1982).

Kan. Stat. Ann. § 2-3201 T0-3203 (1982).

Kan. Stat. Ann. § 45-1505 (1982).

Kan. Stat. Ann § 19-2991 (1982).

Keene, J. C. (1984). Innovative solutions to the age-old problem of agricultural pollution. *Zoning and Planning Law Report, 7*(9), 67–82.

Kemperer, W. D. (1976). The impact of tax alternatives on forest valuations and investments. *Land Economics, 52*, 135–157.

Ky. Rev. Stat. Ann. § 413.072 (Bobbs-Merrill Supp. 1982).

Kerr Tobacco Control Act of 1934, 7 U.S.C. §§ 751–756 (repealed, 1982).

Kirst, L. (1987). American Indian economic development policies. *Journal of Planning Literature, 2*, 101–110.

Klein, S. B. (1980, November). *State soil erosion and sediment control laws.* Paper presented at the National Conference of State Legislatures, Washington, DC.

Knudson, T. J. (1987, August 30). Bullfrog County, Nev. (pop. 0) fights growth. *New York Times*, pp. 1, 16.

Krannich, R. S., Greider, T., & Little, R. L. (1985). Rapid growth and fear of crime: A four-community comparison. *Rural Sociology, 50*, 193–209.

Krmpotich, S. (1980). How regional planners use inventory information. In *Proceedings of the Minnesota Forest Resource Inventory Conference.* St. Paul: U.S. Department of Agriculture, Forest Service, North Central Forest Experiment Station.

Lapping, M. B. (1980). Agricultural land retention: Responses, American and foreign. In A. M. Woodruff (Ed.), *The farm and the city: Rivals or allies?* Englewood Cliffs, NJ: Prentice-Hall.

Lapping, M. B. (1982a). Toward a working rural landscape. In C. H. Reidel (Ed.), *New England prospects.* Hanover: NH: University Press of New England.

Lapping, M. B. (1982b). Beyond the land issue: Farm viability strategies. *GeoJournal, 6*, 519–523.

Lapping, M. B. (1982c). Rural development and land use planning: A forestry perspective. *Journal of Forestry, 80*, 583–584.

Lapping, M. B., & Clemenson, H. A. (1983). The tenure factor in rural land management: A New England perspective. *Landscape Planning, 10*, 255–266.

Lapping, M. B., & Forster, V. D. (1982). Farmland and agricultural policy in Sweden: An integrated approach. *International Regional Science Review, 7*, 293–302.

Lapping, M. B., & Forster, V. D. (1984). From insurgency to policy: Land reform in Prince Edward Island. In C. H. Geigler & F. J. Popper (Eds.), *Land reform, American style.* Totawa, NJ: Rowan & Allenheld.

Lapping, M. B., & Leutwiler, N. (1978). Agriculture in conflict: Right-to-farm laws and the peri-urban milieu for farming. In W. Lockeretz (Ed.), *Sustaining agriculture near cities.* Ankeny, IA: Soil Conservation Society of America.

Lapping, M. B., Penfold, G., & Macpherson, S. (1983). Right to farm laws: Do they reduce land use conflicts? *Journal of Soil and Water Conservation, 38*, 465–467.

Lawton, M. P. (1980). Housing the elderly: Residential quality and residential satisfaction. *Research on Aging, 2*, 309–328.

Leuchtenburg, W. E. (1963). *Franklin Delano Roosevelt and the New Deal, 1932–1940.* New York: Harper & Row Torchbooks.

Levitan, S., & Johnston, W. (1975). *Indian giving: Federal programs for Native Americans.* Baltimore: Johns Hopkins University Press.

Levitan, S., & Shapiro, I. (1987). *Working but poor: America's contradiction.* Baltimore: Johns Hopkins University Press.

Lewis, J. A. (1978). *Landownership in the United States, 1978.* Washington, DC: U.S. Department of Agriculture, Economic Statistics and Cooperative Service.

Lindblom, C. E. (1959). The science of muddling through. *Public Administration Review, 119*, 79–88.

Lovejoy, S. B., & Napier, T. L. (Eds.). (1986). *Conserving soil: Insights from socioeconomic research.* Ankeny, IA: Soil Conservation Society of America.

Lucas, R., & Himelfarb, A. (1977). Some social aspects of medical care in small communities. *Canadian Journal of Public Health, 62*, 6–62.

Maine Act to Promote Orderly Economic Growth and Natural Resource Conservation, Me. Laws Stat. 2317 (1988).

Majchrowicz, T. A., & DeBraal, J. P. (1984). *Foreign ownership of U.S. agricultural land through December 31, 1983.* Washington, DC: U.S. Department of Agriculture, Economic Research Service.

Marans, R. W., Hunt, M. E., & Vakalo, K. L. (1984). Retirement communities. In I. Altman, M. Lawton, & J. Wohlwill (Eds.), *Elderly people and the environment.* New York: Plenum.

Marine Protection Research and Sanctuaries Act, 16 U.S.C. § 1431 *et seq.* (1972).

Market Agreement Act of 1937, 7 U.S.C. §§ 601–674 (1976 & Supp. III 1979).

Marshak, P. (1983). *Green gold: The forestry industry in British Columbia.* Vancouver: University of British Columbia Press.

Max, W. (1981). *The impact of taxation on the nonindustrial private forest sector* (Department of Economics Discussion Paper No. 157). Boulder: University of Colorado, Department of Economics.

Meyers, W. H., Womack, A. W., Johnson, S. R., Brandt, J., & Young, R. E., II. (1986). Impacts of alternative programs indicated by the FAPRI analysis. *American Journal of Agricultural Economics, 69*, 972–979.

Mich. Comp. L. §§ 286.471 (1980).

Milkove, D. L., & Sullivan, P. J. (1987). Financial aid programs as a component of economic development strategy. In D. L. Brown & K. L. Deavers (Eds.), *Rural economic development in the 1980's: Preparing for the future.* Washington, DC: U.S. Department of Agriculture, Economic Research Service.

Miller, D. (1984). Strategies to achieve public land use and forest resource goals. In G. A. Bradley (Ed.), *Land use and forest resources in a changing environment*. Seattle: University of Washington Press.

Minnesota Family Farm Security Act, Minn. Stat. ch. 41 (1976).

Morgan, D. (1980). *The merchants of grain*. New York: Viking Books.

Morrill Act of 1862, 12 Stat. 503, amended 44 Stat. 247 (1926).

Moss, E. (Ed.). (1977). *Land use control in the United States*. New York: Dial Press.

Multiple Use–Sustained Yield Act of 1960, Pub. L. No. 86-517, 16 U.S.C. §§ 528–531 (1976).

Nash, N. C. (1987, December 20). Senate approves farm banking aid. *New York Times*, Sect. 1, p. 33.

National Association of Counties. (1978). *County year book, 1978*. Washington, DC: Author.

National Association of State Departments of Agriculture Research Foundation. (1985, January). *Farmland project*. Washington, DC: Author.

National Environmental Policy Act of 1969, Pub. L. No. 91-190, 42 U.S.C. §§ 4321–4361 (1969).

National Forest Management Act, Pub. L. No. 94-588, 16 U.S.C. §§ 1601–1614 (1976).

National Historic Preservation Act, 16 U.S.C. §§ 470–470n (1966).

National Land Investment Co. v. Easttown, 215 A.2d 597 (1965).

National Trust for Historic Preservation Act, 36 C.F.R. § 68.2 (1949).

Natural Resources Defense Council. (1977). *Land use controls in the United States: A handbook on the legal rights of citizens*. New York: Dial Press.

N.H. Rev. Stat. Ann. § 430c (Supp. 1981).

N.J. Stat. Ann. § 4:1C-1 to 37 (West 1984).

N.Y. Agric. & Mkts. Law §§ 300–309 (McKinney 1971).

Noise Control Act, Pub. L. No. 92-574, 49 U.S.C. § 1431 (1972).

Nordin, D. S. (1974). *Rich harvest: A history of the Grange, 1867–1900*. Jackson: University Press of Mississippi.

N.C. Gen. Stat. § 106-700 to 701 (Supp. 1983).

Nuclear Regulatory Commission Standards for Protection Against Radiation, 10 C.F.R. § 20 app. (1986).

Occupational Safety and Health Act of 1970, Pub. L. No. 91-596, 29 U.S.C. §§ 657–678 (1976 & Supp. III 1979).

O'Hare, W. P. (1988). *The rise of poverty in rural America* (Report No. 15). Washington, DC: Population Reference Bureau.

Ohio Rev. Code Ann. § 3704.01 (page supp. 1982).

Oregon Land Conservation and Development Commission. (1980). *Statewide planning goals and guidelines*. Salem: State of Oregon.

Oregon Land Use Act, Or. Rev. Stat. § 215.505 *et seq*. (1973).

Ostrom, V. (1965). Planning and un-planning. *Public Administration Review*, *125*, 337–338.

Ottauquechee Regional Planning and Development Commission. (1977). *Regional land use plan: Interim report*. Woodstock, VT: Author.

Owens, N. J. (1979). The effects of reservation bordertowns and energy exploitation on American Indian economic development. In G. Dalton (Ed.), *Research in economic anthropology* (Vol. 2). Greenwich, CT: JAI Press.

Page, G. W. (1987). *Planning for groundwater protection*. New York: Academic Press.

Pennsylvania Coal Co. v. Mahon, 260 U.S. 393 (1922).

Pidot, J. R. (1982). Maine's land-use regulation commission. *Journal of Forestry*, *80*, 591–593, 602.

Popper, F. J. (1979). Ownership: The hidden factor in land use regulation. In R. Andrews (Ed.), *Land in America: Commodity or natural resource?* Lexington, MA: Lexington Books.

President's National Advisory Commission on Rural Poverty. (1978). *The people left behind.* Washington, DC: U.S. Government Printing Office.

Protecting our wetlands. (1986). *Environmental Protection Agency Journal, 12*(1), p. 1.

Public Works and Economic Development Act, Pub. L. No. 89-136 (1965).

Rasmussen, W. (1985, October). 90 years of rural development programs. *Rural Development Perspectives*, p. 4.

Raup, P. (1982, December). *The changing determinants of urban–rural competition for land.* Paper presented at the annual meeting of the American Agricultural Economics Association, New York.

Reeves, R. (1974). The battle over land. In L. H. Masotti & J. K. Hadden (Eds.), *Suburbia in transition.* New York: New York Times Books.

Reimund, D., & Petrulis, M. (1987). Performance in the agricultural sector. In D. L. Brown & K. L. Deavers (Eds.), *Rural economic development in the 1980's: Preparing for the future.* Washington, DC: U.S. Department of Agriculture, Economic Research Service.

Reisner, M. (1986). *Cadillac desert: The American West and its disappearing water.* New York: Viking Penguin.

Resource Conservation and Recovery Act, 42 U.S.C. §§ 6901–6991 (1982 & Supp. III, 1985).

Rodgers, H., & Weiher, G. (1986). The rural poor in America: A statistical overview. *Policy Studies Journal, 15*, 279–289.

Roosevelt, T. (1917, October). The farmer, the cornerstone of civilization. *North Dakota Farmer*, p. 13.

Rosenblatt, R., & Moscovice, I. (1978). Establishing new rural family practices: Some lessons from a federal experience. *Journal of Family Practice, 7*, 755–763.

Ross, P. J., & Green, B. L. (1985). *Procedures for developing a policy-oriented classification of nonmetropolitan counties.* Washington, DC: U.S. Department of Agriculture, Economic Research Service.

Row, C. (1978). Economics of tract size in timber growing. *Journal of Forestry, 76*, 575–582.

Rowe v. Walker, No. 81-228769 slip op. (Mich. App. 1982).

Rural Development Act of 1972, 7 U.S.C. §§ 1013a, 1921–1995 (1976 & Supp. III 1979).

Safe Drinking Water Act of 1974, 42 U.S.C. § 300(f) (1976).

Sargent, F. O. (1976). *Rural environmental planning.* South Burlington, VT: Author.

Schenker, A. (1985). Zero employment governments: Survival in the tiniest towns. *Small Town, 16*(2), 4–11.

Schertz, L. P., Martin, J. R., Forste, R. H., Frick, G. E., Rogers, G. B., Van Arsdall, R. N., Gilliam, M. C., McArthur, W. C., Lagrone, W. F., Johnson, S. S., Jesse, E. V., & Reimond, D. A. (1979). *Another revolution in U.S. farming?* Washington, DC: U.S. Department of Agriculture.

Self-Determination Act (1975). P.L. 93-638, 88 Stat. 2203, Title 1 (Jan. 4, 1975).

Seroka, J. (ed.). (1986). *Rural public administration: Problems and prospects.* Wesport, Conn.: Greenwood Press.

Shatto v. McNulty, 509 N.E.2d 897 (Ind. App. 1987).

Sherry, S., & Puring, R. (1983). *Executive summary: Hazardous materials disclosure information systems. A handbook for California communities and their officials.* San Francisco: Golden Health Empire Center.

Shover, J. (1976). *First majority—last minority: The transformation of rural America.* DeKalb: Northern Illinois University Press.

Sibson v. State, 336 A.2d 239 (1975).

Sierra Club v. Morton, 405 U.S. 345 (1970).

Smith, H. N. (1950). *Virgin land.* Cambridge, MA: Harvard University Press.

Smith–Lever Act, 38 Stat. 372 (1914).

Snipp, C. M. (1980, April). Determinants of employment—Wisconsin Native American communities. *Growth and Change*, pp. 39–47.

Soil and Water Resources Conservation Act, 16 U.S.C. §§ 2001–2009 (1977).

Solid Wastes Disposal Act, 42 U.S.C. § 3251 *et seq.* (1970).

South Burlington County NAACP v. Township of Mt. Laurel, 336 A.2d 713 (N.J. 1972).

Spectorsky, A. C. (1955). *The exurbanites.* Philadelphia: J. B. Lippincott.

Sporhase v. Nebraska, 455 U.S. 935 (1982).

Surface Mining Control and Reclamation Act, 30 U.S.C. §§ 1201–1328 (1977).

Superfund Amendments and Reauthorization Act, Pub. L. No. 99-499, 100 Stat. 1613 (1986).

Thompson, R. P., & Jones, J. G. (1981). Classifying nonindustrial private forestland by tract size. *Journal of Forestry, 79*, 288–291.

Thurow, C., Toner, W., & Erley, D. (1975). *Performance controls for sensitive lands.* Chicago: American Society of Planning Officials.

Thurston v. Coche County, 626 P.2d 440 (1981).

Timber Relief Act, Pub. L. No. 98-478 (1984).

Toner, W. (1979). Getting to know the people and place. In J. Getzels & C. Thurow (Eds.), *Rural and small town planning.* Chicago: American Planning Association.

Toxic Substances Control Act, 15 U.S.C. § 2601–2629 (1976).

Traylor, A. (1988). *The Rural Development Program: A historical assessment of rural policy under the Eisenhower administration and its application to the "farm crisis" of the 1980s.* Unpublished master's thesis, Kansas State University.

Tweeten, L., & Brinkman, G. (1976). *Micropolitan development.* Ames: Iowa State University Press.

Tyler, L. S. (1973). *A history of Indian policy.* Washington, DC: U.S. Department of the Interior, Bureau of Indian Affairs.

United States v. Butler (*In re* Hoosac Mills Corp.), 297 U.S. 1 (1937).

U.S. Advisory Commission on Intergovernmental Relations. (1964). *The problem of special districts in American government.* Washington, D.C.: U.S. Government Printing Office.

U.S. Bureau of the Census. (1980). *Census of population.* Washington, DC: U.S. Government Printing Office.

U.S. Bureau of the Census. (1985). *Statistical abstract of the United States, 1986.* Washington, DC: U.S. Government Printing Office.

U.S. Bureau of the Census. (1986). *Statistical abstract of the United States, 1987.* Washington, DC: U.S. Government Printing Office.

U.S. Bureau of the Census. (1987, September). State population and household estimates with age, sex, and components of change: 1981 to 1986. *Current Population Reports*, Series P-25, No. 1010.

U.S. Bureau of the Census. (1988). *Rural and rural farm population: 1987.* Washington, DC: U.S. Government Printing Office.

U.S. Congress, General Accounting Office (GAO). (1978, February 15). *Controls are needed over Indian self-determination contracts, grants, and training and technical assistance activities* (Report No. CED-78-44). Washington, DC: U.S. Government Printing Office.

U.S. Congress, Office of Technology Assessment. (1986). *Technology, public policy, and the changing structure of American agriculture* (Report No. OTA-F-285). Washington, DC: U.S. Government Printing Office.

U.S. Department of Agriculture (USDA). (1955) *Development of agriculture's human resources.* Washington, D.C.: U.S. Government Printing Office.

U.S. Department of Agriculture (USDA). (1980a). *An assessment of the forest and range land situation in the United States.* Washington, D.C.: U.S. Government Printing Office.

U.S. Department of Agriculture (USDA). (1980b). *A recommended renewable resources program–1980 update.* Washington, D.C.: U.S. Government Printing Office.

U.S. Department of Agriculture (USDA). (1982). *1982 census of agriculture.* Washington, DC: U.S. Government Printing Office.

U.S. Department of Agriculture (USDA). (1983a). *Better country: A strategy for rural development in the 1980's.* Washington, DC: U.S. Government Printing Office.

U.S. Department of Agriculture (USDA). (1983b). *Handbook of agricultural charts.* Washington, DC: U.S. Government Printing Office.

U.S. Department of Agriculture (USDA). (1984). *Rural community and the American farm.* Washington, DC: U.S. Government Printing Office.

U.S. Department of Agriculture (USDA). (1986). *Land areas of the National Forest System.* Washington, DC: U.S. Government Printing Office.

U.S. Department of Agriculture (USDA). (1987a). *1987 Resources Conservation Act appraisal.* Washington, DC: Author.

U.S. Department of Agriculture (USDA). (1987b). *The rural resources guide.* Washington, DC: U.S. Government Printing Office.

U.S. Department of Agriculture (USDA), Economic Statistics and Cooperative Service. (1979). *Structure issues of American agriculture* (Agricultural Economic Report No. 438). Washington, DC: U.S. Government Printing Office.

U.S. Department of Agriculture (USDA), Economic Statistics and Cooperative Service. (1981a). *National agricultural lands study: An inventory of state and local programs to protect farmlands.* Washington, DC: U.S. Government Printing Office.

U.S. Department of Agriculture (USDA), Economic Statistics and Cooperative Service. (1981b). *Farm real estate market developments 1972–1980.* Washington, DC: U.S. Government Printing Office.

U.S. Department of Housing and Urban Development. (1976). *Rapid growth from energy products: Ideas for state and local action. A program guide.* Washington, DC: U.S. Government Printing Office.

U.S. Department of Housing and Urban Development. (1980). *Housing needs of the rural elderly and the handicapped.* Washington, DC: U.S. Government Printing Office.

U.S. Department of the Interior, Bureau of Indian Affairs. (1976).

U.S. Department of the Interior, Office of Water Use and Land Planning. (1975).

U.S. General Services Administration. (1985). *The catalogue of federal domestic assistance.* Washington, DC: U.S. Government Printing Office.

Vermont Land Use and Development Law, Act 250 of 1970; Vt. Stat. Ann. tit. 10, §§ 6001–6091 (1973, as amended Supp. 1975).

Vt. Stat. Ann. tit. 10, § 155 (1988).

Village of Euclid v. Ambler Realty Co., 272 U.S. 365 (1926).

Vogeler, I. (1981). *The myth of the family farm: Agribusiness dominance of U.S. agriculture.* Boulder, CO: Westview Press.

Wagner, Paul, (ed.) (1974). *County government across the nation.* Wesport, Conn: Greenwood Press.

Wash. Rev. Code Ann. §§ 814.04, 823.08 (West 1982–83).

Weber, B., Castle, E. N., & Shriver, A. L. (1987). The performance of natural resource industries. In D. L. Brown & K. L. Deavers (Eds.), *Rural economic development in the 1980's: Preparing for the future.* Washington, DC: U:S. Department of Agriculture, Economic Research Service.

Weber, B., & Howell, R. B. (Eds.). (1982). *Coping with rapid growth in rural communities.* Boulder, CO: Westview Press.

Wilderness Act, 16 U.S.C. §§ 1131–1136 (1964).

Williams, N., Jr., Kellogg, E. H., & Gilbert, F. B. (Eds.). (1983). *Readings in historic preservation: Why? What? How?* New Brunswick, NJ: Center for Urban Policy Research.

Wilson & Voss v. McHenry County, 416 N.E.2d 426 (1981).

Windley, P., & Scheidt, R. (1983). Housing satisfaction among rural small-town elderly: A predictive model. *Journal of Housing for the Elderly, 1,* 57–68.

Wisconsin Farmland Preservation Act, Wis. Stat. §§ 71.09(11); 91.11–91.79 (1979).

Wolf, P. (1981). *Land in America: Its value, use, and control.* New York: Pantheon.

Wunderlich, G. (1974). *Who owns America's land: Problems in preserving the rural landscape.* Washington, DC: U.S. Department of Agriculture, Economic Statistics and Cooperative Service.

Wunderlich, G. (1979). Landownership as policy. In H. Davis (Ed.), *Proceedings of the Northeastern Agriculture Leadership Assembly.* Amherst: University of Massachusetts, Center for Environmental Policy Studies.

Yetley, M. J. (1982). Fisheries. In D. A. Dillman & D. J. Hobbs (Ed.), *Rural society in the U.S.: Issues for the 1980's.* Boulder, CO: Westview Press.

Yotopolous, P. A., & Nugent, J. B. (1976). *Economics of development: Empirical investigations.* New York: Harper & Row.

Young, J. A., & Newton, J. A. (1980). *Capitalism and human obsolescence.* Montclair, NJ: Allenheld, Osmun.

Index